# Current Topics in Microbiology 245/I and Immunology

Springer
Berlin
Heidelberg
New York
Barcelona
Hong Kong
London
Milan
Paris
Singapore
Tokyo

# Signal Transduction and the Coordination of B Lymphocyte Development and Function I

## Transduction of BCR Signals from the Cell Membrane to the Nucleus

Edited by L.B. Justement
and K.A. Siminovitch

With 22 Figures and 5 Tables

Springer

Louis B. Justement, Ph.D.
Associate Professor
Division of Developmental & Clinical Immunology
Department of Microbiology
University of Alabama at Birmingham
378 Wallace Tumor Institute
Birmingham, AL 35492-3300
USA
*e-mail*: Louis.Justement@ccc.uab.edu

Professor Katherine A. Siminovitch, M.D.
Department of Medicine
University of Toronto
Mount Sinai Hospital
600 University Ave., Rm. 656A
Toronto, Ontario
CANADA M5G 1X5
*e-mail*: ksimin@mshri.on.ca

*Cover Illustration: Part I – Schematic representation of the major signaling pathways activated in response to antigen-mediated cross-linking of the B cell antigen receptor (BCR) complex. Aggregation of the BCR leads to activation of multiple protein tyrosine complexes that inducibly phosphorylate multiple intracellular effector proteins. Tyrosine phosphorylation of adaptor proteins promotes the formation of multimolecular complexes, which facilitate the activation and/or localization of key signaling proteins. Transcription factor activation is ultimately controlled by $Ca^{2+}$-dependent (calcineurin and $Ca^{2+}$-calmodulin kinase) as well as calcium independent pathways (Ras and Rac1). Differential transcription factor activation determines the nature of the B cell biological response.*

*Cover Design: design & production* GmbH, Heidelberg

ISSN 0070-217X
ISBN 3-540-66002-X Springer-Verlag Berlin Heidelberg New York


Library of Congress Catalog Card Number 15-12910
Printed in Germany



Typesetting: Scientific Publishing Services (P) Ltd, Madras

Production Editor: Angélique Gcouta

SPIN: 10675027 27/3020 – 5 4 3 2 1 0 – Printed on acid-free paper

# Preface

Proper development and differentiation of B lymphocytes is essential to ensure that an organism has the ability to mount an effective humoral immune response against foreign antigens. The immune system must maintain a balance between the deletion of harmful self-reactive B cells and the generation of a diverse repertoire of B cells that has the ability to recognize an almost unlimited array of foreign antigens. The need to delete self-reactive cells is tempered by the need to avoid the generation of large functional holes in the repertoire of foreign antigen-specific B cells that patrol the periphery. To accomplish this, the immune system must reach a compromise by eliminating only the most dangerous autoreactive clones, while allowing less harmful autoreactive B cells to exist in the periphery where they may complement the organism's ability to mount a rapid response against invading micro-organisms. Those autoreactive cells that do enter the peripheral pool are subject to a number of conditional restraints that effectively attenuate their ability to respond to self-antigens. Deleterious alterations in the homeostasis between tolerance induction and recruitment of B cells into the functional repertoire may lead to increased susceptibility to autoimmune disease or infection, respectively. Therefore, delineation of the molecular processes that maintain immunological homeostasis in the B cell compartment is critical.

The balance between tolerance and immunity can be altered by factors that are extrinsic as well as intrinsic to the B cell leading either to anergization and elimination, or to activation and differentiation into antibody-secreting plasma cells or memory cells. Extrinsic factors that regulate the balance between tolerance and immunity include differences in the amount, avidity and timing of antigen presentation, and the expression of molecules by other immune cells or tissues that play a role in elimination, retention or expansion of B cells. Equally important is the regulation of the intrinsic signaling threshold that is required to trigger an antigen-specific B cell clone and which determines whether that cell undergoes tolerance induction versus activation in response to a given antigenic challenge. Regulation of signaling thresholds in individual B cell clones is affected by alterations in the expression or function of molecules that adjust the strength of the signal delivered via the BCR. Triggering thresholds can be determined by inherited genetic polymorphisms that affect the expression or activity of specific intracellular effector proteins, or they can be altered within an individual B cell clone by previous encounters with antigen that lead to anergization or memory cell formation.

Significant progress has been made towards delineation of the intrinsic molecular processes that regulate B lymphocyte immune function. Recent observations have provided a clearer picture of the interactive signaling pathways that emanate from the mature B cell antigen receptor (BCR) complex and the different precursor complexes that are expressed during development. Studies have also revealed that the net functional response to a given antigenic challenge is affected by the combined action of BCR-dependent signaling pathways, as well as those originating from various coreceptors expressed by B cells. The chapters in this volume provide a summary of current findings relating to the molecular control of B cell development and differentiation. Because it is virtually impossible to include a discussion of every aspect related to signal transduction in B lymphocytes, an effort has been made to focus on signaling through the BCR complex and co-receptors that regulate BCR-dependent signaling. Part 1 of the volume deals with the biochemical/molecular aspects of BCR-dependent signal transduction beginning with membrane proximal events and culminating with regulation of gene transcription in the nucleus. Part 1 also covers the molecular function of specific coreceptors that are involved in regulation of BCR signaling.

It is now well established that reversible tyrosine phosphorylation plays an important role in regulating B cell biology. In particular, binding of antigen to the BCR promotes the activation of several protein tyrosine kinases (PTK) that, in conjunction with protein tyrosine phosphatases (PTP), alter the homeostasis of reversible tyrosine phosphorylation in the resting B cell. The net effect is a transient increase in protein tyrosine phosphorylation that facilitates the phosphotyrosine-dependent formation of effector protein complexes, promotes targeting of effector proteins to specific microenvironments within the B cell and initiates the catalytic activation of downstream effector proteins. The role of protein tyrosine kinases and phosphatases in initiation and propagation of signal transduction via the BCR is discussed in the first chapter. Initiation of B lymphocyte activation is dependent on the tyrosine phosphorylation-dependent formation of multi-molecular effector protein complexes that activate downstream signaling pathways. The formation of such complexes was initially hypothesized to occur primarily via effector protein binding to the BCR complex itself. However, recent studies have demonstrated that productive signaling via the BCR is, in fact, dependent on tyrosine phosphorylation of one or more adapter proteins that play a crucial role in recruitment and organization of effector proteins at the plasma membrane. The second chapter provides a discussion of the SLP65/BLNK adapter protein and its role in recruitment and activation of key signal transducing effector proteins in the B cell. After the BCR has been engaged by antigen and the activation response has been initiated, numerous second messengers and intermediate signal transducing proteins are activated. These include the production of lipid second messengers by phosphatidylinositol 3-kinase, the hydrolysis of phosphatidylinositol 4,5-bisphosphate to yield diacylglycerol and 1,4,5-inositoltrisphosphate, and the mobilization of $Ca^{2+}$. Numerous intermediate signaling proteins are also activated, including the Ras and Rap1 GTPases, as well as the Erk, JNK and p38 MAP kinases. The third chapter provides a discussion of the current understanding of how these second messengers and intermediate signaling proteins function in a concerted manner to regulate transcription factor activation and gene transcription in the B cell. In the fourth chapter the role of the cytoskeletal apparatus in B cell activation is discussed. Studies from numerous cell systems indicate that the cytoskeletal apparatus not only provides structural integrity for a cell, but it also provides a cell with the capacity to compartmentalize and redistribute proteins in a dynamic manner in order to modulate signal transduction and, thus, cellular function. It is now becoming clear that the cytoskeletal apparatus plays an important role in B cell activation through its physical/functional interactions with signal transducing effector proteins. Finally, the ability of B lymphocytes to respond to antigen is regulated by the expression of specific genes that play a role in the activation response. The fifth chapter presents a discussion on Pax-5/BSAP, which is a DNA-binding protein that plays a pivotal role in controlling the expression of genes that are required for the B cell response to antigen. A number of potential targets for Pax-5/BSAP have been identified that are directly associated with the BCR, or with signal transduction in B cells. The current understanding of how Pax-5/BSAP function is regulated and its role in gene transcription is covered there.

It is now clear that signal transduction through the BCR can be modulated by a number of coreceptors, in effect to maximize the immune response when the B cell encounters a foreign antigen in the appropriate microenvironment, to attenuate the response in instances where the B cell response to a self-antigen might be detrimental to the host and, finally, to modulate B cell activation under conditions in which an adequate immune response has already been made. In general, it is apparent that B cell coreceptors function as docking structures that recruit specific effector proteins in a phosphotyrosine-dependent manner. Depending of the context in which a B cell encounters antigen, one or more coreceptors can be engaged and, based on their ability to recruit selected effector proteins, either enhance or attenuate signaling through the BCR complex. The final two chapters of Part 1 provide an overview of the CD19, CD22 and FcγRIIb coreceptors describing their molecular function as well as their role in regulation of BCR-dependent signal transduction and B cell immune function.

In Part 2 of this volume the role of BCR- and accessory molecule-dependent signal transduction in regulation of specific physiological processes associated with B cell development and differentiation will be covered. The BCR complex performs an essential function during B cell development and differentiation. In pro-B and pre-B cells, expression of the membrane immunoglobulin heavy chain of the BCR in conjunction with pseudolight chain gene products is essential for establishing allelic exclusion and for regulation of ordered gene recombination at the heavy and light chain loci. Subsequent expression of a mature BCR complex on the surface of immature B cells is important for tolerance induction and selection into the peripheral B cell pool. During the process of tolerance induction, binding of self-antigen to the BCR on B cells can lead to anergization, clonal elimination or receptor editing, depending on the qualitative/quantitative nature of the signal delivered and the developmental state of the B cell. Finally, once B cells have been selected and migrate to the periphery, the BCR complex mediates antigen-dependent activation and selection into the memory B cell pool. In the first chapter of Part 2, the discussion focuses on the well-documented differences in the response of immature and mature stage B cells to antigen. Based on current information it remains controversial as to whether the differential response of these populations reflects processes that are intrinsic to B cells at specific stages of development or, rather, reflect extrinsic processes that determine the functional outcome of antigen-dependent signaling in immature versus mature B cells. In this context, the molecular and cellular characteristics of immature B cells are discussed and an evaluation of whether they support a role for intrinsic versus extrinsic factors in regulating the outcome of BCR signaling presented.

Before the mature BCR complex is expressed on the surface of the B cell, the gene elements that encode the heavy and light chain polypeptides that comprise the receptor must be successfully recombined in a process termed V(D)J recombination. This recombination process is highly regulated during development with respect to the expression of recombinase activity and the susceptibility of specific loci to undergo rearrangement. The high degree of control imposed on V(D)J recombination, which is crucial for successful enforcement of allelic exclusion and for maintenance of the fidelity of rearrangement, is exerted at several levels including: (1) transcription of recombinase genes; (2) accumulation of recombinase gene products; and (3) accesibility of specific loci to recombinase activity. The second chapter summarizes what is currently known about the signaling pathways in the B cell that underlie each of the regulatory mechanisms above. Because V(D)J recombination is an error-prone process in terms of the actual joining of gene elements, it is subject to quality control at many levels, inherently allowing for multiple attempts at rearrangement and correction of the final recombined gene product. Additionally, immune tolerance to self-antigen can be established in emerging B cells by reactivating the capability to undergo V(D)J recombination. Alternatively, self-tolerance can be effectively imposed by triggering B cells to undergo apoptosis. The third chapter focuses on what is currently known about the ability of the BCR to mediate both enhancing and suppressing effects on B cell development, V(D)J recombination and apoptosis.

Once B cells have encountered foreign antigen in the appropriate peripheral microenvironment, numerous soluble and cell-associated factors play a crucial role in driving proliferation and differentiation of the antigen-stimulated cell. The fourth chapter discusses current knowledge pertaining to signal transduction through CD40, which has been shown to play a critical role in regulating B cell growth and differentiation, class switching, germinal center formation and the generation of memory cells. Once B cells have encountered antigen and have received additional signals through CD40, they mature as antigen-presenting cells resulting from the upregulation of CD23, B7.1, B7.2 ICAM1 and CD44. In the fifth chapter evidence is presented indicating that the antigen-processing function of B cells is regulated both developmentally and in response to an array of stimuli that are received by the B cell through cell surface signaling molecules, including the BCR, the CD21/CD19 complex, MHC class II, CD40 and FcγRIIb. Finally, as the B cell response to antigen progresses, it is accompanied by class switching in which the isotype of antibody being synthesized is changed. This process does not affect the inherent specificity for antigen, but it does alter the effector function of the antibody produced. Class switching has been shown to occur in response to signals delivered to the B cell through cytokine receptors, CD40 and the BCR complex itself. The sixth chapter discusses what is currently known about the specificity of isotype switch recombination with regard to the regulation of germline transcripts by signals delivered through cell surface receptors.

The topics covered in this two-part volume are intended to provide information on the biochemical/molecular aspects of signal transduction through the BCR and to relate the mechanistic processes associated with signal transduction in B cells to immunologically relevant functional outcomes. It is also intended as a general overview of the molecular processes that underlie B cell development and differentiation and hopefully will serve as a reference for those who wish to delve further in to the issues discussed herein or related topics.

L.B. Justement
K.A. Siminovitch

# List of Contents

# List of Contents of Companion Volume 245/II

# List of Contributors

(Their addresses can be found at the beginning of their respective chapters.)

# Signal Transduction via the B-cell Antigen Receptor: The Role of Protein Tyrosine Kinases and Protein Tyrosine Phosphatases

L.B. Justement



---

Division of Developmental and Clinical Immunology, Department of Microbiology, University of Alabama at Birmingham, Birmingham, AL 35294-3300, USA

# 1 Introduction

The B-cell antigen receptor (BCR) complex performs an essential function during B-cell development and differentiation. In pro-B and pre-B cells, expression of the membrane immunoglobulin heavy chain of the BCR in conjunction with pseudo-light-chain gene products is essential for establishing allelic exclusion and for promotion of gene recombination at the light-chain loci. Subsequent expression of a mature BCR complex on the surface of immature B cells is important for tolerance induction and selection into the peripheral B-cell pool. During the process of tolerance induction, binding of self-antigen to the BCR on B cells can lead to anergization, clonal elimination or receptor editing, depending on the qualitative/quantitative nature of the signal delivered, and the developmental state of the B cell (Goodnow 1996; Hertz and Nemazee 1998; LeBien 1998). Once B cells have been selected and migrate to the periphery, the BCR complex mediates antigen-dependent activation and selection into the memory B-cell pool (Cambier et al. 1994; Goodnow and Cyster 1997; Reth and Wienands 1997; MacLennan 1998). It has also been shown recently that expression of the BCR is critical because it provides a persistence signal that is required for maintenance of the primary B-cell repertoire (Lam et al. 1997). The persistence signal does not promote entry into the cell cycle; thus, it is likely that the nature and/or strength of the persistence signal delivered via the BCR is distinct from that of an antigen-driven activation signal which causes the B cell to enter the cell cycle.

Clearly then, expression of the BCR complex is crucial for proper B-cell development, immune responsiveness to antigen, and for maintenance of the peripheral B-cell repertoire. Because of its important physiological role in B-cell biology, a great deal of effort has been directed toward the goal of delineating the molecular processes that regulate the function of the BCR complex. A significant amount of progress has been made demonstrating that binding of antigen to the BCR initiates multiple downstream signaling cascades that act in a concerted manner to regulate proper B-cell immune function (Cambier et al. 1994; Gold and DeFranco 1994; Reth and Wienands 1997). BCR cross-linking results in activation of phospholipase C$\gamma$ leading to hydrolysis of phosphatidylinositol 4,5-bisphosphate ($PIP_2$) to yield the second messengers diacylglycerol and 1,4,5-inositol trisphosphate ($IP_3$). These second messengers regulate protein kinase C (PKC) function and $Ca^{2+}$ mobilization, respectively. Changes in the concentration of intracellular $Ca^{2+}$ regulate multiple effectors including calcium/calmodulin-dependent kinase and the serine/threonine phosphatase calcineurin, which dephosphorylates nuclear factor of activation and transcription (NF-AT), causing it to translocate to the nucleus. Additional pathways that are activated following BCR cross-linking involve small-molecular-weight guanosine triphosphate (GTP) binding proteins (G proteins), such as Ras, Rac and Rho, which regulate activation of the ERK, JNK and p38 mitogen activated protein kinase (MAPK) pathways (see Chap. 3). These intermediate effector proteins, in conjunction with signals delivered via PKC and $Ca^{2+}$-dependent effectors, control the activation of numerous transcription factors that are important for entry into the cell cycle and the immune response to foreign antigen.

It is now well established that reversible tyrosine phosphorylation plays an important role in regulating B-cell biology. In particular, binding of antigen to the BCR promotes the activation of several protein tyrosine kinases (PTKs) that, in conjunction with protein tyrosine phosphatases (PTPs), alter the homeostasis of reversible tyrosine phosphorylation in the resting B cell. The net effect is a transient increase in protein tyrosine phosphorylation which facilitates the phosphotyrosine-dependent formation of effector protein complexes, promotes targeting of effector proteins to specific microenvironments within the B cell and initiates the catalytic activation of downstream effector proteins (Fig. 1). Because PTK activation and changes in protein tyrosine phosphorylation are the earliest events detected after BCR cross-linking, it is logical to propose that reversible tyrosine phosphorylation plays an important role in both initiation and propagation of the signal that is transduced.

Indeed, changes in the qualitative/quantitative nature of protein tyrosine phosphorylation can have significant effects on the activation of downstream signal-transducing effector proteins and the production of second messengers. Recent studies have demonstrated that changes in the production of second messengers and the subsequent activation of intermediate effector proteins can alter the pattern of transcription factors that are activated (Dolmetsch et al. 1997; Healy et al. 1997). Differential transcription factor activation results in altered gene transcription, which can profoundly affect the biological response of the B cell. The importance of reversible tyrosine phosphorylation in regulation of B-cell biology underscores the need to delineate the complex interplay that occurs between PTKs and PTPs that affect signal transduction via the BCR complex. Progress has been made towards this goal and the relevant studies will be discussed below.

## 2 The BCR Complex

Although the exact molecular composition of the BCR complex is currently being actively investigated, studies have defined the minimal complex that is required for initiation of signal transduction in response to ligation by antigen. The BCR complex on mature, resting B cells consists of an antigen-recognition structure that is composed of membrane-bound immunoglobulin (mIg) of the μ or δ isotype, non-covalently associated with disulfide-linked heterodimers of CD79a (Igα) and CD79b (Igβ) (Reth and Wienands 1997; Cambier et al. 1994). The specific mIg heavy-chain isotype that is expressed on the surface of the B cell varies depending on its developmental and differentiative state. Immature B cells express mIgM, mature B cells co-express mIgM and mIgD, and memory B cells express BCR complexes that contain mIg of the γ, α or ε isotype. Studies are currently ongoing to determine whether the expression of a specific heavy-chain isotype affects signaling

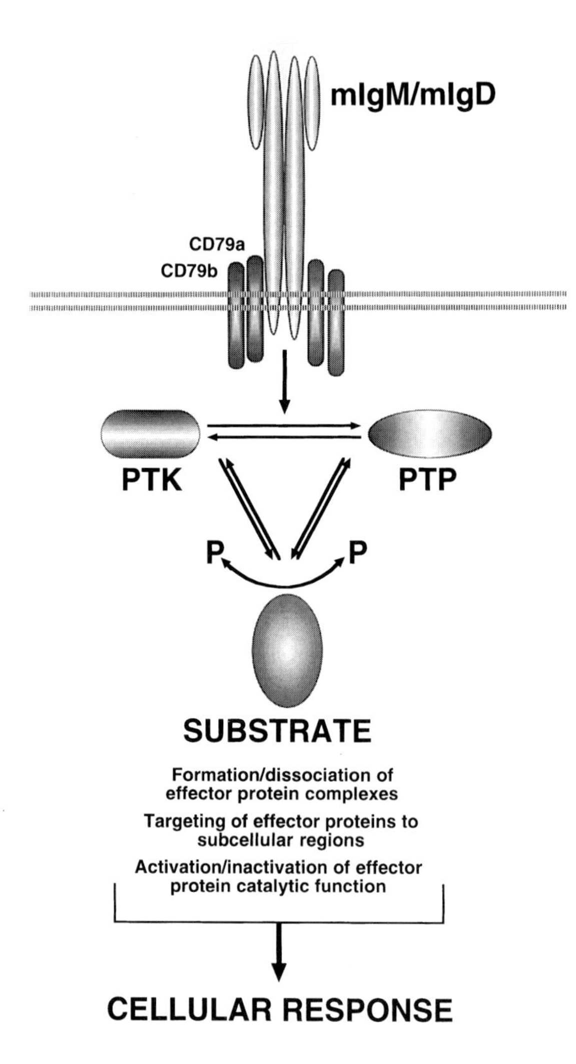


**Fig. 1.** B-cell antigen receptor (BCR)-mediated signal transduction is regulated by reversible tyrosine phosphorylation. Cross-linking of the BCR leads to changes in reversible tyrosine phosphorylation that are regulated by the dynamic interplay between PTKs and PTPs. These enzymes effectively regulate basal and inducible tyrosine phosphorylation of key effector proteins that mediate the production of second messengers and the activation of mitogen-activated protein kinase cascades. Tyrosine phosphorylation of specific effector proteins can result in the formation of multi-molecular complexes through the recruitment of other effector proteins that contain SH2 and phosphotyrosine binding (PTB) domains that recognize and bind to phosphotyrosine motifs. The formation of multi-molecular complexes facilitates subcellular localization of effector proteins, thereby targeting them to regions of the cell where they are efficiently activated and can mediate signal transduction

via the BCR complex. Recent findings suggest that the cytoplasmic domains of mIgG and mIgE isotypic receptor complexes do indeed contain specific information required for the functional immune response of B cells that express these receptors (Achatz et al. 1997; Kaisho et al. 1997; Nussenzweig 1997).

Numerous studies have demonstrated that the mIg-associated CD79a/CD79b heterodimer functions as both a transport structure that is required for expression of mIg on the surface of the B cell and as the primary signal transducing structure that couples the antigen recognition structure, i.e., mIg, to intracellular effector proteins. That CD79a and CD79b function as transport proteins was determined by experiments in which cDNA encoding the heavy and light chain of mIgM were transfected into a B-cell plasmacytoma (J558L). Expression of these cDNAs resulted in the accumulation of mIgM in the cytoplasm but not at the surface of the J558L cell line (Hombach et al. 1988; Williams et al. 1990). When the same cDNAs were transfected into B lymphoma cells (M12) and COS-7 cells, mIgM was expressed on the surface of the B lymphoma cells, whereas it accumulated in the cytoplasm of the COS-7 cells. These findings suggested that B cells, but not differentiated plasma cells or non-B cells, possess an activity that facilitates transport of mIgM to the cell surface.

The retention of mIgM in the endoplasmic reticulum of J558L cells could be overcome by mutations introduced into two polar motifs contained within the cytoplasmic domain of the heavy chains. When either one of these motifs were mutated, mIgM was expressed on the surface of the plasmacytoma (Williams et al. 1990). Thus, it was apparent that the transmembrane region of mIgM contains sequences that mediate retention in the endoplasmic reticulum through the interaction with one or more transport proteins that recognize these polar motifs, thereby facilitating transport to the surface of the cell. Subsequent experiments demonstrated that co-transfection of plasmacytoma cells with cDNA encoding CD79a was sufficient to promote expression of an intact BCR on the surface consisting of mIgM, CD79a and CD79b (Hombach et al. 1990). Similar findings were reported using a murine pituitary cell line. Transfection of mIgM alone into these cells in the absence of CD79a and CD79b led to accumulation of mIgM in the endoplasmic reticulum. In contrast, co-expression of CD79a and CD79b in pituitary cells was observed to promote expression of mIgM on the surface, indicating that these subunits are sufficient to mediate transport of mIgM (Matsuuchi et al. 1992).

The fact that the CD79a/CD79b heterodimer functions as the primary signal transducing structure in the BCR complex has been demonstrated by studies in which the transmembrane domain of mIgM has been mutated to disrupt the association with CD79a and CD79b. The basic strategy used in these experiments has been to replace the transmembrane and cytoplasmic region of mIgM with analogous domains from other transmembrane receptors. This eliminates the endoplasmic retention motifs in the transmembrane domain of native mIg, thereby promoting high level expression of the chimeric mIgM receptors on the surface of the B cell, despite the loss of interaction with the CD79a/CD79b heterodimer. These studies demonstrate that replacement of the transmembrane domain of mIgM with the CD8α (Blum et al. 1993), I-Aα (Parikh et al. 1991, 1992; Michnoff et al. 1994) or major histocompatability complex (MHC) class I (Dubois et al. 1992) transmembrane domain abrogates the interaction with the CD79a/CD79b heterodimer and blocks signal transduction in response to cross-linking of the chimeric receptor. In particular, disruption of the interaction between mIg and the CD79a/CD79b heterodimer abrogates PTK activation in response to BCR cross-linking, indicating that the heterodimer is responsible for physically and functionally coupling mIg to intracellular PTKs (Blum et al. 1993; Michnoff et al. 1994).

## 2.1 Analysis of the Physical Interaction Between the CD79a/CD79b Heterodimer and mIg

Studies have shown that the interaction between mIgM and CD79a/CD79b involves regions within the heavy chain CH3 and/or CH4 domains and the transmembrane spanning region (Reth 1992). An analysis of the transmembrane domain of mIgM reveals that there are two polar motifs (–TTAST– and –YSTTVT–) within this region that may function as points of contact with the CD79a/CD79b heterodimer (Reth 1992). Mutational analysis of these motifs indicates that the –YSTTVT– motif located towards the carboxyl-terminal end of the transmembrane domain does indeed play an important role in mediating the interaction with CD79a and CD79b. Studies have shown that a YS:VV mutation within this motif disrupts the interaction with CD79a and CD79b and also blocks signal transduction via mIgM (Shaw et al. 1990; Grupp et al. 1993; Sanchez et al. 1993). It has been further demonstrated that residues flanking the –YSTTVT– motif, contained within the –SLFYSTTVTLFVK– sequence, form a larger site that is important for signal transfer from the antigen recognition structure to CD79a/CD79b (Pleiman et al. 1994b). Mutation of the outermost serine and lysine residues contained within the larger motif to alanine and isoleucine, respectively, was shown to ablate BCR-mediated signaling in response to antigen but not anti-Ig monoclonal antibody (mAb) or polyclonal anti-Ig. Interestingly, mutation of the outermost tyrosine and threonine contained within the core –YSTTVT– motif resulted in non-responsiveness to antigen or anti-Ig mAb, whereas the cells responded normally to polyclonal anti-Ig (Pleiman et al. 1994b). None of the above mutations were observed to disrupt the interaction between mIg and the CD79a/CD79b heterodimer, suggesting that they clearly were affecting signal transduction in a subtler manner. These results then define a region in the transmembrane domain of mIg that is important for the functional transfer of a signal from mIg to CD79a and CD79b.

In contrast to the results presented above, mutations introduced within the amino-terminal polar motif –TTAST– (TT:AA or TTAST:AAAAA) were not observed to affect either the interaction between mIg and CD79a/CD79b or signaling through the BCR (Shaw et al. 1990; Grupp et al. 1993; Michnoff et al. 1994). Nevertheless, replacement of the amino-terminal third of the transmembrane domain of mIg, which contains the –TTAST– motif, with the comparable region from I-Aα abrogated the association with CD79a/CD79b as well as signal transduction (Michnoff et al. 1994). This suggests that amino acids flanking the –TTAST– motif play an important role in mediating the association with the CD79a/CD79b heterodimer, thus demonstrating once again the critical role that this interaction serves in signal transduction.

## 2.2 CD79a and CD79b are Functional Signal-Transducing Polypeptides

The ability of the CD79a/CD79b heterodimer to mediate signal transduction via a physical/functional interaction with one or more PTKs has been demonstrated by numerous laboratories using a variety of experimental systems. The most commonly used approach involves the generation of chimeric proteins that contain the cytoplasmic domain of either CD79a or CD79b fused to the extracellular/transmembrane domains of another transmembrane receptor. The cytoplasmic domains of CD79a and CD79b have been fused to CD8α (KIM et al. 1993a; TADDIE et al. 1994), platelet-derived growth factor receptor (PDGFR)α or β (LUISIRI et al. 1996; TSENG et al. 1997) and to the heavy chain of mIgM that has been mutated to prevent interaction with native CD79a or CD79b (LAW et al. 1993; PATEL and NEUBERGER 1993; SANCHEZ et al. 1993; BURKHARDT et al. 1994; WILLIAMS et al. 1994; PAPAVASILIOU et al. 1995; TEH et al. 1997).

Through the use of these experimental systems, both CD79a and CD79b chimeric receptors have been shown to possess the ability to transduce a signal when cross-linked with antibodies specific for the heterologous extracellular domain. Moreover, the cytoplasmic domain of both CD79a and CD79b exhibit the ability to transduce signals when expressed in B and T cells as part of a chimeric receptor (COSTA et al. 1992; BURKHARDT et al. 1994; TADDIE et al. 1994). Studies have shown that cross-linking of CD79a and CD79b chimeric proteins leads to activation of PTKs (KIM et al. 1993; LAW et al. 1993; BURKHARDT et al. 1994), production of $IP_3$ (LAW et al. 1993), $Ca^{2+}$ mobilization (KIM et al. 1993a; LAW et al. 1993; BURKHARDT et al. 1994; TADDIE et al. 1994), MAP kinase activation, and interleukin (IL)-2 production (LAW et al. 1993; BURKHARDT et al. 1994; TADDIE et al. 1994). Additional studies involving transgenic mice that express CD79a and CD79b chimeric proteins have revealed that the cytoplasmic domains of these proteins are able to reconstitute B-cell development and to mediate allelic exclusion in mIgM-deficient mice (PAPAVASILIOU et al. 1995; TEH and NEUBERGER 1997). Although the majority of studies support the conclusion that the individual subunits of the CD79a/CD79b heterodimer are functionally competent to regulate similar signaling processes, evidence exists that indicates that these subunits function in a complimentary manner nonetheless. In particular, studies have shown that CD79a chimeric proteins appear to be more efficient in terms of their ability to mediate PTK activation, whereas CD79b chimeras exhibit a greater activity with respect to $Ca^{2+}$ mobilization and IL-2 production (KIM et al. 1993a; BURKHARDT et al. 1994). Thus, it seems that the individual subunits exhibit some degree of selectivity in terms of the downstream processes that they regulate.

This possibility is further supported by studies demonstrating that the cytoplasmic regions of CD79a and CD79b interact with distinct sets of effector molecules (CLARK et al. 1992). Whereas the CD79a tail bound to the PTKs Lyn and Fyn, PI 3-kinase and a 38-kDa phosphoprotein, CD79b was observed to interact with PI 3-kinase and unidentified phosphoproteins of 40kDa and 42kDa. In contrast, other studies have demonstrated equivalent binding of the PTKs Syk, Lyn and Fyn to CD79a and CD79b in response to stimulation (LAW et al. 1993). It is possible then that the nature of the interaction between individual subunits of the heterodimer and downstream effector proteins is regulated by the activation status of the B cell.

It is important to note that the analysis of CD79a- and CD79b-mediated signal transduction has been performed primarily by expressing chimeric receptors containing either the CD79a or the CD79b cytoplasmic domain alone. Thus, the functional analysis of these subunits has largely been based on experimental systems that do not take into account the fact that CD79a and CD79b exist as a disulfide-linked heterodimer, thereby overlooking the possibility that these polypeptides may regulate signal transduction in a concerted manner. However, recent studies employing a modified chimeric expression system have more directly addressed the issue of cooperativity between CD79a and CD79b. In these studies, the extracellular/transmembrane portion of the PDGFR α or β has been fused to the cytoplasmic domain of CD79a or CD79b to construct PDGFRα/CD79b (Igβ) and PDGFRβ/CD79a (Igα) chimeras that can be homo- or hetero-dimerized when expressed in B cells (Luisiri et al. 1996; Tseng et al. 1997). Although cross-linking of either PDGFRα/Igβ or PDGFRβ/Igα alone was sufficient to mediate PTK activation and substrate phosphorylation, it was not possible to induce the full range of substrate phosphorylation seen upon cross-linking of native mIg. Only when PDGFRα/Igβ and PDGFRβ/Igα were expressed in the same cell and were hetero-dimerized was it possible to induce the full range of substrate phosphorylation (Luisiri et al. 1996). Moreover, it was noted that hetero-dimerization of the chimeric receptors decreased the threshold of activation when compared with cross-linking of either chimera alone.

Additional studies demonstrated that co-expression of the CD79b chimera increased the net level of tyrosine phosphorylation of CD79a tenfold, whereas the net level of CD79b phosphorylation decreased significantly in the presence of the CD79a chimera. These findings suggest that the CD79a and CD79b subunits mediate signals that are cooperative in nature. Additional data provide evidence to support the conclusion that the biological function of these polypeptides depends also on their co-expression. Using the WEHI-231 cell line, it was shown that expression and co-ligation of PDGFRα/Igβ and PDGFRβ/Igα is required in order to induce apoptosis. In contrast, cross-linking of either chimeric protein alone was not sufficient to induce an equivalent biological response (Tseng et al. 1997).

In summary, it is clear that signal transduction mediated via mIg is dependent on its interaction with the CD79a/CD79b heterodimer. The distinct subunits of the heterodimer function in a cooperative manner to regulate the activation of multiple PTKs that then phosphorylate a diverse range of substrates in the B cell.

## 3 Recruitment and Activation of PTKs by the BCR Complex

Numerous studies have now shown that BCR cross-linking results in the rapid induction of protein tyrosine phosphorylation (Cambier et al. 1994; Gold and DeFranco 1994; Reth and Wienands 1997). Reversible tyrosine phosphorylation is mediated by the coordinated recruitment and activation of multiple PTKs that function in a concerted manner to regulate downstream signaling cascades leading to gene transcription. Studies have proposed a general temporal relationship for the activation of distinct families of PTKs in response to BCR cross-linking (Saouaf et al. 1994). Specifically, it has been shown that the Src family PTKs (e.g. Lyn, Blk) exhibit increased enzymatic activity within seconds after BCR cross-linking, which precedes significant inducible tyrosine phosphorylation of CD79a and CD79b. The Tec family kinase Bruton's tyrosine kinase (Btk) exhibits a transient activation profile in which it is found to be maximally active within 5min of BCR cross-linking. Finally, Syk activity is observed to increase gradually reaching a maximal level within 10min, paralleling its recruitment and binding to the CD79a/CD79b heterodimer. Thus, these findings provide a temporal framework from which it should be possible to further delineate the interactive and coordinated processes that are regulated by these PTKs.

## 3.1 Recruitment and Activation of Src Family PTKs

Based on the observation that cross-linking of the BCR leads to inducible tyrosine phosphorylation and the fact that the complex does not have inherent kinase activity, it was clear early on that the BCR must physically interact with one or more intracellular PTKs. Initial characterization of the BCR complex and associated proteins revealed that this was indeed the case. Isolation of the BCR complex from resting B cells was observed to result in the co-precipitation of multiple Src family PTKs (Burkhardt et al. 1991; Yamanashi et al. 1991; Campbell and Sefton 1992; Li et al. 1992; Lin and Justement 1992; Burg et al. 1994; Gold et al. 1994). To date, evidence has been presented indicating that the BCR can physically associate with several members of the Src family, including Lyn, Fyn, Blk, Hck, Lck and Fgr, and that cross-linking of the BCR also leads to enzymatic activation of these PTKs (Burkhardt et al. 1991; Li et al. 1992). Characterization of the interaction between the BCR and members of the Src family has revealed that it is mediated through direct binding of Src family PTKs to the CD79a/CD79b heterodimer. The results of co-precipitation experiments suggest that only a small percentage of the available BCR complexes on the surface of resting B cells are associated with kinase (Lin and Justement 1992). It has further been hypothesized that a given heterodimeric structure in the resting BCR complex can interact with a single Src family PTK, such that each intact receptor complex physically associates with a maximum of two kinases (Lin and Justement 1992).

### 3.1.1 Binding of Src Family PTKs to CD79a in Unstimulated B Cells

Characterization of the molecular basis underlying the interaction between Src family PTKs and the CD79a/CD79b heterodimer in unstimulated B cells has implicated regions within the immunoreceptor tyrosine-based activation motif (ITAM) of CD79a as being important. Both CD79a and CD79b possess ITAM motifs within their cytoplasmic domains that contain six precisely spaced conserved amino acids [D/E-$(X)_7$-D/E/$(X)_2$-Y-$(X)_2$-L/I-$(X)_7$-Y-$(X)_2$-L/I] (Cambier 1995). Through the use of glutathione *S*-transferase (GST) fusion proteins containing the CD79a and CD79b cytoplasmic tails, it was found that Src family PTKs preferentially bind to CD79a (Clark et al. 1994; Cassard et al. 1996). Mutational analysis revealed that the specific, low-affinity interaction between Src family PTKs and the heterodimer is mediated by a short sequence, –DCSM–, located between the conserved tyrosine residues within the ITAM motif of CD79a, which is not found in the ITAM motif of CD79b. When the –DCSM– sequence from CD79a is switched with the analogous –QTAT– sequence from CD79b, the specificity of Src family PTK binding is reversed such that the switch mutant of CD79a with the sequence –QTAT– no longer binds PTKs, whereas the CD79b switch mutant with the –DCSM– sequence does (Clark et al. 1994). Interestingly, the ability of CD79a to interact with Src family PTKs in an unstimulated B cell does not require the conserved tyrosine residues contained within the ITAM motif (Clark et al. 1994). Thus, the interaction is not due to recognition of tyrosine-based motifs within the cytoplasmic tail of CD79a, per se.

Analysis of the CD79a binding site contained within members of the Src family has localized the critical region to within the first 30 amino acids of the Src family members Lyn and Fyn (Pleiman et al. 1994a). Studies demonstrated that a Lyn fusion protein containing residues 1–27 has the ability to immunoprecipitate non-phosphorylated CD79a from cell lysates. Further characterization of the specific region involved in the interaction using chimeric recombinant proteins comprised of Fyn, Src and Myc indicates that the first 10 amino acids of Fyn are sufficient to mediate binding to non-phosphorylated CD79a (Pleiman et al. 1994a). In summary, these findings suggest that Src family members are able to bind to non-phosphorylated CD79a via a motif that lies within the SH4 domain and/or the flanking unique region of each member of the family. The specific motif involved has not been identified and it is interesting to note that the SH4/unique regions of the Src family members exhibit the greatest degree of heterogeneity. Thus, it is not clear whether a common motif is employed or whether each PTK has the ability to interact with CD79a via a unique amino acid sequence contained within the first 30 residues. It is also not clear whether all members of the Src family possess the ability to interact directly with non-phosphorylated CD79a. For example, recombinant chimeric proteins containing the first ten amino acids of Src were not observed to interact with non-phosphorylated CD79a (Pleiman et al. 1994a), yet expression of native Src in B cells clearly results in the formation of BCR–Src complexes in unstimulated cells (Lin et al. 1997). Thus, it is possible that multiple mechanisms may be utilized to facilitate the interaction between the BCR and different Src family PTKs in unstimulated B cells to ensure that a functional complex exists that is competent to respond to antigen.

### 3.1.2 Activation of Src Family PTKs in Response to BCR Cross-Linking

In light of numerous studies, it is clear that the concept of the BCR complex expressed on unstimulated B cells can be extended to include the association with

Src family PTKs. However, it is not yet fully understood how these PTKs are activated in response to receptor cross-linking. Ligation of the BCR leads to aggregation of the complex within the plasma membrane, which is a requisite event for initiation of B-cell activation. In conjunction with BCR aggregation, increased tyrosine phosphorylation of the CD79a/CD79b heterodimer is observed (Gold et al. 1991). Interestingly, when mIgM is cross-linked with isotype-specific antibody, it results in selective phosphorylation of CD79a and CD79b subunits associated with mIgM but not mIgD. Similarly, cross-linking of mIgD leads to phosphorylation of mIgD-associated subunits, but not CD79a and CD79b associated with mIgM (Gold et al. 1991). This finding suggests that aggregation of the BCR complex may facilitate the ability of PTKs already associated with the receptor to phosphorylate substrates merely by co-localizing the kinase and substrate (i.e. CD79a and CD79b) within the plasma membrane. Although it is formally possible that Src PTKs associated with the BCR complex in unstimulated B cells are essentially in an active state by virtue of their interaction with CD79a, it is unlikely. Thus, even though receptor aggregation may indeed focus kinase and substrate, it is likely that the process of receptor aggregation also leads to activation of Src family PTKs.

Because BCR cross-linking leads to inducible phosphorylation of CD79a and CD79b, it was initially hypothesized that activation of Src family PTKs is dependent on tyrosine phosphorylation of the ITAM motifs contained within the subunits of the heterodimer. However, recent studies have provided evidence indicating that the initial activation of Src family PTKs is not absolutely dependent on tyrosine phosphorylation of either CD79a or CD79b. This possibility was first suggested by the observation that the kinetics for phosphorylation of CD79a and CD79b are slower than that of other prominent PTK substrates in the B cell (Kim et al. 1993b). Thus, it is unlikely that inducible tyrosine phosphorylation of CD79a and CD79b is required for activation of Src family PTKs, which presumably are responsible for mediating the initial tyrosine phosphorylation of substrates in the B cell.

Two studies have specifically examined the dependence of Src PTK activation on CD79a/b phosphorylation through the use of CD79a and CD79b chimeric proteins in conjunction with mutational analysis of tyrosine residues contained within ITAM motifs (Flaswinkel and Reth 1994; Pao et al. 1998). It was determined that the ITAM motifs in CD79a and CD79b are asymmetrically phosphorylated with the amino-terminal tyrosines in each ITAM motif being preferentially phosphorylated (Pao et al. 1998). Moreover, the amino-terminal tyrosine (Tyr182) in the ITAM of CD79a appears to be the predominant site of phosphorylation for Src family PTKs (Flaswinkel and Reth 1994). This is in keeping with the fact that this tyrosine residue resides within an ideal consensus phosphorylation site for Lyn and Fyn based on phage display analysis (Schmitz et al. 1996).

The fact that Src family PTKs preferentially associate with CD79a in unstimulated B cells is also intriguing because this may play a role in selectively localizing Lyn and Fyn to the amino-terminal tyrosine in the ITAM of CD79a. In keeping with the concept that CD79a is a substrate for Lyn and Fyn, it is logical to hypothesize that the initial activation of these PTKs would not be dependent on phosphorylation of the ITAM in CD79a. Indeed, mutation of the amino-terminal tyrosine in the ITAM of CD79a to phenylalanine diminished, but did not completely abolish, activation of Src family PTKs in response to cross-linking of chimeric CD79a receptors (PAO et al. 1998). Moreover, Src family PTK activation was not abrogated by mutation of any of the tyrosine residues contained within the ITAM motifs of either CD79a or CD79b (PAO et al. 1998). The essential conclusion from these findings is that the physical association of Src family PTKs with the BCR complex, via a tyrosine-independent motif within the ITAM of CD79a, is sufficient to promote the initial activation of Src family PTKs upon cross-linking and aggregation of the BCR. The mechanism by which BCR cross-linking results in the initial activation of Src family PTKs has yet to be elucidated.

#### 3.1.3 Potentiation of Src Family PTK Function by CD79a/b Tyrosine Phosphorylation

Although it is apparent that Src family PTKs can be activated independently of CD79a or CD79b phosphorylation, the net level of PTK activation under these conditions does not reach that seen in conjunction with phosphorylation of CD79a and CD79b (CLARK et al. 1994; PAO et al. 1998). Therefore, tyrosine phosphorylation is likely to play a role in potentiating the activation of Src family PTKs. Presumably, phosphorylation of the ITAM motifs within CD79a and CD79b could promote enhanced recruitment of Src PTKs to the BCR activation complex, thereby facilitating their ability to phosphorylate co-localized substrates. Alternatively, phosphorylation of CD79a or CD79b ITAM motifs could lead to enhanced activation of Src PTK catalytic function. Clearly these possibilities are not mutually exclusive and it is likely that both play a role in optimal activation of Src family PTKs. Analysis of the interaction between Src family PTKs and phosphorylated CD79a or CD79b has revealed that it is mediated through binding of the Src PTK SH2 domain to phosphotyrosine residues contained within the ITAM motifs of the respective subunits (PLEIMAN et al. 1994a). Both CD79a and CD79b have the ability to physically interact with Src family PTKs upon phosphorylation leading to a 20-fold increase in the amount of kinase that is associated with the BCR complex. The enhanced recruitment of Src family PTKs by CD79a or CD79b is dependent on phosphorylation of both tyrosines within the respective ITAM motifs. Subsequent studies demonstrated that binding of Fyn to biphosphorylated fusion proteins containing the cytoplasmic tail of CD79a leads to enhanced activation of its catalytic function (CLARK et al. 1994). Moreover, binding of Src family PTKs to biphosphorylated ITAM motifs from CD79a and CD79b was shown to actually increases the $V_{max}$ of Lyn via a direct interaction with the kinase SH2 domain (JOHNSON et al. 1995). This argues against the possibility that phosphorylation of the heterodimer results in a perceived increase in kinase activity simply through enhanced recruitment of Src PTKs to the BCR complex.

The exact mechanism by which binding of Src family PTKs to biphosphorylated ITAM motifs enhances the $V_{max}$ has not been elucidated. It is formally possible that binding of Src family PTKs to phosphorylated CD79a or CD79b leads to derepression of the kinase. Under conditions in which the carboxyl-terminal tyrosine of a Src PTK is phosphorylated (e.g. Lyn Tyr508), the kinase assumes an inactive conformation caused by an intramolecular association between its SH2 domain and the carboxyl-terminal phosphotyrosine residue (MAYER 1997). Phosphorylation of CD79a and CD79b could theoretically lead to the generation of phosphotyrosine motifs that effectively compete with the carboxyl-terminal phosphotyrosine for binding to the kinase SH2 domain. Thus, SH2-dependent binding of Lyn to phosphorylated CD79a could theoretically cause the kinase to assume an active conformation due to displacement of the carboxyl-terminal phosphotyrosine. Although this is an attractive model, it is not supported by studies demonstrating that Lyn from CD45-deficient B cells, which is hyperphosphorylated on its carboxyl-terminal tyrosine, is not effectively recruited to the phosphorylated BCR complex and is not activated (PAO and CAMBIER 1997). This finding indicates that the carboxyl-terminal tyrosine of Lyn (Tyr508), and possibly the respective inhibitory tyrosines of other Src family PTKs, must be dephosphorylated to permit effective recruitment to phosphorylated CD79a or CD79b. Another possibility is that binding of Src PTKs to biphosphorylated ITAM motifs induces some type of conformational change in the enzyme structure that leads to an increase in $V_{max}$. However, studies have yet to provide definitive evidence that such a conformational change takes place in the Src family PTKs upon SH2 domain binding to phosphotyrosine motifs. Perhaps a more likely scenario is one in which binding of Src family PTKs to biphosphorylated ITAM motifs promotes kinase dimerization, thereby facilitating transphosphorylation of conserved tyrosine residues that promote activation (SORTIRELLIS et al. 1995).

In conclusion, it has been shown that Src family PTKs physically interact with the BCR in resting B cells, thereby constituting a functional receptor complex. Cross-linking of the BCR promotes activation of Src family PTKs that then phosphorylate CD79a and CD79b. The phosphorylated ITAM motifs of CD79a and CD79b effectively recruit additional Src PTKs and promote the maximal activation of these kinases. Additionally, the phosphorylated ITAM motifs act as docking sites for other SH2 domain-containing proteins that are involved in amplification and propagation of signal transduction.

### 3.2 Recruitment and Activation of Syk

Numerous studies have demonstrated that cross-linking of the BCR results in inducible tyrosine phosphorylation and activation of the PTK Syk (HUTCHCROFT et al. 1991, 1992; SAOUAF et al. 1994). The critical role that Syk serves during BCR-mediated signaling has been demonstrated by the analysis of Syk-deficient B cells in which abrogation of Syk expression results in the loss of $IP_3$ generation and $Ca^{2+}$ mobilization (TAKATA et al. 1994). Moreover, studies using bone marrow cells derived from Syk-deficient mice have confirmed that Syk is essential for proper signal transduction via the BCR based on the inability of these cells to develop normally into mature B cells in chimeric mice (Cheng et al. 1995, Turner et al. 1995). Recruitment of Syk to the BCR complex is mediated by tyrosine phosphorylation of the ITAMs in CD79a and CD79b, creating docking sites that are recognized by the dual SH2 domains of Syk (Kurosaki et al. 1995; Rowley et al. 1995). Efficient recruitment of Syk requires that both conserved tyrosines within a given ITAM be present and that they be phosphorylated. Additionally, both SH2 domains of Syk must be present and functional in order to observe optimal association of the kinase with CD79a or CD79b and its subsequent activation. Mutation of either the amino- or carboxyl-terminal SH2 domain of Syk has been shown to abrogate BCR-dependent signal transduction (Kurosaki et al. 1995). Thus, Syk binds to biphosphorylated ITAMs on CD79a or CD79b via its dual SH2 domains and this presumably induces a conformational change in its structure, thereby promoting increased autophosphorylation.

The addition of biphosphorylated ITAM peptides to Syk accelerates the rate at which Syk undergoes autophosphorylation. However, addition of biphosphorylated ITAM peptides to Syk that is already maximally phosphorylated does not further activate its catalytic function (Rowley et al. 1995). These findings support the concept that binding of Syk to biphosphorylated ITAMs promotes a conformational change that leads to enhanced autophosphorylation that is critical for optimal activation. Studies indicate that other PTKs may play an important role in mediating phosphorylation of Syk once it has been recruited to the ITAMs of CD79a or CD79b. In particular, it was noted that the net level of Syk phosphorylation is significantly diminished in B cells that are Lyn deficient, suggesting that this Src family PTK is involved in phosphorylation of Syk (Kurosaki et al. 1994). Additionally, co-transfection of COS cells with Syk and Lyn resulted in significant tyrosine phosphorylation of Syk. Finally, it has been shown that Lyn and Syk physically interact with one another based on co-immunoprecipitation experiments and reassociation of purified proteins in vitro (Sidorenko et al. 1995). Binding of Syk to the ITAMs of CD79a and CD79b may also play a crucial role in its activation by focusing Syk within the BCR activation complex. This effectively brings Syk into close proximity to Lyn and essentially promotes multimerization of Syk itself, both of which would be predicted to facilitate its phosphorylation.

Syk exhibits increased tyrosine phosphorylation in response to B-cell activation and subsequent binding of Syk to biphosphorylated ITAM peptides. Analysis of Syk autophosphorylation has revealed that numerous tyrosines exhibit increased phosphorylation within 5min, although it is not clear whether multiple phosphotyrosine residues are critical for enhanced catalytic activity (Furlong et al. 1997; Keshvara et al. 1997). Mutation of Tyr 518/519 to Phe was observed to produce a mutant Syk enzyme that is defective in terms of its ability to mediate BCR-dependent signal transduction when transfected into Syk-deficient B cells (Kurosaki et al. 1995). Further analysis revealed that the Tyr518/519Phe mutation did not affect recruitment of the mutant kinase to the CD79a/CD79b heterodimer. Therefore, it is apparent that Tyr 518/519 plays a role in regulating the inherent enzymatic activity of Syk (Kurosaki et al. 1995).

Syk possesses a unique linker region between its tandem SH2 domains and the catalytic region. Within this linker is a 23 amino acid insert that is not found either in ZAP-70 or SykB, a naturally occurring splice variant of Syk (LATOUR et al. 1998). Reconstitution of Syk-deficient B cells with SykB revealed that this splice variant is inefficient in terms of its ability to mediate signal transduction via the BCR (LATOUR et al. 1998). The functional defect associated with the loss of the 23-amino-acid insert within the linker region was not due to the loss of Tyr 290 which is inducibly phosphorylated in response to B-cell activation. Moreover, the loss of function was not due to an inherent change in the catalytic activity of SykB (LATOUR et al. 1998). However, studies clearly demonstrated that the 23-amino-acid insert is involved in regulating the ability of Syk to bind to biphosphorylated ITAMs. It is possible that the linker region may function as a molecular hinge that regulates the intramolecular interaction between the amino- and carboxyl-terminal regions of Syk. The 23-amino-acid insert may affect the flexibility of this hinge region, in effect regulating the ability of the SH2 domains to engage biphosphorylated ITAMs. Alternatively, the linker and its insert may stabilize the dual SH2 domains of Syk so that they can bind to biphosphorylated ITAMs with higher avidity.

Although it has been shown that Syk is inducibly recruited to the BCR complex and undergoes tyrosine phosphorylation leading to activation, studies have demonstrated that the bulk of tyrosine phosphorylated and activated Syk is actually localized in soluble cell fractions (PETERS et al. 1996; KESHVARA et al. 1997). The active Syk that is isolated from soluble cell fractions is associated with alpha tubulin as opposed to BCR complex components. Thus, Syk is recruited to the BCR where it is phosphorylated leading to its activation, and then it dissociates from the complex. It has been shown that Tyr 130 located within the inter-SH2 domain region plays a role in regulating the dissociation of Syk from the BCR complex (KESHVARA et al. 1997). Mutation of Tyr 130 to Phe results in increased binding of Syk to the BCR complex and increased tyrosine phosphorylation of intracellular substrates in response to BCR cross-linking. In contrast, mutation of Tyr 130 to Glu decreases binding of Syk to the BCR complex and also attenuates inducible tyrosine phosphorylation after BCR ligation. The effects on overall protein tyrosine phosphorylation were observed despite the fact that the Tyr130Phe mutant exhibits a decrease in inherent activity whereas the Tyr130Glu mutant is hyperactive (KESHVARA et al. 1997). This would suggest that the association of Syk with the ITAMs of CD79a and CD79b and, thus, its localization to the BCR activation complex is crucial for optimal phosphorylation of intracellular substrates. Indeed, localization of Syk to the membrane may be as important as its inherent activation status in terms of promoting phosphorylation of intracellular substrates in response to BCR cross-linking.

## 3.3 Recruitment and Activation of Bruton's Tyrosine Kinase

As discussed previously, in addition to the Src family PTKs and Syk, a third type of kinase is inducibly activated in response to BCR cross-linking and plays an important role in signal transduction (Desiderio 1997; Satterthwaite et al. 1997). Several studies have documented increased tyrosine phosphorylation and activation of Btk a member of the Tec family (Aoki et al. 1994; De Weers et al. 1994). Like the Src family PTKs, Btk contains contiguous SH3, SH2 and SH1 domains, although it does not possess a carboxyl-terminal negative regulatory tyrosine residue or a myristylation site. In addition to the SH domains, Btk contains an amino-terminal pleckstrin homology (PH) domain and an adjacent proline- and cysteine-rich Tec homology (TH) domain (Desiderio 1997). Mutations in the gene encoding Btk are associated with X-linked agammaglobulinemia (XLA) in man and X-linked immunodeficiency (*xid*) in mouse (Rawlings et al. 1993; Thomas et al. 1993; Tsukada et al. 1993 Vetrie et al. 1993; Sideras and Smith 1995). An almost total block in B-cell development at the pre-B cell stage characterizes XLA, whereas the defect in *xid* is much less severe in that the mice have 50% fewer B cells in the periphery and the residual cells exhibit an abnormal $IgM^{high}IgD^{low}$ phenotype. The developmental abnormalities observed in the B-cell compartment are indicative of the fact that Btk serves an important function in BCR-mediated signal transduction. Indeed, B cells from *xid* mice fail to proliferate in response to BCR cross-linking with anti-Ig antibodies (Desiderio 1997). It is likely that the lack of responsiveness is due to the fact that B cells, which are Btk deficient, exhibit several alterations in BCR-mediated signal transduction processes.

Available evidence indicates that the Src family kinases act as positive regulators of Btk activity. This is in keeping with the fact that BCR cross-linking initially leads to activation of Lyn and Fyn followed by activation of Btk (Saouaf et al. 1994). Additionally, co-expression of Btk and Src PTKs results in increased phosphorylation of Btk and enhanced catalytic activity (Mahajan et al. 1995; Rawlings et al. 1996). Finally, co-expression of Csk, a negative regulator of Src family kinases, with the E41K mutant of Btk, has been observed to attenuate tyrosine phosphorylation of the kinase and to decrease its ability to transform fibroblasts, once again indicating that Src PTKs are involved in upregulation of Btk activity (Afar et al. 1996). Activation of Btk is associated with an increase in the phosphorylation of two regulatory tyrosine residues. One tyrosine residue is contained within the SH1 domain (Y551) of Btk while the other is located within the SH3 domain (Y223) (Park et al. 1996; Rawlings et al. 1996). Cross-linking of the BCR results in rapid phosphorylation of Btk at these two sites in a coordinated manner. It has been determined that Y551 is initially transphosphorylated by Src family PTKs and that this promotes autophosphorylation at Y223 (Park et al. 1996). The specific mechanism by which tyrosine phosphorylation leads to Btk activation has not been fully explored; however, it is possible that autophosphorylation of Y223 within the SH3 domain of Btk may affect inter- or intramolecular interactions that involve this domain. Deletion of the Btk SH3 domain or mutation of Y223 to phenylalanine has been observed to synergistically enhance the transforming activity of E41K mutants of Btk indicating that this region does indeed negatively regulate the catalytic activity of Btk (Park et al. 1996). Examination of the related PTK Itk has shown that its SH3 domain is involved in the formation of an intramolecular association via binding to proline-rich sequences in the TH domain (Andreotti et al. 1997). Due to the similarities between Btk and Itk, it is possible that such an interaction may negatively regulate the catalytic activity of the Btk SH1 domain.

Although it does not appear as though Btk is recruited directly to the BCR complex, it presumably must be localized to regions where activated Src family PTKs exist. In this regard, it has been shown that Btk can physically interact with the Src family PTKs Lyn and Fyn which bind to proline-rich sequences within the TH domain of Btk via their respective SH3 domains (CHENG et al. 1994). A point mutation within the TH domain of Btk that abolishes the SH3-dependent interaction with Lyn or Fyn has been shown to impair tyrosine phosphorylation of Btk (CHENG et al. 1994). This finding supports the functional relevance of the interaction between these PTKs and further suggests that Btk may be localized to the BCR activation complex via this association.

Studies have also shown that both the SH2 and PH domains of Btk are essential for optimal activation, presumably due to the fact that these domains are important for targeting Btk to the membrane and/or other effector proteins (DESIDERIO 1997). The molecular basis for the functional importance of the Btk SH2 domain has not been elucidated as of yet. Nevertheless, the SH2 domain could be important for targeting Btk to tyrosine phosphorylated proteins in the cell, thereby facilitating its localization to regions where it can either be inducibly activated or mediate phosphorylation of key substrates. The PH domain of Btk has been shown to mediate the interaction of Btk with both proteins and phospholipids (TSUKADA et al. 1994; YAO et al. 1994; SALIM et al. 1996). In particular, the PH domain of Btk specifically binds to liposomes that contain phosphatidylinositol-3,4,5-trisphosphate (PI(3,4,5)$P_3$), an interaction that requires Arg28, which is mutated in some XLA patients. Thus, the ability of Btk to bind to PI(3,4,5)$P_3$ via its PH domain and the fact that abrogation of this interaction leads to dysregulation of Btk function, clearly indicate that the PH domain plays an important role in localizing Btk to the plasma membrane where it can be activated and efficiently phosphorylate physiological substrates (YAO et al. 1994). Furthermore, the specific interaction of the Btk PH domain with PI(3,4,5)$P_3$ suggests that Btk is a target for 3′-phosphorylated products that are produced as a result of phosphatidylinositol 3-kinase (PI 3-kinase) activation. Indeed, whereas Btk activation is not significantly enhanced by co-expressing a weakly activating form of Src (E378G) with wild-type Btk in fibroblasts, co-expression of the two subunits of PI 3-kinase-$\gamma$ in the same cells results in synergistic activation of Btk, as determined by increased phosphotyrosine content and the ability to mediate cellular transformation (LI et al. 1997). These studies indicate that the production of 3′-phosphorylated products by PI 3-kinase is important for targeting Btk to the membrane where Src activates it.

## 4 Phosphorylation of Intracellular Effector Proteins by PTKs

Significant progress is being made towards elucidation of the physiological role that individual PTKs serve in the B-cell activation response. The ultimate goal of these studies is to identify the specific set of effector proteins that is phosphorylated by each PTK involved in BCR-mediated signaling. Nevertheless, this task has been and continues to be difficult, due to several factors that must be considered when trying to determine whether a particular effector protein is directly phosphorylated by a given PTK. First, it is clear that multiple PTKs are activated in response to cross-linking of the BCR (CAMBIER et al. 1994; RETH and WIENANDS 1997). Perhaps the most confounding issue in this regard is the fact that several Src family PTKs can associate with the BCR complex and are inducibly activated in response to ligand binding (BURKHARDT et al. 1991; LI et al. 1992). It is therefore important to determine whether the individual members of the Src family perform unique functions during B-cell activation or whether they are functionally redundant. Indeed, these two ends of the spectrum are not mutually exclusive. For example, it is quite possible that a given Src family kinase could perform a unique function(s) that is not duplicated by other members of the family while, at the same time, that PTK may perform a redundant function that is common to other members of the family. Similarly, it is possible that PTKs from more than one class might be involved in directly regulating the phosphorylation state and function of a given effector protein. This has been demonstrated to be the case for PLCγ, which is regulated by both Syk and Btk (TAKATA et al. 1994; TAKATA and KUROSAKI 1996).

The second issue that must be considered when establishing the relationship between a PTK and a putative substrate is the fact that one class of PTKs may regulate the catalytic activity of another. As previously discussed, the Src family PTKs as a group are involved in regulating the activation of both Syk and Btk (KUROSAKI et al. 1994; RAWLINGS et al. 1996). Therefore, elimination or alteration of Src family PTK function could exert either a direct or indirect effect, or both, on the phosphorylation of downstream effector proteins.

Clearly, changes in the phosphorylation of effector proteins that are direct substrates for Src kinases would be expected. Alternatively, changes in substrate phosphorylation could also result from an indirect process in which downstream PTKs such as Syk or Btk are dysregulated. This could, in turn, alter the phosphorylation status of effector proteins that are direct substrates of downstream PTKs, but not the Src family kinase. A related scenario could also be envisioned in which alteration of PTK function might affect the function of a PTP, thereby leading indirectly to changes in effector protein phosphorylation. Finally, it is well documented that all PTKs possess structural/functional domains in addition to their catalytic domain. The Src family PTKs contain a unique amino-terminal domain, an SH3 domain and a SH2 domain, whereas Syk contains two SH2 domains. Btk contains PH, TH, SH2 and SH3 domains in addition to its Src-like SH1 (catalytic) domain. Theoretically, these non-catalytic functional domains could be involved in targeting a specific PTK to a subcellular region in order to facilitate its activation and/or ability to phosphorylate substrates. Alternatively, the same functional domains could be involved in targeting effector proteins to the kinase. This might facilitate the ability of the PTK to phosphorylate that effector protein directly, or it could target that protein to a microenvironment in the cell where other PTKs exist. Therefore, each PTK has the ability to regulate the phosphorylation status of effector proteins either directly via its catalytic domain or indirectly by virtue of its ability to interact with a substrate via one or more functional domains. Because of the complexities associated with the interpretation of experiments to identify PTK substrates, it is important to utilize multiple strategies to confirm the direct relationship between a PTK and a putative substrate.

An in-depth discussion of effector protein phosphorylation is clearly beyond the scope of this review. However, a brief discussion of the evidence that is available concerning the role of PTKs in regulating the phosphorylation of key effector proteins responsible for BCR-mediated signal transduction will be discussed. In general, it is now thought that activation of the Src family PTKs is responsible for mediating the activation-dependent phosphorylation of CD79a and CD79b (CAMBIER et al. 1994; RETH and WIENANDS 1997). Additionally, members of this class of PTKs are involved in phosphorylation of Syk and Btk, thereby potentiating their catalytic activity (KUROSAKI et al. 1994; RAWLINGS et al. 1996) (Fig. 2). Currently, studies have not definitively established whether specific Src family PTKs are primarily responsible for mediating CD79a, CD79b, Syk and Btk phosphorylation. Evidence from studies of gene-targeted mice and the DT40 chicken B-cell line that lacks specific Src family kinases suggest that members of the Src family may be functionally redundant in terms of their ability to initiate and then propagate signaling via the BCR (SILLMAN and MONROE 1994; TAKATA et al. 1994; APPLEBY et al. 1995; HIBBS et al. 1995; CHAN et al. 1997). Nevertheless, it should also be noted that Lyn, for example, appears to serve a unique function in the B cell in terms of its role in phosphorylating proteins (e.g. CD22) that are involved in attenuation of BCR signaling (CHAN et al. 1997; NISHIZUMI et al. 1998; SMITH et al. 1998).

The proximal activation of Src family kinases, Syk and Btk results in the activation of at least three major downstream signaling pathways that regulate $Ca^{2+}$ mobilization, Ras/Raf function and Rho-GTPase function through tyrosine phosphorylation of PLCγ, Shc and Vav, respectively (DEFRANCO 1997; KUROSAKI 1997) (Fig. 2). The PLCγ-dependent signaling pathway is responsible for mediating the subsequent production of second messengers (i.e. DAG and IP3) that regulate PKC activity and $Ca^{2+}$ mobilization, leading to nuclear translocation of the transcription factor NF-AT (RAO et al. 1997). The Shc-dependent pathway mediates recruitment of the Grb2/Sos complex to the membrane where Sos acts as a guanine nucleotide exchange factor (GEF) that promotes activation of the Ras/Raf pathway leading to activation of ERK1/2 MAPKs (GOLD et al. 1992; HARWOOD and CAMBIER 1993; LAZARUS et al. 1993; SUTHERLAND et al. 1996). These MAPKs are involved in regulating the function of the transcription factors Elk1 and Sap-1a (SU and KARIN 1996; JANKNECHT and HUNTER 1997). Finally, activation of the GEF Vav is involved in regulating Rac function (CRESPO et al. 1997). This effector plays a role in regulating cytoskeletal rearrangement and the function of JNK and p38 MAPKs. The JNK and p38 MAPKs regulate the activation of Jun and Atf2 transcription factors (SU and KARIN 1996; TREISMAN 1996; KARIN et al. 1997). As previously discussed, the combinatorial activation of

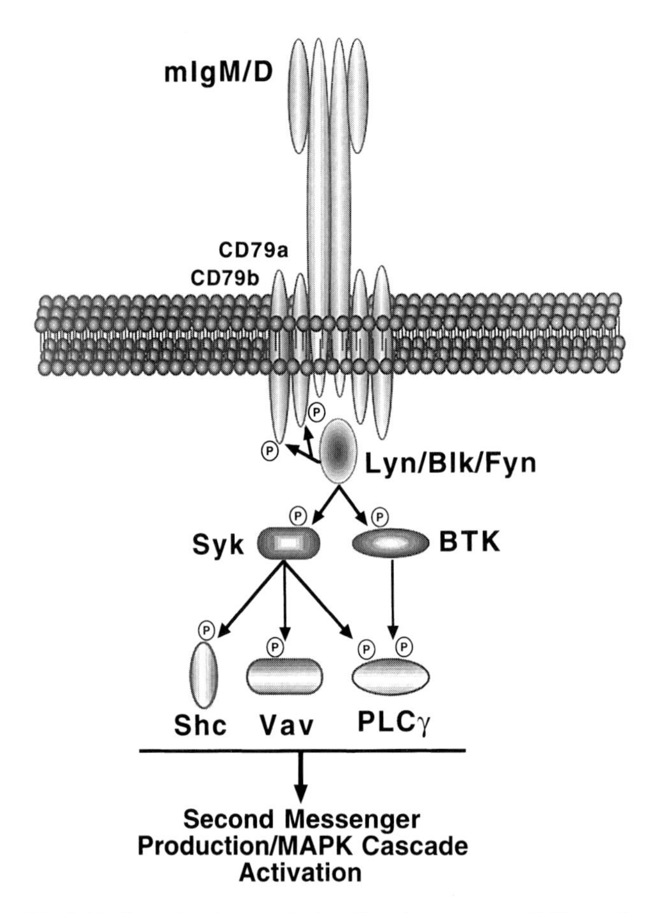


**Fig. 2.** Binding of antigen to the B-cell antigen receptor (BCR) mediates the activation of multiple protein tyrosine kinases (PTKs). Three families of PTK are activated in response to cross-linking of the BCR including the Src family PTKs, Syk and the Tec family kinase Bruton's tyrosine kinase (Btk). Initiation of signal transduction is mediated by the Src family kinases, which phosphorylate the immunoreceptor tyrosine-based activation motifs of CD79a and CD79b promoting recruitment of additional Src kinases, the PTK Syk and other effector proteins such as Shc. The Src family kinases also play an important role in regulating the activation of downstream PTKs, including Syk and Btk, which propagate signal transduction. Syk and BTK act in concert to promote optimal activation of phospholipase C (PLC)γ. In addition, Syk is involved in phosphorylating and regulating the activity of Shc and Vav. Thus, the concerted action of three distinct families of PTK is required to promote optimal phosphorylation of effector proteins that mediate second-messenger production and regulate that activation of mitogen-activated protein kinase cascades

transcription factors has a direct effect on the biological response of the B cell (Dolmetsch et al. 1997; Healy et al. 1997). Thus, it is clear that understanding the way in which tyrosine phosphorylation of PLCγ, Shc and Vav is controlled is critical for elucidating the overall molecular mechanism responsible for determining the biological outcome of BCR cross-linking.

## 4.1 Regulation of PLCγ Phosphorylation and Activation

The initial activation of PTKs is responsible for mediating the production of second messengers that subsequently regulate intermediate signaling processes leading to transcription factor activation. Studies have elegantly demonstrated that PLCγ is activated in response to BCR cross-linking and is responsible for the production of diacylglycerol and $IP_3$ (CARTER et al. 1991; COGGESHALL et al. 1992; HEMPEL et al. 1992). These in turn promote PKC activation and mobilization of $Ca^{2+}$ (DEFRANCO 1997). PLCγ activation is correlated with its phosphorylation on three tyrosine residues (LEE and RHEE 1995) and it has been shown that the PTK Syk is required for the phosphorylation of PLCγ (TAKATA et al. 1994). Studies using DT40 chicken B-cell lines that are genetically deficient for Lyn or Syk demonstrated that tyrosine phosphorylation of PLCγ and the production of $IP_3$ are dependent on Syk, but not Lyn (TAKATA et al. 1994). Moreover, studies have demonstrated that the SH2 domain of PLCγ can bind to phosphotyrosine residues located within the linker domain of Syk, between the tandem SH2 domains of the kinase (SILLMAN and MONROE 1995). This physical interaction between PLCγ and phospho-Syk may be involved in potentiating the ability of Syk to phosphorylate PLCγ in an activation-dependent process.

It is interesting to note that studies using the DT40 chicken B-cell line indicate that Syk activation may be potentiated by, but is not absolutely dependent on, phosphorylation by Src family kinases (TAKATA et al. 1994). Subsequent experiments using additional variants of the DT40 B-cell line have demonstrated that Lyn and Syk are responsible for mediating the majority of inducible tyrosine phosphorylation in response to cross-linking of the BCR (TAKATA et al. 1994). In contrast, cells deficient in Btk exhibited little or no significant alteration in the overall pattern of inducible protein tyrosine phosphorylation (TAKATA and KUROSAKI 1996). These studies essentially demonstrated that the activity of Lyn and Syk are comparable in the absence of Btk and that Btk is not critical for initiation of inducible protein tyrosine phosphorylation after cross-linking of the BCR. Nevertheless, it was noted that inducible phosphorylation of PLCγ in Btk-deficient mice was significantly decreased, approximately threefold, and that this was coincident with a dramatic impairment of the $Ca^{2+}$ mobilization response in these cells (TAKATA and KUROSAKI 1996).

Based on numerous findings, it has been concluded that Syk and Btk act in concert to regulate the phosphorylation and activity of PLCγ (TAKATA and KUROSAKI 1996; DESIDERIO 1997; FLUCKIGER et al. 1998). One possibility is that these PTKs target distinct tyrosine residues on PLCγ and that both Syk and Btk are required to mediate optimal phosphorylation and activation of PLCγ. Studies in the mouse and human are consistent with those described for the chicken in that Btk expression is essential for BCR-mediated $Ca^{2+}$ mobilization (WICKER and SCHER 1986; RIGLEY et al. 1989). Expression of Btk has been shown to restore calcium signaling in Btk-deficient XLA B cells. Moreover, the ability of Btk to reconstitute signaling is dependent on its catalytic function, SH2 and PH domains and the Btk activation loop tyrosine (Y551) that is phosphorylated by Src family

PTKs (Fluckiger et al. 1998). Interestingly, overexpression of Syk was not observed to restore calcium signaling in XLA B cells confirming the hypothesis that these PTKs cannot compensate for one another and that they must act in concert to regulate PLCγ function (Desiderio 1997; Fluckiger et al. 1998). The ability of Btk to restore $Ca^{2+}$ flux in XLA B cells was consistent with its ability to mediate phosphorylation of PLCγ. Mutation of the Src PTK transphosphorylation site, the ATP-binding site or the SH2 domain of Btk abrogated Btk-dependent phosphorylation of PLCγ (Takata and Kurosaki 1996; Fluckiger et al. 1998). The requirement for the SH2 domain of Btk is suggestive of the fact that an interaction between phosphorylated PLCγ and the SH2 domain could stabilize the interaction between these proteins, perhaps facilitating additional transphosphorylation of PLCγ.

## 4.2 Regulation of Shc Phosphorylation and Function

As previously discussed, ligation of the BCR leads to inducible phosphorylation of CD79a and CD79b as well as the accumulation of GTP-bound, activated Ras (Harwood and Cambier 1993; Lazarus et al. 1993). Studies have shown that the mechanism by which the phosphorylated receptor is coupled to Ras activation involves the recruitment of Shc and its subsequent tyrosine phosphorylation (Saxton et al. 1994; Smit et al. 1994). Although Shc is observed to associate with the non-phosphorylated BCR complex via a mechanism that involves the -DCSM- sequence within the ITAM of CD79a, it is also inducibly recruited to the BCR complex (D'Ambrosio et al. 1996). Activation-dependent recruitment of Shc occurs via a phosphotyrosine-based interaction between the SH2 domain of Shc and phosphotyrosine residues contained within the ITAMs of CD79a and CD79b (D'Ambrosio et al. 1996). Binding to CD79a and CD79b in response to BCR cross-linking promotes inducible tyrosine phosphorylation of Shc resulting in the subsequent recruitment of Grb2, which binds to phospho-Shc via its SH2 domain (Saxton et al. 1994; Smit et al. 1994). Grb2 is constitutively associated with Sos, a GEF, through binding of its SH3 domain to proline-rich sequences in Sos (Saxton et al. 1994). Thus, the recruitment of Grb2 to the BCR activation complex localizes Sos at the plasma membrane where it promotes the exchange of GDP bound to Ras with GTP. This leads to accumulation of Ras bound to GTP and activation of the Ras/Raf pathway.

Experiments using the DT40 chicken B-cell line have demonstrated that BCR-mediated phosphorylation of Shc and its association with Grb2 are significantly attenuated in cells that lack either Lyn or Syk (Nagai et al. 1995). Additional studies revealed that the catalytic function of both kinases is essential for optimal Shc phosphorylation. The requirement that both PTKs be catalytically active raises the question of whether one or both kinases are responsible for mediating direct phosphorylation of Shc. Interestingly, Shc was observed to physically interact with Syk in a phosphorylation-independent manner and co-transfection of COS cells with Syk and Shc resulted in significant tyrosine phosphorylation of Shc (Nagai et al. 1995). These results suggest that Src family PTKs, such as Lyn, may be required for optimal phosphorylation of Shc, not because they phosphorylate Shc directly, but because they are important for phosphorylation and activation of Syk, which then mediates tyrosine phosphorylation of Shc. These data further suggest that recruitment of Shc to the BCR complex may be mediated through its association with Syk and/or through binding of Shc to the phosphorylated CD79a/b heterodimer. As previously discussed, Shc does indeed bind to phospho-CD79a/b (D'Ambrosio et al. 1996), indicating that Lyn or other Src family PTKs enhance Shc phosphorylation indirectly through the phosphorylation of CD79a and CD79b, which then effectively co-recruit Shc and Syk to the BCR activation complex.

### 4.3 Regulation of Vav Phosphorylation and Function

Vav expression has been shown to be critical for normal lymphocyte development presumably because it serves an essential role in antigen receptor-mediated signal transduction (Bustelo and Barbacid 1992; Gulbins et al. 1994; Tarakhovsky et al. 1995; Zhang et al. 1995). This has been confirmed by analyzing BCR-dependent signaling using B cells isolated from Vav-deficient mice (Tarakhovsky et al. 1995; Zhang et al. 1995). When compared to B cells isolated from wild-type animals, Vav-deficient B cells exhibit greatly reduced proliferative responses after antigen receptor engagement. The Vav proto-oncogene product contains several functional domains including an SH2 domain, two SH3 domains, a PH domain, a Dbl homology (DH) domain, a leucine-rich region and a cysteine-rich region similar to the lipid-binding domain of PKC (Bonnefoy-Berard et al. 1996). Due to the presence of these functional domains, it is possible that Vav regulates signaling via the BCR through one or more distinct processes.

It has been shown that Vav is inducibly phosphorylated on tyrosine residues in response to BCR cross-linking (Bustelo and Barbacid 1992; Gulbins et al. 1994). The yeast two-hybrid system has been used to demonstrate that Vav physically interacts with the PTK Syk (Deckert et al. 1996). The association between these proteins requires the catalytic activity of Syk, the SH2 domain of Vav and specific tyrosine residues located in the linker domain of Syk. Thus, it is apparent that Vav is recruited to phosphorylated Syk via its SH2 domain (Deckert et al. 1996). Presumably, the physical interaction between Syk and Vav is involved in promoting Vav phosphorylation by Syk. Evidence to support this has been presented in the form of experiments demonstrating that Syk, both in vivo and in vitro, phosphorylates Vav (Deckert et al. 1996). Finally, it has been shown that Vav and Syk cooperate with one another to activate NF-AT in a synergistic manner.

# 5 Regulation of B-Cell Development and Activation by the PTP CD45

As previously discussed, the BCR complex plays an essential role in determining the fate of the B cell throughout its development and differentiation. It is now appreciated that the functional consequences of an encounter with antigen are to a great extent determined by the strength of the signal delivered via the BCR. Alterations in the intrinsic signaling threshold within an individual B cell, in conjunction with extrinsic factors provided by other immune cells or the microenvironment in which the B cell resides, determine whether it is tolerized or activated (Goodnow 1996). Therefore, it is essential to delineate the molecular mechanisms that regulate reversible tyrosine phosphorylation, because the signaling threshold for any given cell is ultimately determined by the homeostasis between phosphorylation and dephosphorylation of key regulatory proteins that control second-messenger production and transcription factor activation (Dolmetsch et al. 1997; Healy et al. 1997). CD45 is a transmembrane PTP that plays a critical role in regulating lymphocyte development and activation (Justement 1997). Analysis of CD45-deficient cell lines and cells derived from gene targeted mice, has demonstrated that CD45 is involved in regulating reversible tyrosine phosphorylation and, thus, the intrinsic sensitivity of the BCR to a given antigenic challenge. Delineation of the specific mechanism(s) by which CD45 does so is an important area of research because the results obtained will yield insight into processes that are associated with regulation of signaling thresholds that maintain the balance between tolerance and immunity.

## 5.1 CD45 Regulates the Threshold of Signaling via the BCR

The generation of CD45-deficient mice through homologous recombination has confirmed the important role that this PTP serves in regulation of lymphocyte development and activation by virtue of its ability to regulate signaling via the BCR (Justement 1997). Initial studies were reported with mice carrying a mutation in CD45 variable exon 6, resulting in a complete loss of CD45 expression by B cells and a significant reduction in its expression on peripheral T cells (Kishihara et al. 1993). Examination of the B-cell compartment in CD45-exon6$^{-/-}$ mice revealed little or no effect on colony formation of bone marrow-derived myeloid or B-cell progenitors. Nor was there any difference in the number of colony forming B cells isolated from spleen, bone marrow or peritoneum (Kishihara et al. 1993). Although the number of B cells was not decreased in CD45-exon6$^{-/-}$ mice, they did exhibit certain phenotypic alterations suggesting that development within the B-cell compartment is affected by the loss of CD45 expression. Both the high- and low-density splenic B-cell populations exhibited a significant decline in the frequency of $IgD^{hi}$, $IgM^{lo}$ cells in addition to a significant decline in the number of B cells expressing CD23 and MHC class II (BENATAR et al. 1996). In contrast, the phenotype of bone marrow B cells was essentially normal. These alterations in B-cell phenotype suggest that the loss of CD45 causes a developmental arrest at the transition from immature to mature B cells. Analysis of the proliferative response following AgR cross-linking demonstrated that CD45-deficient B cells are unable to proliferate when stimulated with either polyclonal or monoclonal anti-IgM, whereas the response to lipopolysaccharide (LPS) is relatively unaffected (KISHIHARA et al. 1993; BENATAR et al. 1996).

More recently CD45-null mice have been generated through the targeted disruption of exon 9 within the CD45 locus (BYTH et al. 1996). In contrast to the CD45-exon6$^{-/-}$ knockout mice, these animals exhibit a complete loss of CD45 expression on both T cells and B cells. The spleens from CD45-exon9$^{-/-}$ mice contained twice the number of B cells and only one fifth as many T cells when compared with wild-type animals. The increased number of B cells in the spleen was due to expansion of two B-cell subpopulations that express high levels of IgM, indicating that CD45 may regulate the generation of specific B-cell subpopulations. As in the CD45-exon6$^{-/-}$ mice, BCR-mediated proliferation of B cells isolated from CD45-exon9$^{-/-}$ mice was absolutely dependent on the expression of CD45 (BYTH et al. 1996). Neither polyclonal anti-IgM or anti-IgD antibody was observed to induce cellular proliferation of B cells, whereas the response to anti-CD40 mAb was relatively unaffected. Moreover, the response to anti-CD40 and IL-4 was normal based on upregulation of MHC class II expression, indicating that CD45 is not involved in regulating signal transduction via these receptors. Finally, proliferation of B cells stimulated with anti-CD38 mAb was analyzed, revealing a significant impairment in responsiveness associated with the loss of CD45 expression. Because signal transduction via CD38 requires the expression of a functional BCR complex (LUND et al. 1996), and may in fact utilize components of the BCR complex, it is not entirely surprising to find that CD38 function is compromised in CD45-deficient cells.

Although intrinsic defects in BCR-mediated signaling in CD45-deficient B cells may be compensated by extrinsic help provided by CD45$^{+/+}$ T cells, there is substantial evidence to support the conclusion that CD45 does indeed play an important role in setting the signaling threshold required to trigger B cells via the BCR. The role of CD45 in regulation of B-cell-positive and -negative selection has been assessed by crossing CD45-exon6$^{-/-}$ mice with mice carrying immunoglobulin transgenes specific for hen egg lysozyme (HEL) (CYSTER et al. 1996). Naive B cells isolated from these animals were examined for their ability to respond to soluble HEL in vitro. Activation of the ERK/RSK/EGR1 signaling cascade was significantly diminished in CD45-deficient B cells, as was the $Ca^{2+}$ mobilization response when compared with control B cells. Antigenic stimulation of CD45-deficient B cells resulted in decreased expression of CD86 and little or no increase in the expression of CD54. The essential conclusion from these experiments is that expression of CD45, although not absolutely required for initiation and propagation of signaling via the BCR, does play an important role in regulating the intensity of the signal and, thus, the threshold of sensitivity of the BCR for antigen.

It was further shown that alterations in BCR sensitivity have practical consequences during the selection of high- and low-affinity self-reactive B cells. Whereas high affinity self-reactive B cells were eliminated from the bone marrow of both CD45$^{+/+}$ and CD45$^{-/-}$ mice, loss of CD45 expression actually potentiated positive selection of low affinity auto-reactive B cells into the recirculating B-cell repertoire (Cyster et al. 1996). At a molecular level, CD45-deficient B cells that have been chronically exposed to self-antigen and are therefore "tolerant" do not exhibit transient $Ca^{2+}$ oscillations characteristically seen in CD45-positive self-reactive B cells, nor do they activate ERK/pp90rsk (Dolmetsch et al. 1997). Presumably, the inability of CD45-deficient B cells to activate ERK/pp90rsk and possibly NF-AT protects them from being eliminated from the bone marrow in response to self-antigen. Thus, it is apparent that CD45 is directly involved in regulating the BCR signal-threshold and therefore determines whether B cells undergo negative or positive selection.

## 5.2 Analysis of Molecular Processes Altered in the Absence of CD45 Expression

The loss of CD45 expression in either B cell lines or in cells derived from genetically deficient mice leads to an alteration in the signal-transduction response that is in many ways similar to that observed in anergic B cells. Thus, an analysis of the role that CD45 plays in regulating signaling should yield important insight into the mechanisms responsible for antigen-mediated tuning of the signaling threshold for the BCR.

### 5.2.1 CD45 Regulates the Ability of B Cells to Respond to T-Dependent Antigen In Vivo

Studies have utilized an anti-CD45 mAb (RA3.6B2) that recognizes a B-cell restricted epitope to analyze the role of this PTP in regulating B-cell activation in vivo. Administration of anti-CD45 mAb was found to abrogate a T-cell-dependent antibody response against the hapten fluorescein isothiocyanate (FITC) (Domiati-Saad et al. 1993). In these studies, both the plaque-forming cell response and serum antibody production were measured and were significantly inhibited by a single dose of anti-CD45 mAb. The effect of antibody administration was transient in that once the mAb had been cleared from the system, anti-FITC responses could be detected. A kinetic analysis of mAb administration provided data to suggest that the anti-CD45 mAb was exerting an effect on B-cell activation, thereby preventing cells from responding to antigen and undergoing subsequent differentiation. Administration of anti-CD45 mAb did not affect the secondary response following re-challenge with antigen (Domiati-Saad et al. 1993). Therefore, it is possible that memory B cells have lost the specific epitope recognized by the anti-CD45 mAb used in this study due to isoform switching. Alternatively, the response to antigen by B cells in the memory pool may involve a qualitatively distinct BCR-mediated signal transduction process that is not regulated by CD45 in a manner analogous to that of resting naive B cells. Finally, based on in vitro studies, it is possible that memory B cells are competent to receive accessory signals that override the negative effects elicited by ligation of CD45 (Gruber et al. 1989).

### 5.2.2 CD45 Exists in a Multi-Molecular Complex with the BCR

To date, the investigation of intermolecular associations between CD45 and other transmembrane proteins in the B cell has been relatively limited in scope. Nevertheless, it has been shown that CD45 interacts with both mIgM and mIgD isolated from resting splenic B cells based on the reciprocal co-precipitation of these proteins (Brown et al. 1994). Furthermore, immunoprecipitation of CD45 from detergent lysates prepared from B cells followed by in vitro radiolabeling of immune-complex proteins revealed the presence of tyrosine phosphorylated proteins with molecular weights corresponding to the CD79a and CD79b subunits of the mIg-associated heterodimer. The presence of the CD79a/CD79b heterodimer in the CD45 immune complex was confirmed by secondary immunoprecipitation with antibodies specific for these subunits (Brown et al. 1994). Although this study did not directly address the specific nature of the interaction between CD45 and the BCR complex, it provides evidence to support the conclusion that CD45 interacts with the BCR complex in an activation-independent manner. The physical association between the BCR complex and CD45 raises the possibility that the basal tyrosine phosphorylation state of the Igα/Igβ heterodimer may be regulated by CD45 in a manner similar to that proposed for the ζ chain in T cells (Furukawa et al. 1994).

In the B cell, evidence has accumulated indicating that there is a physical association between CD45 and the PTK Lyn (Brown et al. 1994; Burg et al. 1994) but not other src family PTKs, such as Fyn or Blk. Isolation of CD45 immune-complex material from detergent lysates of resting splenic B cells followed by radiolabeling of proteins using an in vitro kinase assay revealed that CD45 is constitutively associated with one or more PTKs (Brown et al. 1994). Subsequent identification of CD45-associated PTKs using a panel of anti-Src family kinase antibodies indicated that Lyn co-precipitates with CD45 in a selective manner. Further analysis demonstrated that a GST:Lyn fusion protein could be used to immunoprecipitate CD45 from thymocyte lysates indicating that Lyn does not require components of the BCR complex in order to associate with CD45. Because Lyn constitutively associates with CD45 in lysates prepared from resting B cells or from unstimulated thymocytes, it is apparent that this interaction is not dependent on inducible tyrosine phosphorylation of CD45 or Lyn (Brown et al. 1994). However, the results from another study demonstrate the rapid induction of Lyn binding to CD45 in response to BCR cross-linking (Burg et al. 1994). Thus, it is possible that Lyn may associate with CD45 via more than one mechanism and the association between these proteins may be potentiated by tyrosine phosphorylation of Lyn, CD45 or other intermediate proteins.

### 5.2.3 CD45 Regulates the Phosphorylation State of CD79a and CD79b

Support for the hypothesis that CD45 is involved in regulating the phosphorylation state of the BCR complex has been provided by studies demonstrating that physical sequestration of CD45 within the plasma membrane leads to increased phosphorylation of CD79a and CD79b (LIN et al. 1992). Incubation of splenic B cells with anti-CD45 mAb alone did not induce CD45 to associate with the cytoskeleton or to form caps within the plasma membrane. Under these conditions, little or no change in the phosphorylation of CD79a or CD79b was observed. However, incubation of B cells with anti-CD45 mAb and a secondary cross-linking reagent induced the rapid association of CD45 with the cytoskeleton and caused CD45 to form a cap within the plasma membrane. These events were associated with a significant increase in the tyrosine phosphorylation of CD79a and CD79b. Kinetic analysis of heterodimer phosphorylation revealed a correlation between this event and the formation of an intermolecular association between CD45 and the cytoskeleton (LIN et al. 1992). These results suggest that the activity of CD45 may be regulated by its interaction with cytoskeletal elements and/or physical redistribution within the membrane. Recent findings have suggested an additional mechanism by which CD45 function may be regulated. In these studies, dimerization of an EGF:CD45 chimeric protein results in inhibition of CD45 activity due to the reciprocal insertion of a wedge domain from one catalytic subunit into the substrate binding pocket of another (MAJETI et al. 1998). The reciprocal insertion of wedge domains into catalytic domains within the homodimer presumably results in steric inhibition of substrate binding.

Further support for the hypothesis that CD79a and CD79b are substrates for CD45 comes from studies of the CD45-deficient B-cell plasmacytoma J558μm3 and a CD45-positive transfectant. Analysis of basal as well as inducible phosphorylation of CD79a and CD79b revealed that the heterodimer exhibits elevated basal tyrosine phosphorylation in CD45-deficient cells compared with CD45-positive transfectants (JUSTEMENT et al. 1991; PAO and CAMBIER 1997). Moreover, antigen stimulation resulted in hyperphosphorylation of CD79a and CD79b in CD45-deficient cells, indicating that CD45 may actually regulate basal as well as inducible phosphorylation of these polypeptides. Similar findings were obtained in studies of the K46–17μmλ cell line and CD45-deficient variants (S.F. Greer and L.B. Justement, unpublished observations). In contrast to these observations, B cells isolated from CD45-exon6$^{-/-}$ mice exhibited basal and inducible phosphorylation of the heterodimer that was comparable to B cells isolated from CD45-exon6$^{+/-}$ control mice (BENATAR et al. 1996).

### 5.2.4 CD45 Regulates the Function of Src Family PTKs

Numerous studies have demonstrated that activation and recruitment of the Src family PTKs play an important role in signal initiation and propagation in response to antigen (CAMBIER et al. 1994). Until recently, very little direct evidence was available from studies on the B-cell documenting regulation of Src kinase function by CD45. However, studies have revealed that Lyn isolated from the CD45-negative J558μm3 cell line is hyperphosphorylated in the absence of stimulus and does not exhibit increased tyrosine phosphorylation above this basal level in response to BCR ligation (PAO and CAMBIER 1997). Moreover, Lyn is not recruited to the BCR complex and does not exhibit appreciable catalytic activity, nor were phosphopeptides derived from CD79a observed to bind to Lyn isolated from CD45-negative cells, suggesting that the kinase is in an inactive conformation. This hypothesis was supported by phosphopeptide analysis of Lyn isolated from CD45-deficient cells, which revealed increased phosphorylation of the carboxyl-terminal inhibitory tyrosine residue 508 (PAO and CAMBIER 1997). A more recent study documenting a role for CD45 in regulation of Src kinase function utilized the DT40 chicken B-cell line in which the gene for CD45 has been disrupted by gene-targeting to create a CD45 knockout cell line (YANAGI et al. 1996). Analysis of Lyn from these cells revealed that both autophosphorylation and carboxyl-terminal inhibitory tyrosines are hyperphosphorylated in the absence of CD45. Additionally, Lyn activation was profoundly inhibited in the CD45 knockout B cells. Although previous studies demonstrated a selective interaction between CD45 and the PTK Lyn, the activity of other Src family PTKs appears to be dependent on the expression of CD45. BCR-mediated inducible activation of both Fyn and Blk was significantly decreased in the CD45-deficient J558μm3 cell line when compared to CD45-positive transfectants (PAO et al. 1997). Finally, ligation of either the BCR or MHC class II molecules on CD45-positive K46–17μmλ B lymphoma cells caused a two- to threefold increase in the activity of Lyn based on phosphorylation of the exogenous substrate enolase. In contrast, cross-linking of either the BCR or MHC class II failed to activate Lyn in either of two CD45-deficient clones (3S5 and 35S5) derived from the K46–17μmλ cell line (GREER et al. 1998). These data indicate once again that expression of CD45 is required for activation of Lyn, as well as other Src family kinases, in response to stimulation of B cells via either the BCR or MHC class II (Fig. 3).

### 5.2.5 CD45 Expression Is Not Required for Activation of the PTK Syk

Whereas the requirement for CD45 expression is tightly correlated with successful initiation of signal transduction following T cell AgR cross-linking, the expression of this PTP is not absolutely required to trigger certain early activation events following BCR ligation (KIM et al. 1993b; JUSTEMENT 1997). Cross-linking of the BCR expressed by CD45-deficient J558μm3 cells or 3S5 and 35S5 clones of the K46 B cell lymphoma leads to inducible tyrosine phosphorylation of numerous substrates (KIM et al. 1993b; PAO et al. 1997; GREER et al. 1998). It should be noted that both qualitative and quantitative differences in basal and stimulated tyrosine phosphorylation patterns are observed. Based on these findings and those in other CD45-deficient cell lines, it is apparent that CD45 expression is not absolutely required for coupling of the BCR to its signal-transduction pathway. Moreover, these results suggest that even though CD45 is required for activation of multiple Src family PTKs in the B cell, there must be other PTKs whose function is

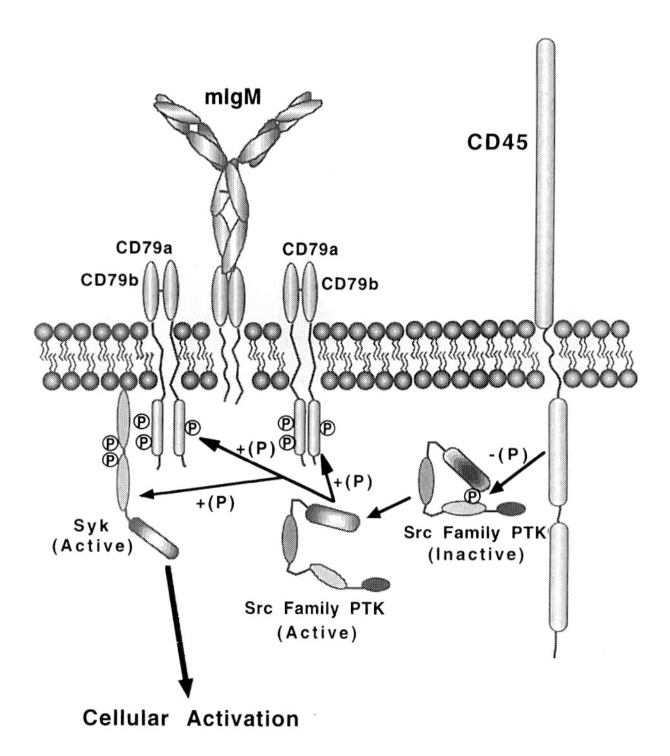


**Fig. 3.** CD45 promotes signal transduction via the B-cell antigen receptor (BCR) by regulating the activation of Src family protein tyrosine kinases. CD45 is responsible for dephosphorylating the carboxyl-terminal inhibitory tyrosine residue of Src kinases. Dephosphorylation of the carboxyl-terminal tyrosine by CD45 disrupts the inhibitory intramolecular association that is formed between the SH2 domain and the carboxyl-terminal phosphotyrosine motif within a given kinase. This in turn promotes recruitment of Src kinases to phosphorylated immunoreceptor tyrosine-based activation motifs within the cytoplasmic domains of CD79a and CD79b. Recruitment of Src kinases to CD79a and CD79b via their SH2 domains focuses the kinases to the activation complex where they are able to transphosphorylate stimulatory tyrosine residues resulting in optimal activation. The localization of Src kinases to the BCR activation complex also facilitates their ability to phosphorylate downstream substrates

essentially normal. Perhaps the most surprising finding from these studies was that CD79a and CD79b are inducibly phosphorylated and recruit Syk, in spite of the fact that Lyn, Fyn and Blk appear to be catalytically inactive in the absence of CD45 (Pao and Cambier 1997; Greer et al. 1998; Yanagi et al. 1996). Thus, it is possible that Syk is able to phosphorylate the BCR complex following receptor ligation in the absence of Src family kinase activation. Alternatively, one or more PTKs distinct from Syk may exist in the B cell that are inducibly activated and phosphorylate CD79a and CD79b in the absence of CD45 expression.

In agreement with the hypothesis that Syk activation can occur in the absence of CD45, Syk phosphorylation and activity are not significantly altered in CD45-deficient cells when compared with CD45-positive cells (Yanagi et al. 1996). Based on an analysis of putative Syk substrate phosphorylation in response to cross-linking of either the BCR or class II in CD45-positive K46–17μmλ B cells and the

CD45-deficient variants 3S5 and 35S5, it appears that this PTK is constitutively active in the absence of CD45. In agreement with this finding, previous studies using the J558μm3 plasmacytoma demonstrated that Syk is approximately three- to fourfold more active in unstimulated cells that lack CD45 than it is in CD45-positive transfectants (PAO and CAMBIER 1997). Therefore, it is possible that the loss of CD45 expression may result in aberrant constitutive activation of the PTK Syk.

Analysis of PLCγ phosphorylation in the K46–17μmλ cell line and its CD45-deficient variants revealed that it is constitutively phosphorylated on tyrosine in the absence of CD45, in agreement with the possibility that Syk may be constitutively active in the absence of this PTP (GREER et al. 1998). Additionally, it has been shown that putative endogenous substrates for Syk are hyperphosphorylated in the CD45 deficient variants of K46–17μmλ (GREER et al. 1998; GREER and JUSTEMENT 1999). This finding is in contrast to previous studies using B cells isolated from CD45-deficient mice, in which apparently normal inducible tyrosine phosphorylation of PLCγ was observed in response to cross-linking of the BCR (BENATAR et al. 1996). Moreover, experiments with the CD45-negative J558μm3 cell line revealed normal inducible tyrosine phosphorylation of PLCγ1 and PLCγ2 even though the basal activity of Syk was elevated in these cells compared with CD45-positive J558μm3 transfectants (PAO and CAMBIER 1997; PAO et al. 1997).

Although PLCγ exhibits high basal phosphorylation in the CD45-deficient 35S5 cells derived from the K46–17μmλ B cell lymphoma, subsequent studies to measure $IP_3$ production demonstrated that PLCγ is indeed functional and that its activity is increased upon cross-linking of the BCR or MHC class II. In agreement with the observation that PLCγ is constitutively hyperphosphorylated in the CD45-negative clones, basal levels of $IP_3$ were 1.5 to 2-fold higher in the 35S5 cells when compared with parental K46–17μmλ cells (GREER et al. 1998). Moreover, the level of $IP_3$ present in CD45-deficient cells stimulated via the BCR or MHC class II was elevated at all time points assayed. Previous studies with the CD45-deficient J558μm3 plasmacytoma have demonstrated that cross-linking of the BCR mediates increased production of $IP_3$ in the absence of CD45 (PAO et al. 1997). However, because the basal level of $IP_3$ in the J558μm3 cells was only a fraction of that in the CD45-positive transfectants, the maximum level of $IP_3$ produced in response to cross-linking of the BCR was not observed to surpass the baseline level of $IP_3$ present in the CD45-positive transfectants (PAO et al. 1997). Although there is general agreement that cross-linking of the BCR, as well as MHC class II, leads to increased production of $IP_3$ in the absence of CD45, there is a discrepancy between the 35S5 and J558μm3 cell lines as to whether the net level of $IP_3$ produced upon stimulation is sufficient to mediate downstream signaling. Similar studies have not been performed with B cells isolated from CD45exon6$^{-/-}$ or exon9$^{-/-}$ mice.

### 5.2.6 Calcium Mobilization Is Altered in the Absence of CD45 Expression

Although BCR and class II-mediated activation of the Src family PTK Lyn is abrogated in the absence of CD45 expression, experiments suggest that both Syk and PLCγ are functional, whereas there is still some question as to whether $IP_3$ production is significantly affected by the loss of CD45 expression. Regardless of this, numerous studies have demonstrated that loss of CD45 expression causes significant alterations in $Ca^{2+}$ mobilization in response to BCR cross-linking. Analysis of $Ca^{2+}$ mobilization in the CD45-negative J558μm3 cell line reveals little or no detectable $Ca^{2+}$ flux after stimulation with anti-Ig mAb (Justement et al. 1991). In contrast, the CD45-deficient DT40 cell line exhibits a slow, gradual increase in the concentration of free intracellular $Ca^{2+}$ in response to BCR cross-linking (Yanagi et al. 1996). Although this response is significantly different from that observed in the J558μm3 cell line, it is consistent with the phenotypic response observed in Lyn-deficient DT40 B cells (Takata et al. 1994).

Most recently, studies were conducted to determine whether $Ca^{2+}$ mobilization is altered in the K46–17μm λ cell line and its CD45-deficient variants. Analysis of $Ca^{2+}$ mobilization in response to either anti-Ig or anti-class II mAbs revealed that there are profound alterations in the ability of cells to mobilize $Ca^{2+}$ in the absence of CD45. Whereas cross-linking of the BCR elicited a transient and somewhat diminished $Ca^{2+}$ flux response, there was little or no detectable increase in the intracellular concentration of free $Ca^{2+}$ after stimulation through class II (Greer et al. 1998). Experiments were performed in which it was determined that the decreased $Ca^{2+}$ mobilization response after cross-linking of the BCR was due to lack of extracellular influx of $Ca^{2+}$. These results indicated that there is a significant defect in both BCR- and MHC class II-mediated $Ca^{2+}$ mobilization in the absence of CD45, even though Syk activation and the production of $IP_3$ appear to occur (Greer et al. 1998). The selective abrogation of the extracellular influx of $Ca^{2+}$ in the 35S5 and 3S5 clones in response to BCR cross-linking is in agreement with results from previous studies in which B cells from CD45-deficient mice were examined (Benatar et al. 1996).

The underlying defect that is responsible for loss of $Ca^{2+}$ influx from the extracellular space in the absence of CD45 has not been elucidated. Recent studies have demonstrated that depletion of intracellular $Ca^{2+}$ stores leads to the generation of a signal that activates $Ca^{2+}$ influx across the plasma membrane through a process termed capacitative $Ca^{2+}$ entry (Clapham 1995). The specific mechanism by which depletion-activated $Ca^{2+}$ channels are regulated is currently not well understood. However, there is evidence to support the hypothesis that depletion of intracellular stores initiates the production of a second messenger(s) from the microsomal fraction of the cell that diffuses through the cytoplasm to the plasma membrane where it opens the $Ca^{2+}$-release-activated $Ca^{2+}$ (CRAC) channel (Lewis and Cahalan 1995). The identity of the calcium influx factor (CIF) is not well established at present. Nevertheless, the direct involvement of CD45 in regulating the production of CIF might provide an explanation for the lack of BCR-mediated $Ca^{2+}$ influx in CD45-negative cells. Alternatively, studies have suggested that depletion of intracellular $Ca^{2+}$ stores may regulate CRAC channel function via a PTK-dependent mechanism (Vostal et al. 1991; Lee et al. 1993). Thus, it is possible that CD45 expression is required to enable a member(s) of the Src family, or an as yet undefined PTK, to regulate CRAC channel function.

### 5.3 Summary of CD45-Dependent Regulation of BCR Signal Transduction

In general, it is now widely accepted that CD45 regulates the activation of Src family PTKs that are responsible for initiating signal transduction via the BCR (YANAGI et al. 1996; PAO and CAMBIER 1997; GREER et al. 1998). Although activation of these PTKs is abrogated in the absence of CD45, inducible tyrosine phosphorylation of numerous substrates is still observed (YANAGI et al. 1996; PAO and CAMBIER 1997; GREER et al. 1998). Recent studies in the B cell have revealed that the downstream PTK Syk is inducibly activated even in the absence of CD45 expression or Src kinase function. Likewise, PLCγ is inducibly phosphorylated and hydrolyzes $PIP_2$ to yield variable amounts of the second messenger $IP_3$, depending on the cell lines examined (TAKATA et al. 1994). Nevertheless, in numerous studies, profound alterations in the $Ca^{2+}$ mobilization response elicited by cross-linking the BCR have been documented in the absence of CD45 expression. Thus, the differences seen in the response of CD45-positive versus CD45-deficient B cells are in many ways similar to those seen between naive versus anergized B cells (HEALY et al. 1997).

In agreement with these results, antibody-mediated cross-linking of the BCR on CD45exon6$^{-/-}$ B cells mediates inducible tyrosine phosphorylation of intracellular substrates involved in cellular activation, including CD79a and PLCγ2 (BENATAR et al. 1996). However, the calcium mobilization response in the CD45exon6$^{-/-}$ B cells is not normal and is characterized by a rapid and transient increase in the concentration of intracellular $Ca^{2+}$. The transient rise in $Ca^{2+}$ was determined to be due to the release of $Ca^{2+}$ from intracellular stores (BENATAR et al. 1996). These results suggest that CD45 may regulate the influx of $Ca^{2+}$ from the extracellular environment into the B cell, perhaps by controlling the function of CRAC channels in the plasma membrane either directly or indirectly. Finally, CD45 expression is important for Ras activation and downstream signaling via the ERK2-dependent pathway (KAWAUCHI et al. 1994; CYSTER et al. 1996). Therefore, it is evident that CD45 regulates at least two downstream signaling pathways that lead to transcription factor activation in the B cell.

## 6 Regulation of BCR-Mediated Signal Transduction by SHP-1

Analysis of the spontaneous mutant motheaten (*me*) and motheaten viable (*me*$^{v}$) mouse strains has provided convincing evidence that supports the important role that SHP-1 plays in regulation of B-cell development and activation (SIDMAN et al. 1989; KOZLOWSKI et al. 1993; TSUI and TSUI 1994; CYSTER and GOODNOW 1995; PANI et al. 1995). Both strains of mice exhibit severe defects in the B-cell compartment resulting in B-cell immune deficiency and the breakdown of self-tolerance, leading to the production of autoantibodies. Specifically, these mice lack mature conventional B cells and, therefore, mount a weak humoral response when challenged with antigens. Nevertheless, these mice also develop autoantibodies by 4 weeks of age. Presumably, the increased prevalence of autoantibodies is due to the abnormal peripheral expansion of $CD5^+$ B cells, which constitute a putative autoreactive B-cell subpopulation that is normally localized in the peritoneal and pleural cavities (Kozlowski et al. 1993). The mutant *me* alleles have been shown to result in the disruption of the gene that encodes SHP-1, an SH2 domain-containing PTP, which has been shown to negatively regulate signal transduction via growth factor and cytokine receptors (Yi et al. 1993; Klingmuller et al. 1995; Vambutas et al. 1995). The mutation in the *me* mutant allele completely ablates the production of SHP-1, whereas the mutation in the $me^v$ allele disrupts a splice site within the PTP domain of SHP-1 leading to the production of aberrantly sized products that exhibit only 10–20% of wild-type catalytic activity (Kozlowski et al. 1993; Tsui and Tsui 1994). In either instance, the loss of SHP-1 activity severely alters normal B-cell function due to dysregulation of BCR-dependent signal transduction.

## 6.1 SHP-1 Negatively Regulates Signal Transduction via the BCR

The abnormal expansion of $CD5^+$ B cells in *me* and $me^v$ mice and the associated high levels of serum immunoglobulins, autoantibodies and B-cell lymphokines provided an initial indication that SHP-1 function is associated with attenuation of B-cell activation, presumably through its ability to attenuate BCR-mediated signal transduction. Indeed, this was found to be the case based on stimulation of B cells isolated from me and $me^v$ mice through the BCR-complex. B cells deficient in the expression of functional SHP-1 were maximally stimulated by concentrations of anti-Ig antibody that were five to ten times lower than those that induced peak responses in normal B cells (Pani et al. 1995). In contrast, the response of SHP-1-deficient and normal B cells to LPS was comparable, indicating that loss of SHP-1 expression does not result in a general change in the functional status of the B cell (Pani et al. 1995). These studies were extended by work providing genetic evidence that SHP-1 is an important negative regulator of BCR signal transduction. In these studies, crossing $me^v$ mice with mice expressing a transgenic BCR specific for HEL was employed to assess the role of SHP-1 (Cyster and Goodnow 1995). In SHP-1-deficient B cells, it was observed that antigen triggers a greater and more rapid elevation of intracellular $Ca^{2+}$, again indicating that this PTP negatively regulates BCR-mediated signaling. Subsequent studies demonstrated that the increased responsiveness to antigen has a direct effect on the process of B-cell selection. Binding a form of HEL with a lower valency than is normally required to mediate clonal deletion of wild-type B cells efficiently eliminated SHP-1-deficient self-reactive B cells in this system (Cyster and Goodnow 1995). This finding provided convincing evidence that SHP-1 plays an important role in regulating the threshold of signaling through the BCR that is necessary to elicit a given biological response.

## 6.2 Recruitment and Activation of SHP-1 in Response to BCR Cross-Linking

SHP-1 is a non-transmembrane PTP that contains two tandem SH2 domains (MATTHEWS et al. 1992; PLUTZKY et al. 1992; YI et al. 1992). Studies have demonstrated that the SH2 domains of SHP-1 interact with the PTP domain in a phosphotyrosine-independent manner causing the PTP domain to assume an inactive conformation (PEI et al. 1994). It has been shown that the tandem SH2 domains of SHP-1 actually perform different functions (PEI et al. 1996). Whereas the amino-terminal SH2 domain is primarily responsible for mediating autoinhibition of SHP-1 catalytic activity, the carboxyl-terminal domain is not involved in this process. In contrast, both the amino- and the carboxyl-terminal domains are important for binding to phosphotyrosine motifs within the cytoplasmic domains of transmembrane proteins to promote recruitment of SHP-1 to the plasma membrane (PEI et al. 1996). Upon cross-linking of the BCR, several transmembrane co-receptors, including CD22, PIR-B and FcγRIIb, are inducibly phosphorylated on tyrosine residues, thereby generating phosphotyrosine-based docking sites that effectively recruit SHP-1 to the membrane (DOODY et al. 1995; D'AMBROSIO et al. 1995; MAEDA et al. 1998) (Fig. 4). These transmembrane proteins all possess one or more immunoreceptor tyrosine-based inhibitory motif(s) (ITIM) in their cytoplasmic domains that are responsible for recruitment of SHP-1 via its tandem SH2 domains (THOMAS 1995). Moreover, these transmembrane proteins are known to play a role in negatively regulating signal transduction through the BCR (DOODY et al. 1995; D'AMBROSIO et al. 1995; MAEDA et al. 1998). Binding of SHP-1 to ITIMs within the cytoplasmic domains of these co-receptors promotes activation of its catalytic function by disrupting the interaction between its SH2 and PTP domains and targets the PTP to critical regions at the membrane where it can effectively dephosphorylate specific substrates (DOODY et al. 1995; D'AMBROSIO et al. 1995).

### 6.2.1 Recruitment of SHP-1 to FcγRIIb

Studies have demonstrated that the transmembrane protein FcγRIIb negatively regulates signaling through the BCR complex (COGGESHALL 1998). Until recently, the mechanism by which FcγRIIb attenuates signaling was unknown; however, experiments revealed that BCR-mediated tyrosine phosphorylation of FcγRIIb promotes binding of SHP-1 to the ITIM contained within its cytoplasmic domain (D'AMBROSIO et al. 1995). The interaction between SHP-1 and FcγRIIb is mediated primarily by the carboxyl-terminal SH2 domain of SHP-1, which interacts with the sole ITIM of native FcγRIIb (D'AMBROSIO et al. 1995). Although SHP-1 does appear to be recruited to FcγRIIb in response to stimulation of B cells with intact anti-Ig antibody, several studies have demonstrated that the inhibitory function of FcγRIIb is not dependent on the expression of SHP-1 (ONO et al. 1996, 1997). Indeed, the inhibitory activity of FcγRIIb appears to be dependent on a lipid phosphatase called SHIP. The fact that SHP-1 does not play an obligatory role in

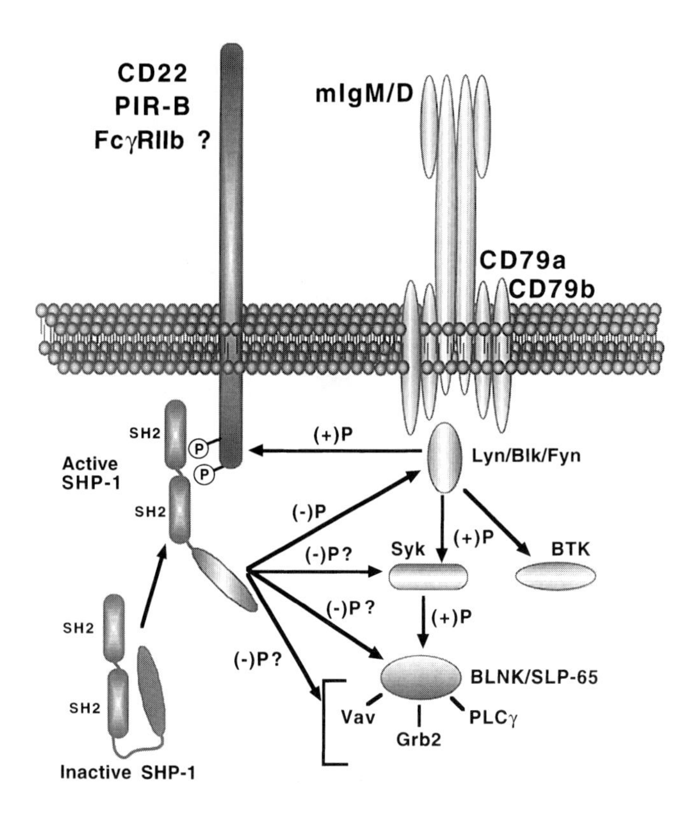


**Fig. 4.** The protein tyrosine phosphatase SHP-1 attenuates signal transduction via the B-cell antigen receptor (BCR). Cross-linking of the BCR mediates activation of Src family protein tyrosine kinases (PTKs), which in turn phosphorylate one or more inhibitory receptors that contain immunoreceptor tyrosine-based inhibition motifs (ITIMs) within their respective cytoplasmic domains. Tyrosine phosphorylation of the ITIMs promotes the recruitment and activation of SHP-1, which binds to the phosphorylated inhibitory molecule via its dual SH2 domains. This effectively targets active SHP-1 to the BCR complex under specific conditions, depending on the particular inhibitory receptor involved. SHP-1 negatively regulates the catalytic activity of Src family PTKs, thereby attenuating signal transduction via the BCR. Based on studies in the T cell, additional SHP-1 substrates have been proposed including Syk and BLNK/SLP-65. SHP-1 has been shown to physically interact with additional effector proteins including Vav and Grb2. Such interactions may be important for recruitment of SHP-1 into activation complexes where it can dephosphorylate key effector proteins. Alternatively, SHP-1 may directly dephosphorylate one or more of the intracellular effector proteins with which it interacts (e.g. Vav)

mediating the inhibitory function of FcγRIIb may be due to the fact that there is only one ITIM in the cytoplasmic domain of this receptor (Muta et al. 1994).

Although studies have shown that the phosphorylated ITIM peptide from FcγRIIb binds to both the amino- and carboxyl-terminal SH2 domains of SHP-1, and can in fact activate its PTP function (Pei et al. 1996), two issues must be considered. First, there is only one ITIM in the cytoplasmic domain of FcγRIIb and, second, the carboxyl-terminal SH2 domain of SHP-1 exhibits a higher affinity for that ITIM than does the amino-terminal SH2 domain (Pei et al. 1996). Thus, when the amount of phosphorylated ITIM peptide is not limiting, both SH2 domains are engaged and the autoinhibition of SHP-1 catalytic function is released. It should be noted that the phospho-ITIM peptide from FcγRIIb is several fold less efficient in terms of activating SHP-1 than phosphopeptides from other receptors with which SHP-1 interacts (PEI et al. 1996). However, when SHP-1 binds to the single ITIM in native FcγRIIb, it would be predicted to do so primarily via its carboxyl-terminal SH2 domain, which has a lower $K_d$ and, therefore, would outcompete the amino-terminal SH2 domain. This would lead to the effective recruitment of SHP-1 to FcγRIIb, but it would not necessarily lead to its activation because the amino-terminal SH2 domain would still be engaged with the PTP domain. In support of this hypothesis, recent studies have shown that superclustering of FcγRIIb activates SHP-1 (SATO and OCHI 1998). This may be due to the fact that under conditions in which excess phosphorylated FcγRIIb receptors are recruited into the complex, binding of the two SH2 domains of a given SHP-1 molecule to the ITIMs of two distinct FcγRIIb receptors may occur in-trans.

### 6.2.2 Recruitment of SHP-1 to Paired Ig-Like Receptors-B

Recently, a novel family of proteins has been identified that includes the paired Ig-like receptors (PIRs) which are expressed on B cells and myeloid-lineage cells (KUBAGAWA et al. 1997). Two forms of PIR have been identified including PIR-A molecules, with a short cytoplasmic domain, and PIR-B molecules, which contain three potential ITIMs within the cytoplasmic domain. Because PIR-B contains ITIMs, it has been hypothesized that this receptor may function as an inhibitory co-receptor that could attenuate signal transduction via the BCR. Subsequent studies have indeed demonstrated that PIR-B is able to negatively regulate signaling through the BCR (BLERY et al. 1998; MAEDA et al. 1998). Upon co-ligation of mIg and PIR-B, it was demonstrated that there is a significant decrease in the $Ca^{2+}$ mobilization response when compared with the level of $Ca^{2+}$ mobilization in cells that were stimulated through the BCR alone. Moreover, it was observed that co-ligation of PIR-B with the BCR completely abrogated mobilization of $Ca^{2+}$ from intracellular stores. These findings were corroborated by experiments in which a chimeric FcγRIIb/PIR-B receptor was transfected into IIA1.6 cells. The chimeric FcγRIIB/PIR-B receptor containing the extracellular/transmembrane regions of FcγRIIB and the wild-type cytoplasmic domain of PIR-B was observed to inhibit $Ca^{2+}$ mobilization mediated by cross-linking of the BCR (MAEDA et al. 1998). Again, it was observed that co-ligation of the chimeric receptor significantly inhibited intracellular mobilization of $Ca^{2+}$.

The mechanism through which PIR-B mediates its negative effect on BCR signaling is dependent on SHP-1, which is recruited to phosphorylated ITIMs within the cytoplasmic domain of PIR-B. Two groups have independently confirmed that SHP-1 associates with phosphorylated PIR-B via an SH2 domain-mediated interaction (BLERY et al. 1998; MAEDA et al. 1998). In particular, it was found that SHP-1 binding is mediated by the fourth and fifth tyrosine motif in the carboxyl-terminus of PIR-B, both of which are ITIMs. This was determined by site-specific mutation of individual tyrosine residues within the cytoplasmic domain of

PIR-B, demonstrating that mutation of the fourth or fifth tyrosine residue to phenylalanine decreased the inhibitory activity of PIR-B (BLERY et al. 1998; MAEDA et al. 1998). It was further noted that the inhibitory activity of PIR-B was decreased by 60–80% only when both carboxyl-terminal tyrosines were mutated. Presumably both tyrosine residues are required to facilitate binding of SHP-1 via its tandem SH2 domains, thereby disrupting the autoinhibitory association between the amino-terminal SH2 and PTP domains of the PTP. Thus, phosphorylation of PIR-B leads to recruitment of SHP-1 to the plasma membrane and promotes activation of its catalytic function.

Additional studies using plasmon resonance demonstrated that the fourth and fifth tyrosine motifs from PIR-B preferentially bind to the PTPs SHP-1 and SHP-2, whereas they exhibit little ability to interact with the lipid phosphatase SHIP (BLERY et al. 1998). This was in contrast to the sole ITIM from FcγRIIb that was found to bind SHP-2, SHP-1 and SHIP, with approximately the same affinity. These findings further supported the hypothesis that the inhibitory function of PIR-B is dependent on its association with SHP-1 and possibly SHP-2, but not SHIP. This was confirmed by studies using chicken DT40 B-cell lines that are deficient in SHIP, SHP-1, SHP-2 or both SHP-1 and SHP-2 (MAEDA et al. 1998). A chimeric FcγRIIb/PIR-B receptor was transfected into these cell lines and assayed for its ability to inhibit signaling via the BCR. It was shown that loss of SHIP expression had little or no effect on PIR-B inhibitory function, whereas loss of either SHP-1 or SHP-2 expression decreased PIR-B inhibitory function slightly. PIR-B inhibitory function was significantly affected only in cells that lacked both SHP-1 and SHP-2, suggesting that these PTPs serve a redundant function in relation to PIR-B (MAEDA et al. 1998).

### 6.2.3 Recruitment of SHP-1 to CD22

A third type of inhibitory receptor is expressed in B lineage cells that has been shown to attenuate signal transduction via the BCR. The transmembrane glycoprotein CD22 contains three ITIMs in its cytoplasmic domain similar to other inhibitory co-receptors (CAMBIER 1997). Not surprisingly, studies demonstrated that tyrosine phosphorylation of CD22 promotes the recruitment of the PTP SHP-1 (CAMPBELL and KLINMAN 1995; DOODY et al. 1995; LANKESTER et al. 1995). Moreover, the SH2 domain-mediated binding of SHP-1 to CD22 presumably results in activation of its PTP function, based on studies demonstrating that phosphopeptides representing the ITIMs from CD22 have the ability to activate recombinant SHP-1 in vitro (DOODY et al. 1995). Additional studies demonstrated that these phosphopeptides also have the ability to block binding of SHP-1 to native CD22 and can immunoprecipitate SHP-1 from cell lysates (DOODY et al. 1995; YOHANNAN et al. 1999a). However, analysis of peptide binding to GST fusion proteins containing the dual SH2 domains of SHP-1 using reverse Far-Western blotting revealed that only two of the three ITIMs from CD22, the fifth and sixth tyrosine motifs located at the carboxyl-terminus of the cytoplasmic domain, bind directly to CD22 (YOHANNAN et al. 1999b). Mutational analysis of the tyrosine residues in the cytoplasmic domain of CD22 has shown that mutation of either the fifth or sixth tyrosine, both of which are contained in ITIMs, to phenylalanine significantly decreases SHP-1 binding (YOHANNAN et al. 1999b). In contrast, mutation of the second tyrosine in the cytoplasmic domain of CD22, which is in the third ITIM, does not significantly decrease binding of SHP-1. These results are in agreement with the reverse Far-Western analysis and suggest that the carboxyl-terminal ITIMs of CD22 containing tyrosine 5 and 6 are important for optimal recruitment of SHP-1.

It was therefore hypothesized that recruitment of SHP-1 to the CD22/BCR complex is involved in attenuating signal transduction. Independent ligation of CD22 using immobilized anti-CD22 mAb was observed to potentiate B-cell proliferation in response to anti-Ig and IL-4, and actually decreased the threshold of stimulus required more than tenfold (DOODY et al. 1995). This finding was interpreted as providing evidence that sequestration of CD22 away from the BCR enhances signal transduction. In light of these findings, previous antibody cross-linking experiments have been interpreted as providing proof that sequestration of CD22 promotes signaling via the BCR (PEZZUTTO et al. 1997). In support of this, co-cross-linking experiments in which CD22 is localized in the membrane with the BCR complex leads to suppression of MAPK activation (TOOZE et al. 1997). More definitive proof that CD22 negatively regulates signal transduction via the BCR has been provided by a series of studies examining the B-cell compartment in $CD22^{-/-}$ mice (O'KEEFE et al. 1996; OTIPOBY et al. 1996; SATO et al. 1996; NITSCHKE et al. 1997).

Loss of CD22 expression does not significantly affect B-cell development in the bone marrow, despite the fact that CD22 is expressed on preB cells after initiation of V-D-J recombination. However, a significant decrease in the number of $B220^{HI}$, $IgM^{+}$, $IgD^{+}$ mature recirculating B cells was observed in the bone marrow of $CD22^{-/-}$ mice. The decrease in recirculating B cells was shown to be due, in part, to an inability of $CD22^{-/-}$ B cells to home to the bone marrow (NITSCHKE et al. 1997). The overall number of B cells observed in the periphery was essentially normal in $CD22^{-/-}$ mice, although there was a decrease in the number of transitional B cells in one study (NITSCHKE et al. 1997) and a decrease in marginal zone B cells in another (SATO et al. 1996). In all cases, B cells in the periphery exhibited a significant decrease in membrane IgM expression (40–50%) and had elevated MHC class II expression. The decrease in membrane IgM expression is similar to that observed for $mIg^{HEL}$ transgenic B cells that have become anergized in the presence of soluble HEL (CYSTER and GOODNOW 1997). This finding suggests that loss of CD22 expression leads to chronic BCR-dependent stimulation of B cells that results in the acquisition of phenotypic characteristics commonly associated with anergic B cells.

Examination of B cells from $CD22^{-/-}$ mice revealed that the loss of CD22 expression causes them to become hyper-responsive to acute stimulation through the BCR (O'KEEFE et al. 1996; OTIPOBY et al. 1996; SATO et al. 1996; NITSCHKE et al. 1997). $CD22^{-/-}$ B cells exhibited a tenfold decrease in their threshold of sensitivity to anti-IgM or anti-IgD antibodies based on an analysis of $Ca^{2+}$ mobilization. Additionally, the $Ca^{2+}$ response in $CD22^{-/-}$ B cells was elevated and prolonged when compared with that of wild-type B cells. Signal transduction via CD19 was significantly enhanced in a similar manner (SATO et al. 1996). Most recently, studies have shown that BCR cross-linking on $CD22^{-/-}$ B cells leads to hyper phosphorylation of Vav, indicating that this effector protein is a potential target for the PTP SHP-1 following its recruitment to CD22 (SATO et al. 1997).

Despite being hyper-responsive to acute stimulation via the BCR, $CD22^{-/-}$ B cells exhibit decreased proliferative responses to anti-Ig either in the presence or absence of IL-4 (OTIPOBY et al. 1996; SATO et al. 1996). In contrast, the proliferative response to anti-CD40 or LPS is either normal or enhanced, respectively (OTIPOBY et al, 1996; NITSCHKE et al. 1997). The decreased proliferative response in $CD22^{-/-}$ mice may be explained by the finding that anti-Ig stimulation does not induce entry into the cell cycle as efficiently and actually promotes increased apoptosis when compared with wild-type B cells (OTIPOBY et al. 1996; SATO et al. 1996; NITSCHKE et al. 1997). These results suggest that $CD22^{-/-}$ B cells have a decreased lifespan due to chronic stimulation via the BCR. This was confirmed by BrdU labeling of B cells demonstrating that $CD22^{-/-}$ mice exhibit increased labeling of B cells at all time points examined as well as an accelerated loss of short-term B cells using pulse/chase labeling experiments (OTIPOBY et al. 1996; SATO et al. 1996; NITSCHKE et al. 1997). Based on the results obtained, it is apparent that the loss of CD22 expression results in a decreased lifespan and increased rate of turnover in the B-cell compartment. This, in conjunction with an inability to properly home to the bone marrow, accounts for the decrease in recirculating B cells observed in this compartment.

The loss of either Lyn, CD22 or SHP-1 expression is associated with hyper-responsive signaling via the BCR, spontaneous downregulation of membrane IgM, diminished numbers of recirculating B cells and increased expression of MHC class II (CYSTER and GOODNOW 1995; HIBBS et al. 1995; NISHIZUMI et al. 1995; PANI et al. 1995; CYSTER and GOODNOW 1997). This suggests that these effector proteins comprise a common pathway that regulates the threshold of signaling via the BCR. Mice deficient in any one of these effector proteins also exhibit an expansion of their B-1 B cell population and have elevated levels of serum IgM and IgA (CYSTER and GOODNOW 1997). Therefore, it is likely that genetic polymorphisms in these regulatory molecules could affect the balance between tolerance and immunity potentially leading to autoimmune disease. Recent studies support the conclusion that CD22 is functionally linked with the PTK Lyn and the PTP SHP-1. In Lyn-deficient ($Lyn^{-/-}$) mice, basal and inducible phosphorylation of CD22 is virtually abrogated and SHP-1 is not recruited to CD22 (CHAN et al. 1998; CORNALL et al. 1998; NISHIZUMI et al. 1998; SMITH et al, 1998). Analysis of CD22 function in the $Lyn^{-/-}$ mice demonstrates that CD22 no longer has the ability to attenuate the $Ca^{2+}$ mobilization response when it is co-cross-linked with the BCR (CHAN et al. 1998). Thus, the failure to recruit SHP-1 to the CD22/BCR complex may result in hyper-stimulation of B cells, leading to autoantibody production. Indeed, $Lyn^{-/-}$ mice secrete elevated levels of autoantibodies and develop autoimmune glomerulonephritis resembling systemic lupus erythematosus (HIBBS et al. 1995; NISHIZUMI et al. 1995; CHAN et al. 1997). Similarly, $CD22^{-/-}$ mice have elevated levels of serum IgM and, in one study, have been found to produce autoantibodies against double and single stranded DNA (O'KEEFE et al. 1996; OTIPOBY et al. 1996; SATO et al. 1996; NITSCHKE et al. 1997). Finally, several autoimmune strains of mice are known to express a specific allele of CD22 that does not effectively couple to SHP-1 (NADLER et al. 1997).

## 6.3 Identification of SHP-1 Substrates Involved in Signal Transduction via the BCR

Although it is well documented that SHP-1 negatively regulates signal transduction mediated by the BCR, studies have not yet definitively identified the intracellular substrates that are targeted by this PTP. Because SHP-1-deficient B cells exhibit increased $Ca^{2+}$ mobilization in response to BCR cross-linking, it is likely that SHP-1 regulates the phosphorylation of effector proteins that are upstream of this point in the signal transduction cascade. Studies of thymocytes from *me* mice have revealed that a lack of SHP-1 expression results in hyperphosphorylation of numerous substrates on tyrosine residues that correlated with increased activation of the Src family PTKs Lck and Fyn (LORENZ et al. 1996). These data indicate that SHP-1 is an important negative regulator of Src family kinases. Similarly, studies of B-cell lines derived from *me* mice reveal heightened Lyn expression as well as increased activity (YANG et al. 1998). Together, these studies suggest that SHP-1 regulates Src kinase function, but it has not been determined whether SHP-1 does so directly. In other studies, it has been shown that SHP-1 physically interacts with ZAP-70 and regulates its catalytic activity in T cells (PLAS et al. 1996). Thus, it is logical to predict that a similar regulatory interaction may take place between SHP-1 and Syk in the B cell. It is interesting to note that Syk and SHP-1 are both recruited to the inhibitory receptor CD22 after it has been tyrosine phosphorylated (TEDDER et al. 1997). Theoretically, this could facilitate dephosphorylation of Syk by SHP-1 resulting in inhibition of Syk function, however, this has yet to be demonstrated.

In addition to regulating the phosphorylation of PTKs and thus their function, it is equally possible that SHP-1 targets downstream effector proteins that are substrates for these kinases. In this regard, recent studies have demonstrated that SLP-76 physically interacts with SHP-1 and is a direct substrate of SHP-1 in T cells and natural killer cells (MIZUNO et al. 1996; BINSTADT et al. 1998). SLP-76 plays an important role as a multifunctional adapter protein in these cells that recruits key effector proteins into multi-molecular complexes (PETERSON et al. 1998). Phosphorylation of SLP-76 was shown to be essential for optimal activation of cytotoxic T cells, suggesting that targeted dephosphorylation of SLP-76 by SHP-1 is an important mechanism for negative regulation of immune cell activation by inhibitory receptors (BINSTADT et al. 1998). Recently a functionally homologous protein has been cloned in B cells called BLNK or SLP-65 (FU et al. 1998; WIENANDS et al. 1998). This protein, like SLP-76, acts as an adapter molecule and physically interacts with PLC$\gamma$, Grb2 and Vav, thus bringing together into one multi-molecular complex key effector proteins that are involved in regulating three essential signaling pathways in the B cell. It is possible then that SHP-1 could dephosphorylate SLP-65/BLNK, thereby abrogating its ability to recruit one or more of the above effector proteins. This would presumably attenuate signal transduction mediated by the BCR. It is interesting to note that SHP-1 has been shown to physically interact with both Vav and Grb2 raising the possibility that one or both of these effectors could recruit SHP-1 to SLP65/BLNK (Kon-Kozlowski et al. 1996). Alternatively, the physical interaction between SHP-1 and either Vav or Grb2 might be indicative of the fact that these proteins are substrates for this PTP.

## 7 Conclusions

Significant progress has been made in terms of delineating the molecular mechanisms that underlie regulation of reversible protein tyrosine phosphorylation in the B cell. It is now clear that several classes of PTK are involved in promoting tyrosine phosphorylation of key effector proteins in response to BCR cross-linking. The Src family kinases are responsible for initiation and propagation of signal transduction after ligand binding to the BCR. These PTKs play a role in phosphorylating the BCR complex (i.e, CD79a and CD79b) and regulatory co-receptors (i.e., CD22 and CD19). Additionally, these kinases phosphorylate downstream PTKs, such as Syk and Btk, resulting in enhanced catalytic function. Once Syk and Btk are activated, they phosphorylate PLCγ, which enhances its catalytic function leading to $Ca^{2+}$ mobilization. Additionally, Syk is important for phosphorylating Shc and Vav, thereby promoting activation of additional signaling pathways in the cell. The function of these PTKs is in turn regulated by the activity of at least two PTPs, CD45 and SHP-1. CD45 is involved in promoting BCR-dependent signal transduction by virtue of its ability to regulate the activity of Src family kinases. Studies further suggest that CD45 may regulate $Ca^{2+}$ influx from the extracellular environment either directly by regulating the phosphorylation of $Ca^{2+}$ channels or the production of CIF, or indirectly through an effect on other PTKs or PTPs. Finally, SHP-1 has been shown to negatively regulate signaling mediated by cross-linking of the BCR through its recruitment to multiple inhibitory molecules. SHP-1 is thought to negatively regulate the catalytic activity of Src family kinases and Syk. Additionally, SHP-1 may target specific substrates of these kinases, such as BLNK/SLP-65.

Although significant progress has been made, there are still numerous issues that must be resolved before a complete understanding of the intricate regulatory network that regulates reversible tyrosine phosphorylation can be attained. Additional work must be performed to characterize specific substrates for each of the kinases and phosphatases that are involved in BCR-dependent signaling. Moreover, studies have not yet fully delineated the functional interrelationships that exist between various PTKs and PTPs. This will be important for elucidating the way in which homeostasis of reversible tyrosine phosphorylation is maintained in the B cell. Ultimately, it will be important to define how changes in the quantitative and/or qualitative nature of reversible tyrosine phosphorylation affects the production of second messengers, the activation of intermediate signaling pathways and the activation of transcription factors that regulate gene transcription and the biologic function of the B cell.

*Acknowledgements.* The work performed in the author's laboratory and described herein was supported by grants R01 GM46524 and R01 AI36401 from the National Institutes of Health.

## References

Achatz G, Nitschke L, Lamers MC (1997) Effect of transmembrane and cytoplasmic domains of IgE on the IgE response. Science 276:409–411

Afar DEH, Park H, Howell BW, Rawlings DJ, Cooper J, Witte ON (1996) Regulation of Btk by Src family tyrosine kinases. Mol Cell Biol 16:3465–3471

Andreotti AH, Bunnell SC, Feng S, Berg LJ, Schreiber SL (1997) Regulatory intramolecular association in a tyrosine kinase of the Tec family. Nature 385:93–97

Aoki Y, Isselbacher KJ, Pillai S (1994) Bruton tyrosine kinase is tyrosine phosphorylated and activated in pre-B lymphocytes and receptor-ligated B cells. Proc Natl Acad Sci USA 91:10606–10609

Appelby MW, Kerner JD, Chien S, Malizewski CR, Bondadaa S, Perlmutter RM (1995) Involvement of p59fyn$^T$ in interleukin -5 receptor signaling. J Exp Med 182:811–820

Benatar T, Carsetti R, Furlonger C, Kamalia N, Mak T, Paige CJ (1996) Immunoglobulin-mediated signal transduction in B cells from CD45-deficient mice. J Exp Med 183:329–334

Binstadt BA, Billadeau DD, Jeremovic D, Williams BL, Fang N, Yi T, Koretzky GA, Abraham RT, Leibson PJ (1998) SLP-76 is a direct substrate of SHP-1 recruited to killer cell inhibitory receptors. J Biol Chem 273:27518–27523

Blery M, Kubagawa H, Chen C-C, Vely F, Cooper MD, Vivier E (1998) The paired Ig-like receptor PIR-B is an inhibitory receptor that recruits the protein-tyrosine phosphatase SHP-1. Proc Natl Acad Sci USA 95:2446–2451

Blum JH, Stevens TL, DeFranco AL (1993) Role of the μ immunoglobulin heavy chain transmembrane and cytoplasmic domains in B cell antigen receptor expression and signal transduction. J Biol Chem 268:27236–27245

Bonnefoy-Berard N, Munshi A, Yon I, Wu S, Collins TL, Deckert M, Shalom-Barak T, Giampa L, Herbert E, Hernandez J, Meller N, Couture C, Altman A (1996) Vav: function and regulation in hematopoietic cell signaling. Stem Cells 14:250–268

Brown VK, Ogle EW, Burkhardt AL, Rowley RB, Bolen JB, Justement LB (1994) Multiple components of the B cell antigen receptor complex associate with the protein tyrosine phosphatase, CD45. J Biol Chem 269:17238–17244

Burg DL, Furlong MT, Harrison ML, Geahlen RL (1994) Interactions of Lyn with the antigen receptor during B cell activation. J Biol Chem 269:28136–28142

Burkhardt AL, Brunswick M, Bolen JB, Mond JJ (1991) Anti-immunoglobulin stimulation of B lymphocytes activates src-related protein-tyrosine kinases. Proc Natl Acad Sci USA 88:7410–7414

Burkhardt AL, Costa T, Misulovin Z, Stealy B, Bolen JB, Nussenzweig MC (1994) Iga and Igb are functionally homologous to the signaling proteins of the T-cell receptor. Mol Cell Biol 14:1095–1103

Bustelo XR, Barbacid M (1992) Tyrosine phosphorylation of the vav proto-oncogene product in activated B cells. Science 256:1196–1199

Byth KF, Conroy LA, Howlett S, Smith AJH, May J, Alexander DR, Holmes N (1996) CD45-null transgenic mice reveal a positive regulatory role for CD45 in early thymocyte development, in the selection of $CD^+CD8^+$ thymocytes, and in B cell maturation. J Exp Med 183:1701–1718

Cambier JC (1995) Antigen and Fc receptor signaling: the awesome power of the immunoreceptor tyrosine-based activation motif (ITAM). J Immunol 155:3281–3285

Cambier JC (1997) Inhibitory receptors abound? Proc Natl Acad Sci USA 94:5993–5995
Cambier JC, Pleiman CM, Clark MR (1994) Signal transduction by the B cell antigen receptor and its coreceptors. Annu Rev Immunol 12:457–486
Campbell M-A, Sefton BM (1992) Association between B-lymphocyte membrane immunoglobulin and multiple members of the Src family of protein tyrosine kinases. Mol Cell Biol 12:2315–2321
Campbell MA, Klinman NR (1995) Phosphotyrosine-dependent association between CD22 and protein tyrosine phosphatase 1 C. Eur J Immunol 25:1573–1579
Carter RH, Park DJ, Rhee SG, Fearon DT (1991) Tyrosine phosphorylation of phospholipase C induced by membrane immunoglobulin in B lymphocytes. Proc Natl Acad Sci USA 88:2745–2749
Cassard S, Choquet D, Fridman WH, Bonnerot C (1996) Regulation of ITAM signaling by specific sequences in Ig-β B cell antigen receptor subunit. J Biol Chem 271:23786–23791
Chan VWF, Meng F, Soriano P, DeFranco AL, Lowell CA (1997) Characterization of the B lymphocyte populations in Lyn-deficient mice and the role of Lyn in signal initiation and down-regulation. Immunity 7:69–81
Chan VWF, Lowell CA, DeFranco AL (1998) Defective negative regulation of antigen receptor signaling in Lyn-deficient B lymphocytes. Curr Biol 8:545–553
Cheng AM, Rowley B, Pao W, Hayday A, Bolen JB, Pawson T (1995) Syk tyrosine kinase required for mouse viability and B-cell development. Nature 378:303–306
Cheng G, Ye Z-S, Baltimore D (1994) Binding of Bruton's tyrosine kinase to Fyn, Lyn or Hck through a Src homology 3 domain-mediated interaction. Proc Natl Acad Sci USA 91:8152–8155
Clapham DE (1995) Calcium signaling. Cell 80:259–268
Clark MR, Campbell KS, Kazlauskas A, Johnson SA, Hertz M, Potter TA, Pleiman C, Cambier JC (1992) The B cell antigen receptor complex: association of Ig-α and Ig-β with distinct cytoplasmic effectors. Science 258:123–126
Clark MR, Johnson SA, Cambier JC (1994) Analysis of Ig-α-tyrosine kinase interaction reveals two levels of binding specificity and tyrosine phosphorylated Ig-α stimulation of Fyn activity. EMBO J 13:1911–1919
Coggeshall KM (1998) Inhibitory signaling by B cell FcγRIIb. Curr Opin Immunol 10:306–312
Coggeshall KM, McHugh JC, Altman A (1992) Predominant expression and activation-induced tyrosine phosphorylation of phospholipase C-γ2 in B lymphocytes. Proc Natl Acad Sci USA 89:5660–5664
Cornall RJ, Cyster JG, Hibbs ML, Dunn AR, Otipoby KL, Clark EA, Goodnow CC (1998) Polygenic autoimmune traits: Lyn, CD22and SHP-1 are limiting elements of a biochemical pathway regulating BCR signaling and selection. Immunity 8:497–508
Costa T, Franke RR, Sanchez M, Misulovin Z, Nussenzweig MC (1992) Functional reconstitution of an immunoglobulin antigen receptor in T cells. J Exp Med 175:1669–1676
Crespo P, Schuebel KE, Ostrom AA, Gutkind JS, Bustelo XR (1997) Phosphorylation-dependent activation of rac-1 GDP/GTP exchange by the vav proto-oncogene product. Nature 385:169–172
Cyster JG, Goodnow, CC (1995) Protein tyrosine phosphatase 1C negatively regulates antigen receptor signaling in B lymphocytes and determines thresholds for negative selection. Immunity 2:13–24
Cyster JG, Goodnow CC (1997) Tuning Antigen receptor signaling by CD22: Integrating cues from antigens and the microenvironment. Immunity 6:509–517
Cyster JG, Healy JI, Kishihara K, Mak TW, Thomas ML, Goodnow CC (1996) Regulation of B-lymphocyte negative and positive selection by tyrosine phosphatase CD45. Nature 381:325–328
D'Ambrosio D, Hippen KL, Minskoff SA, Mellman I, Pani G, Siminovitch KA, Cambier JC (1995) Recruitment and regulation of PTP1C in negative regulation of antigen receptor signaling by FcgRIIb. Science 268:293–297
D'Ambrosio D, Hippen KL, Cambier JC (1996) Distinct mechanisms mediate SHC association with the activated and resting B cell antigen receptor. Eur J Immunol 26:1960–1965
DeFranco AL (1997) The complexity of signaling pathways activated by the BCR. Curr Opin Immunol 9:296–308
Desiderio S (1997) Role of Btk in B cell development and signaling. Curr Opin Immunol 9:534–540
deWeers M, Brouns GS, Hinshelwood S, Kinnon C, Schuurman RKB, Hendriks RW, Borst J (1994) B-cell antigen receptor stimulation activates the human Bruton's tyrosine kinase, which is deficient in X-linked agammaglobulinemia. J Biol Chem 269:23857–23860
Dolmetsch RE, Lewis RS, Goodnow CC, Healy JI (1997) Differential activation of transcription factors induced by $Ca^{2+}$ response amplitude and duration. Nature 386:855–858
Domitati-Saad R, Ogle EW, Justement LB (1993) Administration of anti-CD45 mAb specific for a B cell-restricted epitope abrogates the B cell response to a T-dependent antigen in vivo. J Immunol 151:5936–5947

Doody GM, Justement LB, Delibrias CC, Matthews RJ, Lin J, Thomas ML, Fearon DT (1995) A role in B cell activation for CD22 and the protein tyrosine phosphatase SHP. Science 269:242–244

Dubois PM, Stepinski J, Urbain J, Sibley CH (1992) Role of the transmembrane and cytoplasmic domains of surface IgM in endocytosis and signal transduction. Eur J Immunol 22:851–857

Flaswinkel H, Reth M (1994) Dual role of the tyrosine activation motif of the Ig-a protein during signal transduction via the B cell antigen receptor. EMBO J 13:83–89

Fluckiger A-C, Li Z, Kato RM, Wahl MI, Ochs HD, Longnecker R, Kinet J-P, Witte ON, Scharenberg AM, Rawlings DJ (1998) Btk/Tec kinases regulate sustained increases in intracellular $Ca^{2+}$ following B-cell receptor activation. EMBO J 17:1973–1985

Fu C, Turck CW, Kurosaki T, Chan AC (1998) BLNK: a central linker protein in B cell activation. Immunity 9:93–103

Furlong MT, Mahrenholz AM, Kim K-H, Ashendel CL, Harrison ML, and Geahlen RL (1997) Identification of the major site of autophosphorylation of the murine protein-tyrosine kinase Syk. Biochim Biophys Acta 1355:177–190

Furukawa T, Itoh M, Krugger NX, Streuli M, Saito H (1994) Specific interaction of the CD45 protein-tyrosine phosphatase with tyrosine-phosphorylated CD3 ζ chain. Proc Natl Acad Sci USA 91:10928–10932

Gold MR, DeFranco AL (1994) Biochemistry of B lymphocyte activation. Adv Immunol 55:221–295

Gold MR, Matsuuchi L, Kelly RB, DeFranco AL (1991) Tyrosine phosphorylation of components of the B-cell antigen receptors following receptor crosslinking. Proc Natl Acad Sci USA 88:3436–3440

Gold MR, Sanghera JS, Stewart J, Pelech SL (1992) Selective activation of p42 MAP kinase in murine B lymphoma cell lines by membrane immunoglobulin crosslinking. Biochem J 287:269–276

Gold MR, Chiu R, Ingham RJ, Saxton TM, van Oostveen I, Watts JD, Affolter M, Aebersold R (1994) Activation and serine phosphorylation of the $p56^{lck}$ protein tyrosine kinase in response to antigen receptor cross-linking in B lymphocytes. J Immunol 153:2369–2380

Goodnow CC (1996) Balancing immunity and tolerance: deleting and tuning lymphocyte repertoires. Proc Natl Acad Sci USA 93:2264–2271

Goodnow CC, Cyster JG (1997) Lymphocyte homing: the scent of a follicle. Curr Biol 7:R219-R222

Greer SF, Lin J, Clarke CH, Justement LB (1998) Major histocompatibility class II-mediated signal transduction is regulated by the protein-tyrosine phosphatase CD45. J Biol Chem 273:11970–11979

Greer SF, Justement LB (1999) CD45 regulates tyrosine phosphorylation of CD22 and its association with the protein tyrosine phosphatase SHP-1. J Immunol 162:5278–5286

Gruber MF, Bjorndahl JM, Nakamura S, Fu SM (1989) Anti-CD45 inhibition of human B cell proliferation depends on the nature of activation signals and the state of B cell activation. J Immunol 142:4144–4152

Grupp SA, Campbell K, Mitchell RN, Cambier JC, Abbas AK (1993) Signaling-defective mutants of the B lymphocyte antigen receptor fail to associate with Ig-α and Ig-β/γ. J Biol Chem 268:25776–25779

Gulbins E, Langlet C, Baier G, Bonnefoy-Berard N, Herbert E, Altman A, Coggeshall, KM (1994) Tyrosine phosphorylation and activation of Vav GTP/GDP exchange activity in antigen receptor-triggered B cells. J Immunol 152:2123–2129

Harwood AE, Cambier JC (1993) B cell antigen receptor cross-linking triggers rapid PKC independent activation of $p21^{ras}$. J Immunol 151:4513–4522

Healy JI, Dolmetsch RE, Timmerman LA, Cyster JG, Thomas ML, Crabtree GR, Lewis RS, Goodnow CC (1997) Different nuclear signals are activated by the B cell receptor during positive versus negative signaling. Immunity 6:419–428

Hempel WM, Schatzman RC, DeFranco AL (1992) Tyrosine phosphorylation of phospholipase C-γ2 upon cross-linking of membrane Ig on murine B lymphocytes. J Immunol 148:3021–3027

Hertz M, Nemazee D (1998) Receptor editing and commitment in B lymphocytes. Curr Opin Immunol 10:208–213

Hibbs ML, Tarlinton DM, Armes J, Grail D, Hodgson G, Maglitto R, Stacker SA, Dunn AR (1995) Multiple defect in the immune system of Lyn-deficient mice, culminating in autoimmune disease. Cell 83:301–311

Hombach J, Sablitzky F, Rajewski K, Reth M (1988) Transfected plasmacytoma cells do not transport the membrane form of IgM to the cell surface. J Exp Med 167:652–657

Hombach J, Tsubata T, Leclerq, Stappert H, Reth M (1990) Molecular components of the B-cell antigen receptor complex of the IgM class. Nature 343:760–762

Hutchcroft JE, Harrison ML, Geahlen RL (1991) B lymphocyte activation is accompanied by phosphorylation of a 72-kDa protein-tyrosine kinase. J Biol Chem 266:14846–14849

Hutchcroft JE, Harrison ML, Geahlen RL (1992) Association of the 72-kDa protein tyrosine kinase PTK72 with the B cell antigen receptor. J Biol Chem 267:8613–8619
Janknecht R, Hunter T (1997) Convergence of MAP kinase pathways on the ternary complex factor Sap-1a. EMBO J 16:1620–1627
Johnson SA, Pleiman CM, Pao L, Schneringer J, Hippen K, Cambier JC (1995) Phosphorylated immunoreceptor signaling motifs (ITAMS) exhibit unique abilities to bind and activate Lyn and Syk tyrosine kinases. J Immunol 155:4596–4603
Justement LB (1997) The role of CD45 in signal transduction. Adv Immunol 66:1–65
Justement LB, Campbell KS, Chien NC, Cambier JC (1991) Regulation of B cell antigen receptor signal transduction and phosphorylation by CD45. Science 252:1839–1842
Kaisho T, Schwenk F, Rajewski K (1997) The roles of $\gamma 1$ heavy chain membrane expression and cytoplasmic tail in IgG1 responses. Science 276:412–414
Karin M, Liu Z-G, Zandi E (1997) AP-1 function and regulation. Curr Opin Biol 9:240–246
Kawauchi K, Lazarus AH, Rapoport MJ, Harwood A, Cambier JC, Delovitch TL (1994) Tyrosine kinase and CD45 tyrosine phosphatase activity mediate $p21^{ras}$ activation in B cells stimulated through the antigen receptor. J Immunol 152:3306–3316
Keshvara LM, Isaacson C, Harrison ML, Geahlen RL (1997) Syk activation and dissociation from the B-cell antigen receptor is mediated by phosphorylation of tyrosine 130. J Biol Chem 272:10377–10381
Kim K-M, Alber G, Weiser P, Reth M (1993a) Differential signaling through the Ig-a and Ig-b components of the B cell antigen receptor. Eur J Immunol 23:911–916
Kim K-M, Alber G, Weiser P, Reth M (1993b) Signaling function of the B-cell antigen receptors. Immunol Rev 132:125–146
Kishihara K, Penninger J, Wallace VA, Kundig TM, Kawai K, Wakeham A, Timms E, Pferrer K, Ohashi PS, Thomas ML, Furlonger C, Paige CJ, Mak TW (1993) Normal B lymphocyte development but impaired T cell maturation in CD45-exon6 protein tyrosine phosphatase-deficient mice. Cell 74: 143–156
Klingmuller U, Lorenz U, Cantley LC, Neel BG, Lodish HF (1995) Specific recruitment of SH-PTP to the erythropoietin receptor causes inactivation of JAK2 and termination of proliferative signals. Cell 80:729–738
Kon-Kozlowski M, Pani G, Pawson T, Siminovitch KA (1996) The tyrosine phosphatase PTP1C associates with Vav, Grb2 and mSos1 in hematopoietic cells. J Biol Chem 271:3856–3862
Kozlowski M, Mlinaric-Rascan I, Feng G-S, Shen R, Pawson T, Siminovitch KA (1993) Expression and catalytic activity of the tyrosine phosphatase PTP1C is severely impaired in motheaten and viable motheaten mice. J Exp Med 178:2157–2163
Kubagawa H, Burrows PD, Cooper MD (1997) A novel pair of immunoglobulin-like receptors expressed by B cells and myeloid cells. Proc Natl Acad Sci USA 94:5261–5266
Kurosaki T (1997) Molecular mechanisms in B cell antigen receptor signaling. Curr Opin Immunol 9: 309–318
Kurosaki T, Takata M, Yamanashi Y, Inazu T, Taniguchi T, Yamamoto T, Yamamura H (1994) Syk activation by the Src-family tyrosine kinase in the B cell receptor signaling. J Exp Med 179:1725–1729
Kurosaki T, Johnson SA, Pao L, Sada K, Yamamura H, Cambier JC (1995) Role of the Syk autophosphorylation site and SH2 domains in B cell antigen receptor signaling. J Exp Med 182:1815–1823
Lam K.-P., Kuhn R, Rajewski K (1997) In vivo ablation of surface immunoglobulin on mature B cells by inducible gene targeting results in rapid cell death. Cell 90:1073–1083
Lankester AC, van Schijndel GMW, van Lier RAW (1995) Hematopoietic cell phosphatase is recruited to CD22 following B cell antigen receptor ligation. J Biol Chem 270:20305–20308
Law DA, Chan VWF, Datta SK, DeFranco AL (1993) B-cell antigen receptor motifs have redundant signalling capabilities and bind the tyrosine kinases PTK72, Lyn and Fyn. Curr Biol 3:645–657
Lazarus AH, Kawauchi K, Rapoport MJ, Delovitch TJ (1993) Antigen-induced B lymphocyte activation involves the p21 and ras.GAP signaling pathway. J Exp Med 178:1765–1769
LeBien TW (1998) B-cell lymphopoiesis in mouse and man. Curr Opin Immunol 10:188–195
Lee K-M, Toscas K, Villereal ML (1993) Inhibition of bradykinin- and thapsigargin-induced $Ca^{2+}$ entry by tyrosine kinase inhibitors. J Biol Chem 268:9945–9948
Lee SB, Rhee SG (1995) Significance of $PIP_2$ hydrolysis and regulation of phospholipase C isozymes. Curr Opin Cell Biol 7:183–189
Lewis RS, Cahalan MD (1995) Potassium and calcium channels in lymphocytes. Annu Rev Immunol 13:623–653

Li Z, Wahl MI, Eguinoa A, Stephens LR, Hawkins PT, Witte ON (1997) Phosphatidylinositol 3-kinase-γ activates Bruton's tyrosine kinase in concert with Src family kinases. Proc Natl Acad Sci USA 94:13820–13825

Li Z-H, Mahajan S, Prendergast MM, Fargnoli J, Zhu X, Klages S, Adam D, Schieven GL, Blake J, Bolen JB, Burkhardt AL (1992) Cross-linking surface immunoglobulin activates src-related tyrosine kinases in WEHI-231 cells. Biochem Biophys Res Commun 187:1536–1544

Lin J, Justement LB (1992) The MB-1/B29 heterodimer couples the B cell antigen receptor to multiple src family protein tyrosine kinases. J Immunol 149:1548–1555

Lin J, Brown VK, Justement LB (1992) Regulation of basal tyrosine phosphorylation of the B cell antigen receptor complex by the protein tyrosine phosphatase, CD45. J Immunol 149:3182–3190

Lin J, Tao J, Dyer RB, Herzog NK, Justement LB (1997) Kinase-independent potentiation of B cell antigen receptor-mediated signal transduction by the protein tyrosine kinase Src. J Immunol 159:4823–4833

Lorenz U, Ravichandran KS, Burakoff SJ, Neel BG (1996) Lack of SHPTP1 results in src-family kinase hyperactivation and thymocyte hyperresponsiveness. Proc Natl Acad Sci USA 93:9624–9629

Luisiri P, Lee YJ, Eisfelder BJ, Clark MR (1996) Cooperativity and segregation of functions within the Ig-α/β heterodimer of the B cell antigen receptor complex. J Biol Chem 271:5158–5163

Lund FE, Yu N, Kim KM, Reth Howard MC (1996) Signaling through CD38 augments B cell antigen receptor (BCR) responses and is dependent on BCR expression. J Immunol 157:1455–1467

MacLennan ICM (1998) B-cell receptor regulation of peripheral B cells. Curr Opin Immunol 10:220–225

Maeda A, Kurosaki M, Ono M, Takai T, Kurosaki T (1998) Requirement of SH2-containing protein tyrosine phosphatases SHP-1 and SHP-2 for paired immunoglobulin-like receptor B (PIR-B)-mediated inhibitory signal. J Exp Med 187:1355–1360

Mahajan S, Fargnoli J, Burkhardt AL, Kut SA, Saouaf SJ, Bolen JB (1995) Src family protein tyrosine kinases induce autoactivation of Bruton's tyrosine kinase. Mol Cell Biol 15:5304–5311

Majeti R, Bilwes AM, Noel JP, Hunter T, Weiss A (1998) Dimerization-induced inhibition of receptor protein tyrosine phosphatase function through an inhibitory wedge. Science 279:88–91

Matsuuchi L, Gold MR, Travis A, Grosschedl R, DeFranco, AL Kelly RB (1992) The membrane IgM-associated proteins MB-1 and Ig-b are sufficient to promote surface expression of a partially functional B-cell antigen receptor in a nonlymphoid cell line. Proc Natl Acad Sci USA 89:3404–3408

Matthews RJ, Bowne DB, Flores E, Thomas ML (1992) Characterization of hematopoietic intracellular protein tyrosine phosphatases: description of a phosphatase containing an SH2 domain and another enriched in proline-, glutamic acid-, serine-, and threonine-rich sequences. Mol Cell Biol 12:2396–2405

Mayer BJ (1997) Signal transduction: clamping down on Src activity. Curr Biol 7:R295–R298

Michnoff CH, Parikh VS, Lelsz DL, Tucker PW (1994) Mutations within the $NH_2$-terminal transmembrane domain of membrane immunoglobulin (Ig) M alters Igα and Igβ association and signal transduction. J Biol Chem 269:24237–24244

Mizuno K, Katagiri T, Hasegawa K, Ogimoto M, Yakura H (1996) Hematopoietic cell phosphatase, SHP-1, is constitutively associated with the SH2 domain-containing leukocyte protein, SLP-76, in B cells. J Exp Med 184:457–463

Muta T, Kurosaki T, Misulovin Z, Sanchez M, Nussenzweig MC, Ravetch JV (1994) A 13-amino-acid motif in the cytoplasmic domain of Fc gamma RIIB modulates B-cell receptor signaling. Nature 368:70–73

Nadler MJS, McLean PA, Neel BG, Wortis HH (1997) B cell antigen receptor-evoked calcium influx is enhanced in CD22-deficient B cell lines. J Immunol 159:4233–4243

Nagai K, Takata M, Yamamura H, Kurosaki T (1995) Tyrosine phosphorylation of Shc is mediated through Lyn and Syk in B cell receptor signaling. J Biol Chem 270:6824–6829

Nishiumi H, Taniguchi I, Yamanashi Y, Kitamura D, Ilic D, Mori S, Watanabe T, Yamamoto T (1995) Impaired proliferation of peripheral B cells and indication of autoimmune disease in lyn-deficient mice. Immunity 3:549–60

Nishizumi H, Horikawa K, Mlinaric-Rascan I, Yamamoto T (1998) A double-edged kinase Lyn: a positive and negative regulator for antigen receptor-mediated signals. J Exp Med 187:1343–1348

Nitschke L, Carsetti R, Ocker B, Kohler G, Lamers MC (1997) CD22 is a negative regulator of B-cell receptor signalling. Curr Biol 7:133–143

Nussenzweig MC (1997) Immune responses: tails to teach a B cell. Curr Biol 7:R355–R357

O'Keefe TL, Williams GT, Davies SL, Neuberger MS (1996) Hyperresponsive B cells in CD22-deficient mice. Science 274:798–801

Ono M, Bolland S, Tempst P, Ravetch JV (1996) Role of the inositol phosphatase SHIP in negative regulation of the immune system by the Fc-gamma-RIIB. Nature 383:263–266

Ono M, Okada H, Bolland S,Yanagi S, Kurosaki T, Ravetch JV (1997) Deletion of SHIP or SHP-1 reveals two distinct pathways for inhibitory signaling. Cell 90:293–301

Otipoby KL, Andersson KB, Draves KE, Klaus SJ, Farr AG, Kerner JD, Perlmutter RM, Law C-L, Clark EA (1996) CD22 regulates thymus-independent responses and the lifespan of B cells. Nature 384:634–637

Pani G, Kozlowski M, Cambier JC, Mills GB, Siminovitch KA (1995) Identification of the tyrosine phosphatase PTP1C as a B cell antigen receptor-associated protein involved in the regulation of B cell signaling. J Exp Med 181:2077–2084

Pao LI, Cambier JC (1997) Syk but not Lyn recruitment to BCR and activation following stimulation of CD45$^{-}$ B cells. J Immunol 158:2663–2669

Pao LI, Bedzyk WD, Persin C, Cambier JC (1997) Molecular targets of CD45 in B cell antigen receptor signal transduction. J Immunol 158:1116–1124

Pao LI, Famiglietti, Cambier JC (1998) Asymmetrical phosphorylation and function of immunoreceptor tyrosine-based activation motif tyrosines in B cell antigen receptor signal transduction. J Immunol 160:3305–3314

Papavasiliou F, Jankovic M, Suh H, Nussenzweig MC (1995) The cytoplasmic domains of immunoglobulin (Ig) a and Igb can independently induce the precursor B cell transition and allelic exclusion. J Exp Med 182:1389–1394

Park H, Wahl MI, Afar DEH, Turck CW, Rawlings DJ, Tam C, Scharenberg AM, Kinet J-P, Witte ON (1996) Regulation of Btk function by a major autophosphorylation site within the SH3 domain. Immunity 4:515–525

Parikh VS, Nakai C, Yokota SJ, Bankert, RB, Tucker PW (1991) COOH terminus of membrane IgM is essential for an antigen-specific induction of some but not all early activation events in mature B cells. J Exp Med 174:1103–1109

Parikh VS, Bishop GA, Liu K-J, Do BT, Ghosh MR, Kim BS, Tucker PW (1992) Differential structure-function requirements of the transmembranal domain of the B cell antigen receptor. J Exp Med 176:1025–1031

Patel KJ, Neuberger MS (1993) Antigen presentation by the B cell antigen receptor is driven by the $\alpha/\beta$ sheath and occurs independently of its cytoplasmic tyrosines. Cell 74:939–946

Pei D, Lorenz U, Klingmuller U, Neel BG, Walsh CT (1994) Intramolecular regulation of protein tyrosine phosphatase SH-PTP1: a new function for Src homology 2 domains. Biochem 33:15483–15493

Pei D, Wang J, Walsh CT (1996) Differential functions of the two Src homology 2 domains in protein tyrosine phosphatase SH-PTP1. Proc Natl Acad Sci USA 93:1141–1145

Peters JD, Furlong MT, Asai DJ, Harrison ML, Geahlen RL (1996) Syk, activated by cross-linking the B-cell antigen receptor, localizes to the cytosol where it interacts with and phosphorylates $\alpha$-tubulin on tyrosine. J Biol Chem 271:4755–4762

Peterson EJ, Clements JL, Fang N, Koretzky GA (1998) Adaptor proteins in lymphocyte antigen-receptor signaling. Curr Opin Immunol 10:337–344

Pezzutto A, Dorken B, Moldenhauer G, Clark EA (1987) Amplification of human B cell activation by a monoclonal antibody to the B cell-specific antigen CD22, Bp 130/140. J Immunol 138:98–103

Plas DR, Johnson R, Pingel JT, Matthews RJ, Dalton M, Roy G, Chan AC, Thomas ML (1996) Direct regulation of ZAP-70 by SHP-1 in T cell antigen receptor signaling. Science 272:1173–1176

Pleiman CM, Abrams C, Gauen LT, Bedzyk W, Jongstra J, Shaw AS, Cambier JC (1994a) Distinct p53/p56$^{lyn}$ and p59$^{fyn}$ domains associate with nonphosphorylated and phosphorylated Ig-$\alpha$. Proc Natl Acad Sci USA 91:4268–4272

Pleiman CM, Chien NC, Cambier JC (1994b) Point mutations define a mIgM transmembrane region motif that determines intersubunit signal transduction in the antigen receptor. J Immunol 152:2837–2844

Plutzky J, Neel BG, Rosenberg RD (1992) Isolation of a src homology 2-containing tyrosine phosphatase. Proc Natl Acad Sci USA 89:1123–1127

Rao A, Luo C, Hogan PG (1997) Transcription factors of the NFAT family: regulation and function. Ann Rev Immunol 15:707–747

Rawlings DJ, Saffran DC, Tsukada S, Largaespada DA, Grimaldi JC, Cohen L, Mohr RN, Bazan JF, Howard M, Copeland NG, Jenkins NA, Witte ON (1993) Mutation of unique region of Bruton's tyrosine kinase in immunodeficient XID mice. Science 261:358–361

Rawlings DJ, Scharenberg AM, Park H, Wahl MI, Lin S, Kato RM, Fluckiger A-C, Witte ON, Kinet J-P (1996) Activation of BTK by a phosphorylation mechanism initiated by SRC family kinases. Science 271:822–825

Reth M (1992) Antigen receptors on B lymphocytes. Annu Rev Immunol 10:97–121
Reth M, Wienands J (1997) Initiation and processing of signals from the B cell antigen receptor. Annu Rev Immunol 15:453–479
Rigley KP, Harnett MM, Phillips RJ, Klaus GGB (1989) Analysis of signaling via surface immunoglobulin receptors on B cells from CBA/N mice. Eur J Immunol 19:2081–2086
Rowley RB, Burkhardt AL, Chao H-G, Matsueda GR, Bolen JB (1995) Syk protein-tyrosine kinase is regulated by tyrosine-phosphorylated Iga/Igb immunoreceptor tyrosine activation motif binding and autophosphorylation. J Biol Chem 270:11590–11594
Salim K, Bottomley MJ, Querfurth E, Zvelebil MJ, Gout I, ScaifeR, Margolis RL, Gigg R, Smith CIE, Driscoll PC, Waterfield MD, Panayotou G (1996) Distinct specificity in the recognition of phosphoinositides by the pleckstrin homology domains of dynamin and Bruton's tyrosine kinase. EMBO J 15:6241–6250
Sanchez M, Misulovin Z, Burkhardt AL, Mahajan S, Costa T, Framke R, Bolen JB, Nussenzweig M (1993) Signal transduction by immunoglobulin is mediated through Ig$\alpha$ and Ig$\beta$. J Exp Med 178:1049–1055
Saouaf SJ, Mahajan S, Rowley RB, Kut SA, Fargnoli J, Burkhardt AL, Tsukada S, Witte ON, Bolen JB (1994) Temporal differences in the activation of three classes of non-transmembrane protein tyrosine kinases following B-cell antigen receptor surface engagement. Proc Natl Acad Sci USA 91:9524–9528
Sato K, Ochi A (1998) Superclustering of B cell receptor and Fc gamma RIIB activates Src homology 2-containing protein tyrosine phosphatase-1. J Immunol 161:2716–2722
Sato S, Miller AS, Inaoki M, Bock CB, Jansen PJ, Tang MLK and Tedder TF (1996) CD22 is both a positive and negative regulator of B lymphocyte antigen receptor signal transduction: altered signaling in CD22-deficient mice. Immunity 5:551–562
Sato S, Jansen PJ, Tedder TF (1997) CD19 and CD22 expression reciprocally regulates tyrosine phosphorylation of Vav protein during B lymphocyte signaling. Proc Natl Acad Sci USA 94:13158–13162
Satterthwaite AB, Cheroutre H, Khan WN, Sideras P, Witte ON (1997) Btk dosage determines sensitivity to B cell antigen receptor cross-linking. Proc Natl Acad Sci USA 94:13152–13157
Saxton TM, van Oostveen I, Bowtell D, Aebersold R, Gold MR (1994) B cell antigen receptor cross-linking induces phosphorylation of the p21$^{ras}$ oncoprotein activators SHC and mSOS1 as well as assembly of complexes containing SHC, GRB-2, mSOS1, and a 145-kDa tyrosine-phosphorylated protein. J Immunol 153:623–636
Schmitz R, Baumann G, Gram H (1996) Catalytic specificity of phosphotyrosine kinases BLK, Lyn, c-Src and Syk as assessed by phage display. J Mol Biol 260:664–677
Shaw AC, Mitchell RN, Weaver YK, Campos-Torres J, Abbas AK, Leder P (1990) Mutations of immunoglobulin transmembrane and cytoplasmic domains:effects on intracellular signaling and antigen presentation Cell 63:381–392
Sideras P, Smith CIE (1995) Molecular and cellular aspects of X-linked agammaglobulinemia. Adv Immunol 59:135–223
Sidman CL, Marshall JD, Allen RD (1989) Murine "viable motheaten" mutation reveals a gene critical to the development of both B and T lymphocytes. Proc Natl Acad Sci USA 86:6279–6282
Sidorenko SP, Law C-L, Chandran KA, Clark EA (1995) Human spleen tyrosine kinase p72Syk associates with the Src-family kinase p53/p56Lyn and a 120-kDa phosphoprotein. Proc Natl Acad Sci USA 92:359–363
Sillman AL, Monroe JG (1994) Surface IgM-stimulated proliferation, inositol phospholipid hydrolysis, $Ca^{2+}$ flux, and tyrosine phosphorylation are not altered in B cells from $p59^{fyn-/-}$ mice. J Leuk Biol 56:812–816
Sillman AL, Monroe JG (1995) Association of p72syk with the src homology-2 (SH2) domains of PLC gamma 1 in B lymphocytes. J Biol Chem 270:11806–11811
Smit L, de Vries-Smits AMM, Bos JL, Boorst J (1994) B cell antigen receptor stimulation induces formation of a Shc-Grb2 complex containing multiple tyrosine-phosphorylated proteins. J Biol Chem 269:20209–20212
Smith KGC, Tarlinton DM, Doody GM, Hibbs ML Fearon DT (1998) Inhibition of the B cell by CD22: a requirement for Lyn. J Exp Med 187:807–811
Sortirellis N, Johnson TM, Hibbs ML, Stanley IJ, Stanley E, Dunn AR, Cheng H-C (1995) Autophosphorylation induces autoactivation and a decrease in the Src homology 2 domain accessibility of the Lyn protein kinase. J Biol Chem 270:29773–29780
Su B, Karin M (1996) Mitogen-activated protein kinase cascades and regulation of gene expression. Curr Opin Immunol 8:402–411

Sutherland CL, Heath AW, Pelech SL, Young PR, Gold MR (1996) Differential activation of the ERK, JNK and p38 mitogen-activated protein kinases by CD40 and the B cell antigen receptor. J Immunol 157:3381–3390

Taddie JA, Hurley TR, Hardwick BS, Bartholomew MS (1994) Activation of B- and T-cells by the cytoplasmic domains of the B-cell antigen receptor proteins Ig-α and Ig-β. J Biol Chem 269:13529–13535

Takata M, Kurosaki T (1996) A role for Bruton's tyrosine kinase in B cell antigen receptor-mediated activation of phospholipase C-γ2. J Exp Med 184:31–40

Takata M, Sabe H, Hata A, Inazu T, Homma Y, Nukada T, Yamamura H, Kurosaki T (1994) Tyrosine kinases Lyn and Syk regulate B cell receptor-coupled Ca2+ mobilization through distinct pathways. EMBO J 13:1341–1349

Tarakhovsky A, Turner M, Schaal S, Mee PJ, Duddy LP, Rajewsky K, Tybulewicz VLJ (1995) 374: 467–470

Tedder TF, Tuscano J, Sato S, Kehrl JH (1997) CD22, a B lymphocyte-specific adhesion molecule that regulates antigen receptor signaling. Annu Rev Immunol 15:481–504

Teh Y-M, Neuberger MS (1997) The immunoglobulin (Ig)a and Igb cytoplasmic domains are independently sufficient to signal B cell maturation and activation in transgenic mice. J Exp Med 185:1753–1758

Thomas JD, Sideras P, Smith CIE, Vorechovsky I, Chapman V, Paul WE (1993) Colocalization of X-linked agammaglobulinemia and X-linked immunodeficiency genes. Science 261:355–358

Thomas ML (1995) Of ITAMs and ITIMs: turning on and off the B cell antigen receptor. J Exp Med 181:1953–1956

Tooze RM, Doody GM, Fearon DT (1997) Counterregulation by the coreceptors CD19 and CD22 of MAP kinase activation by membrane immunoglobulin. Immunity 7:59–67

Treisman R (1996) Regulation of transcription by MAP kinase cascades. Curr Opin Cell Biol 8:205–215

Tseng J, Eisfelder BJ, Clark MR (1997) B-cell antigen receptor-induced apoptosis requires both Igα and Igβ. Blood 89:1513–1520

Tsui FW, Tsui HW (1994) Molecular basis of the motheaten phenotype. Immunol Rev 138:187–206

Tsukada S, Saffran DC, Rawlings DJ, Parolini O, Allen RC, Klisak I, Sparkes RS, Kubagawa H, Mohandas T, Quan S, Belmont JW, Cooper MD, Conley ME, Witte ON (1993) Deficient expression of a B cell cytoplasmic tyrosine kinase in human X-linked agammaglobulinemia. Cell 72:279–290

Tsukada S, Simon MI, Witte ON, Katz A (1994) Binding of βγ subunits of heterotrimeric G proteins to the PH domain of Bruton's tyrosine kinase. Proc Natl Acad Sci USA 91:11256–11260

Turner M, Mee JP, Costello PS, Williams O, Price AA, Duddy LP, Furlong MT, Geahlen RL, Tybulewicz VLJ (1995) Perinatal lethality and blocked B-cell development in mice lacking the tyrosine kinase Syk. Nature 378:298–302

Vambutas V, Kaplan DR, Sells MA, Chernoff J (1995) Nerve growth factor stimulates tyrosine phosphorylation and activation of Src homology-containing protein-tyrosine phosphatase 1 in PC12 cells. J Biol Chem 270:25629–25633

Vetrie DI, Vorechovsky I, Sideras P, Holland J, Davies A, Flinter F, Hammarstrom L, Kinnon C, Levinsky R, Bobrow M (1993) The gene involved in X-linked agammaglobulinemia is a member of the src family of protein tyrosine kinases. Nature 361:226–233

Vostal JG, Jackson WL, Shulman NR (1991) Cytosolic and stored calcium antagonistically control tyrosine phosphorylation of specific platelet proteins. J Biol Chem 266:16911–16916

Wickler LS, Scher I (1986) X-linked immune deficiency (xid) of CBA/N mice. Curr Top Microbiol Immunol 124:87–101

Wienands J, Schweikert J, Wollscheid B, Jumaa H, Nielsen PJ, Reth M (1998) SLP-65: a new signaling component in B lymphocytes which requires expression of the antigen receptor for phosphorylation. J Exp Med 188:791–795

Williams GT, Venkitaraman AR, Gilmore DJ, Neuberger MS (1990) The sequence of the μ transmembrane segment determines the tissue specificity of the transport of immunoglobulin M to the cell surface. J Exp Med 171:947–952

Williams GT, Peaker CJG, Patel KJ, Neuberger MS (1994) The α/β sheath and its cytoplasmic tyrosines are required for signaling by the B-cell antigen receptor but not for capping or for serine/threonine-kinase recruitment. Proc Natl Acad Sci USA 91:474–478

Yamanashi Y, Kakiuchi T, Mizuguchi J, Yamamoto T, Toyoshima K (1991) Association of B cell antigen receptor with protein tyrosine kinase Lyn. Science 251:192–194

Yanagi S, Sugawara H, Kurosaki M, Sabe H, Yamamura H, Kurosaki T (1996) CD45 modulates phosphorylation of both autophosphorylation and negative regulatory tyrosines of Lyn in B cells. J Biol Chem 271:30487–30492

Yang W, McKenna SD, Jiao H, Tabrizi M, Lynes MA, Schultz LD, Yi T (1998) SHP-1 deficiency in B-lineage cells is associated with heightened lyn protein expression and increased lyn kinase activity. Exp Hematol 26:1126–1132

Yao L, Kawakami Y, Kawakami T (1994) The pleckstrin homology domain of Bruton tyrosine kinase interacts with protein kinase C. Proc Natl Acad Sci USA 91:9175–9179

Yi T, Cleveland JL, Ihle JN (1992) Protein tyrosine phosphatase containing SH2 domains: characterization, preferential expression in hematopoietic cells, and localization to human chromosome 12p12-p13. Mol Cell Biol 12:836–846

Yi T, Mui AL-F, Krystal G, Ihle JN (1993) Hematopoietic cell phosphatase associates with the interleukin-3 (IL-3) receptor β chain and down-regulates IL-3-induced tyrosine phosphorylation and mitogenesis. Mol Cell Biol 13:7577–7586

Yohannan BJ, Wienands J, Coggeshall KM, Justement LB (1999a) Analysis of tyrosine phosphorylation-dependent interactions between stimulatory effector proteins and the B cell co-receptor CD22. J Biol Chem (in press)

Yohannan BJ, Brody BA, Justement LB (1999b) Identification of specific ITIM motifs in the cytoplasmic domain of CD22 that mediate binding of the protein tyrosine phosphatase SHP-1 and are required for inhibitory function. Submitted for publication

Zhang R, Alt FW, Davidson L, Orkin SH, Swat W (1995) Defective signalling through the T- and B-cell antigen receptors in lymphoid cells lacking the vav proto-oncogene. Nature 374:470–473

# The B-Cell Antigen Receptor: Formation of Signaling Complexes and the Function of Adaptor Proteins

J. WIENANDS



## 1 Introduction

To initiate the humoral immune response, pathogens or foreign substances are recognized and bound by antigen-specific receptors expressed on the cell surface of mature, resting B lymphocytes. Engagement of the B-cell antigen receptor (BCR) triggers the transition of the B cell from the $G_o$ into the $G_1$ phase of the cell cycle. Further, BCR-antigen complexes become endocytosed, processed, and peptides of

---

Department for Molecular Immunology, Biology III, University of Freiburg and Max-Planck-Institute for Immunobiology, Stübeweg 51, D-79108 Freiburg, Germany
e-mail: wienands@immunbio.mpg.de

protein antigens are presented on the B-cell surface in association with major histocompatibility (MHC) class-II proteins. This enables the B cell to collaborate with antigen-specific T lymphocytes. Such a *cognate* B-cell/T-cell interaction can provide both cell types with appropriate proliferation and differentiation signals. The now fully activated B cells can differentiate, for instance, into plasma cells, which help to neutralize the antigen by secreting large amounts of soluble, antigen-specific antibodies into the body's *humor* (Greek for fluid). The basics of such a scenario were first postulated by Paul Ehrlich in his *Seiten-Ketten* (side chain) theory about one century ago. His idea was developed further by Nils Jerne into the *Natural Selection Theory* (1955) and by Sir MacFarlane Burnet into the *Clonal Selection Theory of Aquired Immunity* (1959). Membrane-bound antibody molecules (immunoglobulins, Ig), as a central element of all three theories, were detected on the cell surface of lymphocytes in the early 1970s, but their signaling and activation function for the cell was neglected until the mid 1980s. Since then, much has been learned about the molecular mechanisms utilized by the BCR to activate the B cell.

In this review, I summarize our current picture of the molecular mechanisms that are involved in the initiation of BCR signal transduction. Some related work on the T-cell antigen receptor (TCR) and Fc receptors (FcRs) will also be described. I will focus on more recent progress in our understanding of the coupling of intracellular effector proteins and how this relates to the specificity of BCR signaling pathways and to the fine tuning of the BCR response.

## 2 The Structure of the B-Cell Antigen Receptor and its Function for the Development of B Lymphocytes

### 2.1 The BCR is a Multiprotein Complex

The BCR and TCR, together with some members of the FcR families for IgG, IgE and IgA, belong to the class of multi subunit immunoreceptors. These receptors share two characteristic features. First, the ligand-binding structure and the signaling function are located on different polypeptides. In all cases, the signaling subunits are non-covalently associated with the ligand-binding subunits. Second, immunoreceptors do not posses an intrinsic enzymatic activity but their signaling components contain a conserved peptide motif, the immunoreceptor tyrosine-based activation motif (ITAM), which couples the receptor to related sets of cytoplasmic effector proteins.

Antigen recognition by the BCR is mediated by transmembrane Ig molecules, which are disulfide-linked, tetrameric proteins of two Ig-heavy and two Ig-light chains and which belong to five different classes: IgM, IgD, IgG, IgE and IgA (Reth 1992). As a result of somatic recombination of the gene segments encoding the variable region of the heavy and light chains, Ig molecules are clonotypic, and a diversity can be generated that enables the organism to acquire immune specificity to any possible antigen, including synthetic antigens that probably do not occur in a natural environment. The signaling function of the BCR is achieved by two transmembrane proteins, Ig-α (CD79a) and Ig-β (CD79b), which form a disulfide-linked heterodimer (Campbell and Cambier 1990; Hombach et al. 1990b; Wienands et al. 1990; Venkitaraman et al. 1991). Ig-α and Ig-β are encoded by the *mb-1* and the *B29* gene, respectively (Hombach et al. 1990a; Campbell et al. 1991), which are expressed in B cells and in some precursor T cells. As deduced from the *mb-1* and *B29* cDNAs, both proteins contain one extracellular Ig-domain, a membrane-spanning α-helix and an intracellular part of 61 and 48 amino acids, respectively. None of the BCR components can be expressed independently on the cell surface but require complex formation.

## 2.2 Expression of Functional BCR Components Is Controlled During B-Cell Development

In Ig-negative pro-B and pre-B cells, the Ig-αβ heterodimer is expressed and may associate with surrogate Ig chains to form the pro-BCR and pre-BCR, respectively (Melchers et al. 1995). Although the exact structure of the pro-BCR is unclear, the functional relevance of these receptors is demonstrated by gene-targeting experiments. In Ig-β-deficient mice, B-cell development is arrested at an early stage, even before V-exon rearrangement on the Ig-heavy-chain locus occurs (Gong and Nussenzweig 1996). The generation of pre-B cells is severely compromised in the absence of the surrogate light chain component λ5, and is completely blocked in mouse mutants that lack the transmembrane form of the μ-heavy chain (Kitamura et al. 1991, 1992). The developmental block of the latter could be overcome by introducing a transgene encoding an IgM fusion protein which contains the cytoplasmic region of either Ig-α or Ig-β (Teh and Neuberger 1997). Deletion of the BCR on mature B cells by inducible targeting of the rearrangement V-exon segment leads to rapid death of the cells by apoptosis (Lam et al. 1997). Together, these results show that expression of a BCR permits precursor B cells to proceed to the next developmental stage and is required for the survival of mature, immunocompetent B cells in the periphery as well. Since BCR expression can couple to the activation of intracellular signaling pathways, it is conceivable that Ig-αβ-mediated signaling is required in all stages of B-cell development. It is still an open question whether a specific ligand triggers signal transduction through the pro- and pre-BCR (for example, by binding to the extracelluar part of Ig-α and/or Ig-β) or whether surface expression of the receptor per se results in a low but constant default signal flow that is sufficient to drive B-cell development and to maintain naive B cells. Alternatively, but less likely, progenitor B cells and mature B cells devoid of a BCR could be recognized by other cellular systems and are actively eliminated.

The functional potential of the BCR is tested not only for reaction with non-self antigens but also for its ability to recognize self antigens, which could lead to autoreactivity and destruction of its own tissues. A number of systems have evolved to prevent this *horror autotoxicus* ("Paul Ehrlich"). One includes the ability of the

BCR to deliver a "negative signal" when expressed on immature B cells ($IgM^+$/$IgD^-$) that are at the exit to leave the bone marrow. Engagement of the BCR on these cells is interpreted as an encounter with an auto antigen and is translated into a suicide signal that triggers apoptotic cell death.

In conclusion, the BCR does not only function as a *Seiten-Kette* that selects the antigen-specific B cell and promotes its maturation into an Ig-secreting plasma cell. The BCR also provides the critical check point signal that, on one hand, supports the development and survival of those lymphocytes that are potentially useful because they have a functional receptor and that, on the other hand, eliminates useless or harmful lymphocytes which either failed to produce a complete BCR or which express a BCR with specificity directed against an auto antigen. How the BCR signaling machinery can accomplish these diverse biological outcomes is poorly understood. Most of our current knowledge – as it is outlined in the following sections – is obtained from experiments with cells or cell lines that are activated upon BCR triggering. Only recently have clues been provided that may help to explain the BCR-mediated inhibition of B-cell growth.

## 3 The ITAM – 10 Years After

The short cytoplasmic tail of the transmembrane Ig molecules made it difficult to understand how they could participate in signal transduction. The discovery of Ig-α and Ig-β as integral components of the BCR resolved this problem because their cytoplasmic parts could provide the necessary link that is required to transduce the initial extracellular signal (i.e., antigen binding) across the plasma membrane into the cell interior. For their communication with intracellular effector proteins, the Ig-α and Ig-β cytoplasmic part contain one copy of the ITAM found by MICHAEL RETH (1989). The ITAM is characterized by six conserved amino acids with the consensus sequence $D/Ex_7D/ExxYxxI/Lx_7YxxI/L$ (in single-letter code for amino acids, where X can be any residue). Beside Ig-α and Ig-β, ITAMs are present with one or three copies in the signaling subunits of the TCR, in the common FcR-γ subunit of FcRs for IgG (FcγRI, FcγRIIA/C, FcγRIIIA), IgE (FcεRI) and IgA (FcαR) and in the FcεRI-β subunit (for review of FcR structures see, DAERON 1997). Interestingly, certain membrane proteins encoded by the Epstein-Barr virus and the bovine leukemia virus also posses one ITAM. In these cases, the activation function of ITAM may be misused by the virus for neoplastic transformation of the host cell.

That the ITAM is the basic signaling module of the BCR and other immunoreceptors was demonstrated by the analysis of chimeric transmembrane proteins containing the cytoplasmic domain of Ig-α or Ig-β (KIM et al. 1993; SANCHEZ et al. 1993; FLASWINKEL and RETH 1994; TADDIE et al. 1994; WILLIAMS et al. 1994) or of the signaling subunits of the TCR. The activation function of the receptor chimeras was dependent on the conserved tyrosine and leucine/isoleucine residues as well as on the precise spacing between them. Moreover, the ITAM was identified as the minimally functional segment that is necessary and sufficient to transmit an activation signal into the cell. Finally, aggregation of ITAM-containing chimeras initiated intracellular responses that were similar to those observed upon antigen-stimulation of wild-type antigen receptors. These events include activation of protein tyrosine kinases (PTKs), hydrolysis of phosphoinositides, mobilization of $Ca^{2+}$ ions, activation of the ras/mitogen-activated protein (MAP) kinase pathway and phosphatidyl 3-OH-kinase (PI3K) activation (CAMPBELL and SEFTON 1990; GOLD et al. 1990, 1992; HARNETT 1994). Prominent PTK substrates in BCR-stimulated cells are the ITAM tyrosine residues of Ig-α and to a lesser extent of Ig-β (CAMPBELL and CAMBIER 1990; GOLD et al. 1991). The functional elucidation of ITAM phosphorylation was tightly connected to the discovery of Src homology (SH) 2 domains. SH2 domains are non-catalytic protein domains that are found in a variety of cytoplasmic proteins and that mediate specific binding to tyrosine-phosphorylated proteins (for review see: MAYER and GUPTA 1998). Thus, phosphorylated ITAMs can provide the activated BCR with a docking site for selective communication with SH2-domain-containing signaling proteins.

## 4 Cytoplasmic PTKs Are Early BCR Effector Proteins

Protein tyrosine phosphorylation can be prevented by treatment of cells with pharmacological agents that specifically inhibit or lead to degradation of cytoplasmic PTKs (CARTER 1991; LANE 1991). These experiments revealed that the above-mentioned BCR signaling pathways are critically dependent on the induction of PTK substrate phosphorylation. Therefore, PTK activation is a receptor-proximal event. BCR engagement activates three classes of cytoplasmic PTKs: the Src family members Lyn, Blk, Fyn, Lck and Fgr; the PTK Syk; and the Tec family member BTK. The PTKs interact in a complicated network not only during the onset of BCR signaling; they also perform an inhibitory function that is required to modulate and to finally switch off the BCR signal.

### 4.1 Src Family PTKs

Eight mammalian members of Src-related PTKs have been identified to date. They are composed of multiple protein domains which include an N-terminal unique region followed by one SH3, SH2 and kinase (SH1) domain. In contrast to other cytoplasmic PTKs and of critical importance for their function, Src family PTKs are constitutively attached to the inner leaflet of the plasma membrane by virtue of N-terminal myristylation and/or palmitylation. Lyn, Blk, Fyn and Lck can associate directly, although at low stoichiometry, with the unligated

BCR via their unique region, and together with Fgr, they become activated upon BCR ligation (Burkhardt et al. 1991; Yamanashi et al. 1991; Campbell and Sefton 1992; Clark et al. 1992; Lin and Justement 1992; Clark et al. 1994). So far no direct evidence suggests a role for Hck, Yes and Src in BCR signal transduction.

Lyn is the dominantly expressed Src family member in the chicken B-cell line DT40. Targeted disruption of the *lyn* gene in these cells resulted in a profound decrease of protein tyrosine phosphorylation and slow $Ca^{2+}$ mobilization upon BCR stimulation (Takata et al. 1994). The parental phenotype could be reconstituted with expression constructs for either wild-type Lyn, Fyn and Lck, but not Src. In B cells from Lyn-deficient mice, PTK substrate phosphorylation was delayed but not drastically impaired (Chan et al. 1997), which is probably due to the compensation of Lyn function by other Src family members. Unexpectedly, in these cells, the peak $Ca^{2+}$ flux, the MAP kinase activation and the proliferative response to BCR engagement increased compared with wild-type B cells. This correlated with the enhanced negative selection of B cells in the presence of antigen and with spontaneous hyperactivity and autoreactivity in the absence of antigen (Hibbs et al. 1995; Nishizumi et al. 1995; Chan et al. 1997; Cornall et al. 1998). Collectively, these results show that Lyn can initiate BCR signal transduction, but this function can also be accomplished by other Src family members. Nevertheless, the presence of at least one of the BCR-regulated Src family PTKs seems to be required for efficient BCR signaling transduction. Lyn itself is an essential component for the inhibition of the BCR. As will be described, this function involves the surface protein CD22, the phosphatase SHP-1 and perhaps Syk.

To keep Src family PTKs in an inactive state, they need to become phosphorylated by the PTK Csk (for C-terminal Src kinase) on a regulatory tyrosine residue at the C-terminal end. Once phosphorylated, the phosphotyrosine residue binds back to the SH2 domain of the same molecule. In concert with the intramolecular binding of the SH3 domain to a proline-rich region, these interactions convey Src family PTKs in a closed conformation that does not allow substrate binding (Sicheri et al. 1997; Xu et al. 1997). Lyn from Csk-negative DT40 B cells is constitutively activated, but BCR-stimulation is still necessary to evoke full PTK substrate phosphorylation in these cells (Hata et al. 1994). This may reflect the requirement of Src-like PTKs for transphosphorylation at a conserved tyrosine residue in the catalytic domain in order to become fully active. This process may be dependent on BCR aggregation to bring the PTKs in close proximity to each other. Transition from the closed to the open conformation requires dephosphorylation of the C-terminal tyrosine residue. The reaction is thought to be catalyzed by the transmembrane phosphatase CD45 (Yanagi et al. 1996), suggesting a positive and BCR-proximal regulatory role for this enzyme. This is consistent with the numerous signaling deficits observed in CD45-negative B-cell lines and in B cells from genetically engineered mice (Justement et al. 1990, 1991; Benatar et al. 1996; Cyster et al. 1996).

## 4.2 Syk

Syk is expressed in most hematopoietic cell types and comprises two tandemly arranged SH2 domains at the N-terminus and a catalytic domain at the C-terminus. The second Syk family member, ZAP-70, is specifically expressed in T and NK cells. A number of experiments suggest that Syk is a downstream effector of Src-like PTKs. (1) Syk phosphorylation and kinase activity is strongly reduced in the Lyn-negative DT40 B cells (KUROSAKI et al. 1994; TAKATA et al. 1994). (2) A direct activation of Syk by Src-like PTKs through a phosphorylation-dependent mechanism is suggested by co-transfection experiments in COS cells (KUROSAKI et al. 1994). (3) The kinetics of Syk activation is delayed compared with that of Src-like PTKs (SAOUAF et al. 1994).

These experiments, together with the analysis of Syk transmembrane chimeras (KOLANUS et al. 1993; RIVERA and BRUGGE 1995), revealed that Syk can also be activated – at least to some extent – in the absence of Src-like PTKs. This may account for the residual $Ca^{2+}$ mobilization observed upon BCR-stimulation of Lyn-negative DT40 cells. In contrast, Syk-negative DT40 cells do not hydrolyze phosphoinositides and do not mobilize $Ca^{2+}$ ions in response to BCR stimulation, demonstrating that Syk is indispensable for these events (TAKATA et al. 1994). Reconstitution of the signaling defects requires not only the catalytic domain but also both Syk SH2 domains (KUROSAKI et al. 1995). The tandem SH2 domains of Syk bind specifically to doubly-phosphorylated ITAMs of Ig-α and/or Ig-β (WIENANDS et al. 1995), and this is reported to increase Syk activity (ROWLEY et al. 1995). It may also bring Syk into proximity of Src family PTKs allowing phosphorylation and further activation of Syk. The current model, therefore, is that, upon ITAM phosphorylation of Ig-α and/or Ig-β by Src-like PTKs and probably also by Syk, a concerted action of these PTK families activates downstream BCR signaling cascades. Indeed, a complete block of protein tyrosine phosphorylation, phosphoinositide hydrolysis and $Ca^{2+}$ mobilization is observed in Lyn/Syk-double negative DT40 B cells (TAKATA and KUROSAKI 1996). An important role for Syk in B-cell development and in signal transduction from the pro- and pre-BCR is demonstrated by tracing the fate of Syk-deficient stem cells in irradiated or recombination-defective mice. The development of Syk-negative B cells is blocked at the transition from the pro- to pre-B cell stage. A few immature, but no mature Syk-negative B cells, accumulate in the periphery (CHENG et al. 1995; TURNER et al. 1995).

## 4.3 Btk

Various mutations in the gene encoding the Bruton's tyrosine kinase (Btk) are responsible for the X-linked agammaglobulinemia (XLA) in humans and the X chromosome-linked immunodeficiency (xid) of CBA/N mice (RAWLINGS and WITTE 1994). The XLA disease is characterized by a severe reduction of BCR-positive B cells due to a developmental arrest at the pre-B-cell stage. This leads to very low levels of all Ig classes in the serum of XLA patients. The phenotype of xid is less severe since some B cells can be detected in the periphery. However, these B cells are immature ($IgM^{high}/IgD^{low}$) and fail to respond to thymus-independent type-II antigens and to lipopolysaccharides. Complete ablation of Btk function by gene targeting in mice leads to a phenotype that is more reminiscent of xid than XLA (KERNER et al. 1995; KHAN et al. 1995), supporting the interesting notion that the development of mature B cells ($IgM^{low}/IgD^{high}$) is less dependent on Btk in mice than in humans.

In addition to Btk, the Tec family of cytoplasmic PTKs includes Tec I and II, Itk, Bmx and Txk. They consist of an N-terminal pleckstrin homology (PH) domain of about 140 amino acids, followed by a Tec homology region (TH), one SH3, one SH2 and a catalytic domain at the C-terminus. Mutations causing XLA are found in all of the Btk domains, while xid is caused by a point mutation in the PH domain. PH domains bind to phosphoinositide phospholipids and thus can tether cytoplasmic proteins to the plasma membrane (LEMMON and FERGUSON 1998). The Btk PH domain binds preferentially to phosphatidylinositiol 3,4,5-triphosphate (PtdIns 3,4,5-$P_3$), which is the product of activated PI3K (SALIM et al. 1996; TOKER and CANTLEY 1997). Thus, in contrast to Src family PTKs, BTK, and perhaps also other Tec family members, are associated with the plasma membrane in a stimulation-dependent manner; however, similar to Src-like PTKs, membrane anchoring is required for BTK to function.

Following BCR stimulation, Btk-deficient DT40 B cells exhibited an almost normal pattern of phosphotyrosine-containing proteins and normal activation of Lyn, Syk and MAP kinase (TAKATA and KUROSAKI 1996). Yet, phosphoinositide hydrolysis and $Ca^{2+}$ mobilization were completely abrogated in these cells. Similarly, $Ca^{2+}$ signaling was severely compromised in B cells from CBA/N mice and XLA patients (RIGLEY et al. 1989; FLUCKIGER et al. 1998). Btk mutants in the PH, SH2 and kinase domains were unable to fully compensate the defects (TAKATA and KUROSAKI 1996; FLUCKIGER et al. 1998). Thus, not only Syk but also Btk is indispensable for the above-mentioned signaling pathways and this requires membrane attachment of Btk and association with other phosphotyrosine-containing proteins via the Btk SH2 domain. Moreover, while activation of Lyn and Syk is independent of Btk (TAKATA and KUROSAKI 1996), recent experiments show that Btk is activated by Src-like PTKs but not by Syk (MAHAJAN et al. 1995; AFAR et al. 1996; RAWLINGS et al. 1996). This involves a two-step mechanism in which Src-like PTKs first phosphorylate Btk, allowing Btk to autophosphorylate at a second site to achieve full activity. Indeed, Btk is phosphorylated and activated with slower kinetics than Src-like PTKs, but before Syk (SAOUAF et al. 1994). Collectively, the data implicate that Btk and Syk act in concert to achieve phosphoinositide hydrolysis and $Ca^{2+}$ mobilization. They seem, however, not to work synergistically, but to fulfill quite distinct functions, both of which are required for the efficient activation of these signaling events.

## 5 The Adaptor Protein SLP-65: Linking PTK Activation to Downstream BCR Signaling Pathways

Key questions in understanding BCR signal transduction are (1) how do activated PTKs find and interact with their substrate proteins, (2) what are these substrate proteins, and (3) how do they achieve coordinated and specific activation of BCR signaling pathways, even in those cases where some signaling elements are common to the BCR and other receptor types? The recent identification of the protein adaptor family SLP (for SH2-containing Leukocyte Protein) provided important insight into some aspects of the molecular mechanisms underlying these processes.

### 5.1 SLP-65 and SLP-76

Kinetic studies of BCR-dependent protein tyrosine phosphorylation in the myeloma cell line J558Lμm3 revealed the very early and prominent phosphorylation of a 65-kDa protein (WIENANDS et al. 1996). This protein was not recognized by antibodies against known PTK substrates and had an unusually high pI of 8.5–9. In the absence of a BCR, treatment of J558L cells with the phosphatase inhibitor pervanadate did not induce phosphorylation of p65, indicating that this protein is involved in coupling intracellular signaling elements specifically to the BCR. Based on peptide sequences, the p65 cDNA was cloned and found to encode a B-cell specific protein of 457 amino acids (WIENANDS et al. 1998). The protein was named SLP-65 because its overall structure and amino acid sequence is most similar to that of the adaptor protein SLP-76, a PTK substrate in TCR-stimulated T cells (JACKMAN et al. 1995). The N-terminal portion of SLP-65 contains seven consensus tyrosine phosphorylation sites, followed by a proline-rich part with several SH3 domain-binding motives and a C-terminal SH2 domain (Fig. 1). The mouse and human cDNAs for this protein were independently cloned by FU et al. (1998) and called BLNK (for B cell linker protein). These authors demonstrated two isoforms of the protein in human B cells, which are generated by alternative splicing.

One common feature of SLP-65 and SLP-76 is the presence of multiple consensus tyrosine phosphorylation sites of the YxxP type, otherwise found in the PTK substrates Cas, p62dok and Cbl (LANGDON et al. 1989; SAKAI et al. 1994; YAMANASHI and BALTIMORE 1997). In TCR-stimulated T cells, SLP-76 is phosphorylated by ZAP-70, the T- and NK-cell specific Syk family member. SLP-65/BLNK is phosphorylated by Syk but not by Lyn or Btk, as demonstrated by coexpression in insect cells and by the analysis of PTK-deficient DT 40 B cells (FU et al. 1998). In vitro kinase assays indicate that SLP-65 phosphorylation requires the presence of the growth receptor-binding protein 2 (Grb2), perhaps to keep SLP-65 in a conformation that is accessible to the kinase (Wienands et al., unpublished results). Indeed, co-precipitation experiments showed that an association between Grb2 and SLP-65 already exists in unstimulated cells (FU et al. 1998; WIENANDS

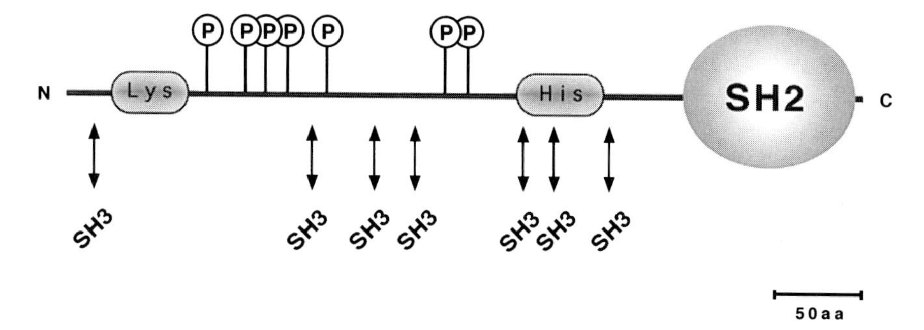


**Fig. 1.** Schematic representation of the SLP-65 protein structure. Consensus peptide motifs for protein tyrosine phosphorylation (*P*) and SH3 domain binding (*SH3*) are indicated. *Lys* and *His* are basic regions with a very high content of lysine and histidine residues, respectively. *SH2*, Src homology 2 domain

et al. 1998) and that this complex contains a PTK activity (probably Syk) that is able to phosphorylate SLP-65 in vitro (Wienands et al., unpublished results). Such a preformed BCR-regulated signaling complex had been previously postulated, and it explains the rapid phosphorylation of SLP-65 and the specific onset of downstream signaling pathways following BCR engagement (Wienands et al. 1996).

## 5.2 SLP-65 and Ras Activation

Biochemical studies demonstrated the constitutive interaction of SLP-65 and SLP-76 with Grb2 and, furthermore, that this association is increased in antigen receptor-stimulated cells (Fu et al. 1998; Wienands et al. 1998). The Grb2 protein comprises two SH3 domains separated by a central SH2 domain. The N-terminal SH3 domain is sufficient to bind SLP-65 from unstimulated cells, while the additional presence of the Grb2 SH2 domain is required for the increased SLP-65 binding in stimulated cells (Wienands et al. 1998). Grb2 has long been implicated in the activation of the ras pathway by virtue of its constitutive association to the guanine nucleotide exchange factor SOS, a positive regulator of ras. A tri-molecular complex of SLP-65/BLNK, Grb2 and SOS was isolated and becomes associated with the membrane fraction following BCR ligation (Fu et al. 1998). The translocation of SOS to its membrane-located effector protein ras may be one way the BCR regulates ras activity. Alternative and/or additional mechanisms may exist as is illustrated by the BCR-induced tyrosine phosphorylation of the adaptor protein SHC (Lankester et al. 1994; Saxton et al. 1994; Smit et al. 1994) and the ras GTPase-activating protein, p120GAP (Gold et al. 1993). Following BCR stimulation, tyrosine phoshorylation of SHC augments assembly of Grb2 and SOS, and a multi-protein complex encompassing SHC and Grb2, but not SLP-65, is translocated to the plasma membrane. The SH2 domain of SHC binds to the tyrosine-phosphorylated Ig-α/Ig-β heterodimer in vitro (Baumann et al. 1994), and an in vivo interaction between Ig-α and SHC was found upon expression of the proteins in COS cells (NAGAI et al. 1995). As p120GAP is a negative regulator of ras, the observed decrease of its activity upon phosphorylation could increase ras activity. One ras effector protein is the serine/threonine kinase Raf, which activates the MAP kinase pathway. Further investigations are needed to convincingly solve the long-standing question of BCR-induced ras/MAPK activation.

## 5.3 Vav and the Reorganization of the Actin Cytoskeleton

A second signaling protein that associates with SLP-65 in a constitutive and stimulation-dependent manner is Vav (FU et al. 1998; WIENANDS et al. 1998). Similar results are reported for SLP-76 in T cells. Tyrosine-phosphorylated Vav is a guanine nucleotide exchange factor for the Rho family of small G proteins (CRESPO et al. 1997) and plays a central role in the antigen receptor-induced reorganization of the actin cytoskeleton. T cells from Vav-negative mutant mice are deficient for this process and show severe defects in TCR cap formation, cytokine production, $Ca^{2+}$ mobilization and cell-cycle progression (TARAKHOVSKY et al. 1995; ZHANG et al. 1995; FISCHER et al. 1998; HOLSINGER et al. 1998). A functional actin cytoskeleton is also thought to be a prerequisite for the endocytosis of BCR-antigen complexes leading to antigen presentation. Coupling of Vav to the activated BCR and TCR and the concomitant reorganization of the cytoskeleton seems to involve SLP-65 and SLP-76, respectively (Fig. 2). A second signaling element that may contribute to this process is the SH3P7 protein (SPARKS et al. 1996). Recent studies identified SH3P7 as a cytoskeleton adaptor protein that becomes phosphorylated in activated B and T cells on two tyrosine residues (LARBOLETTE et al. 1998). Notably, both tyrosines are located within YxxP motifs, also present in SLP-65 in SLP-76.

## 5.4 The Complexity of $Ca^{2+}$ Mobilization

The SLP adaptor proteins posses a key regulatory function for the mobilization of $Ca^{2+}$ in antigen-receptor-stimulated lymphocytes. The TCR-induced $Ca^{2+}$ flux is nearly abrogated in a SLP-76-deficient T-cell line (YABLONSKI et al. 1998) and overexpression of SLP-65 in B cells leads to an enhanced response (FU et al. 1998), which directly correlates with the amount of SLP-65 in the cells (Wollscheid, Reth and Wienands, unpublished results). Mutant SLP-65/BLNK, carrying four Y > F substitutions, has a dominant negative effect, demonstrating the requirement of SLP-65 tyrosine phosphorylation (FU et al. 1998).

How does SLP influence $Ca^{2+}$ signaling? $Ca^{2+}$ mobilization is initiated by tyrosine phosphorylation and subsequent activation of the two isoforms of phospholipase C, PLC-$\gamma$1 and 2, which convert phosphatidylinositol 4,5-diphosphate (PtdIns 4,5-$P_2$) into diacylglycerol (DAG) and Inositol 1,4,5-triphosphate (IP3). These two metabolites act as second messengers to activate protein kinase C (PKC) and to induce the release of $Ca^{2+}$ ions from specialized parts of the endoplasmic reticulum, respectively (Fig. 2a). The transient mobilization of intracellular $Ca^{2+}$ is followed by a sustained capacitative entry of extracellular $Ca^{2+}$ through plasma membrane channels. Tyrosine-phosphorylated SLP-65 and SLP-76 regulate PLC-γ activation by binding the PLC-γ SH2 domain (Fu et al. 1998; Yablonski et al. 1998), hence facilitating PLC-γ phosphorylation by Syk and ZAP-70, respectively. Following overexpression of SLP-65/BLNK, phosphorylation of PLC-γ is augmented while it is reduced in B cells expressing the Y > F mutant and it is uncoupled from TCR stimulation in the SLP-76-deficient T-cell line. This explains the different $Ca^{2+}$ responses of those cells mentioned above. An efficient $Ca^{2+}$ response to BCR stimulation not only requires Syk but also Btk. This reflects the fact that PLC-γ is a direct substrate for both Syk and Btk (Takata et al. 1994; Takata and Kurosaki 1996). Accordingly, IP3 generation and PLC-γ phosphorylation are substantially reduced or abolished in some but not all B cells that are deficient of one of these PTKs. Nonetheless, both PTKs exert a non-redundant function on $Ca^{2+}$ mobilization because activation of Syk and Btk are mutually independent and overexpression of one cannot compensate for the defect of the other. As recently demonstrated, Syk function is responsible for the initial $Ca^{2+}$ release from internal stores, while Btk function specifically affects the second sustained phase of $Ca^{2+}$ mobilization in slope, peak and duration (Fluckiger et al. 1998; Scharenberg et al. 1998). A similar function is described for Itk in T cells (Liu et al. 1998). A likely explanation for these phenomena is that Syk and Tec family members have to phosphorylate the same PLC-γ molecule but on distinct sites to achieve sufficient PLC-γ activation and to evoke a full $Ca^{2+}$ response.

The analysis of the $Ca^{2+}$ signal in stimulated B cells overexpressing wild-type SLP-65 revealed that not only the magnitude but also the duration of the response is strongly enhanced (Fu et al. 1998; Wollscheid et al. unpublished results). The data suggest a positive effect of SLP-65 overexpression on the function of Btk. This observation, together with the fact that Btk needs a wild type SH2 domain to promote sustained $Ca^{2+}$ elevation (see above), could be explained by the recruitment of Btk to tyrosine-phosphorylated SLP-65 (Fig. 2b). This would bring Btk into the vicinity of SLP-65-bound PLC-γ and facilitates phosphorylation of PLC-γ by Btk. The hypothesis is supported by the specific binding of the SH2 domain of either Btk or Itk to tyrosine-phosphorylated SLP-65 in vitro (and to some extent to phosphorylated Ig-α/Ig-β) (Wienands et al., unpublished results). Consistent with this, Itk can substitute for Btk in BCR-induced $Ca^{2+}$ mobilization (Fluckiger et al. 1998; Scharenberg et al. 1998). Since overexpression of SLP-65/BLNK directly augments the Syk-mediated phosphorylation of PLC-γ, it is possible that Syk, Btk and PLC-γ exist in one complex, allowing Syk and Btk to phosphorylate the same PLC-γ molecule (Fig. 2b). So far, we have been unable to detect Btk in anti-SLP-65 immune complexes prepared from BCR-stimulated B cells. This may be due to a more transient interaction of these proteins or to disruption of the complex by our polyclonal antibodies. The pH domain of Btk is required for the Btk activation process (Scharenberg et al. 1998), but may also play a role for the association of Btk with a PLC-γ molecule that has been attached to the plasma membrane via its own PH domain. A PI3K-induced membrane targeting and activation of PLC-γ has been reported in fibroblasts (Rhee and Bae 1997; Bae et al.

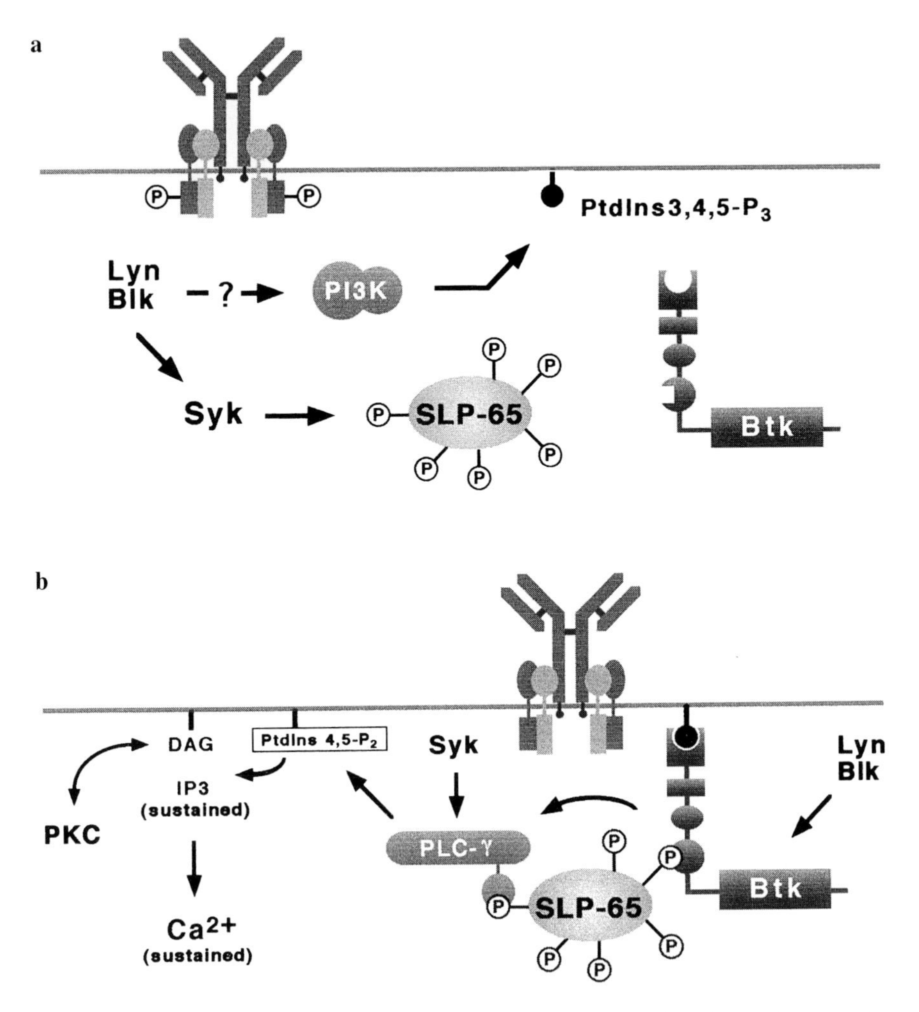


**Fig. 2a–d.** Model for the activation of BCR signaling cascades. **a** Following B-cell antigen receptor (BCR) ligation, Src-like protein tyrosine kinases and Syk mediate early signaling events, including tyrosine phosphorylation of SLP-65 and activation of PI3 kinase. **b** The src homology (SH)2-mediated binding of phospholipase C (PLC)-γ isoforms to phosporylated SLP-65 facilitates phosphorylation of PLC-γ by Syk and probably also by Bruton's tyrosine kinase (Btk), allowing sustained generation of inositol 1,4,5-triphosphate and mobilization of $Ca^{2+}$ ions. **c** Other SLP-65 binding proteins are Grb2, Vav and Nck, which may couple BCR activation to distinct signaling pathways, including ras activation and reorganization of the actin cytoskeleton. **d** Inhibition and termination of the BCR signal involves activation of the protein tyrosine phosphatase SHP-1 and the phosphoinositol 5′-phosphatase SHIP. See text for details

1998; Falasca et al. 1998). Conversely, the PH domain of BTK could also serve to bring the SLP-65/PLC-γ complex to the plasma membrane (Fig. 2b). The proposed mechanisms are not mutually exclusive and future investigations are needed to

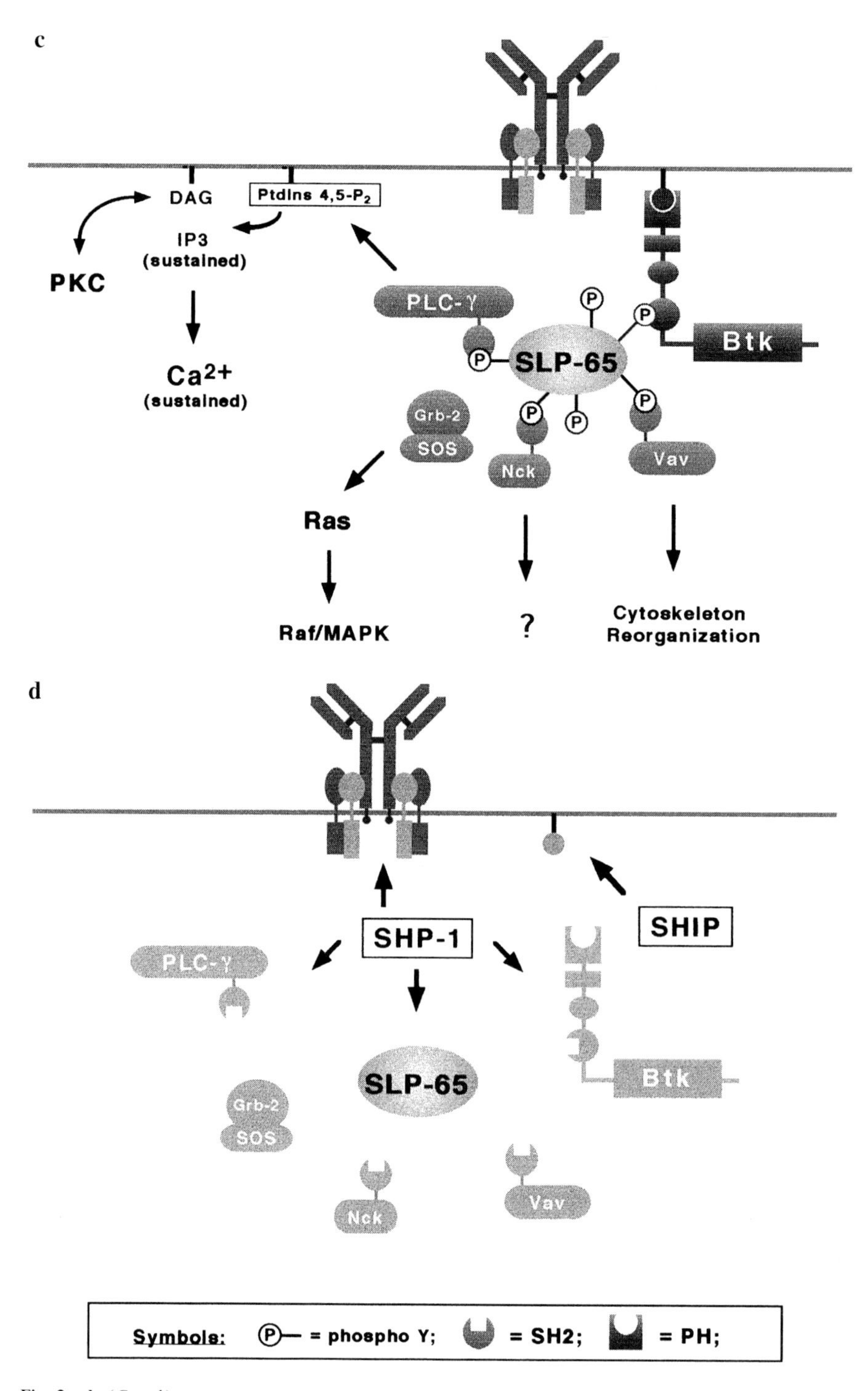


**Fig. 2a–d.** (*Contd.*)

better understand the function of Btk and its different protein modules in $Ca^{2+}$ mobilization and other signaling pathways.

## 5.5 SLP Adaptors and the Organization of Signaling Complexes

SLP adaptors have a scaffolding function before and after BCR stimulation. The reported association of tyrosine-phosphorylated SLP-65/BLNK with the SH2/SH3 adaptor protein Nck (Fu et al. 1998) provides the complex with additional protein–protein interaction sites (Fig. 2c). Moreover, a ligand for the SLP-65 SH2 domain has not yet been identified. In T cells, the SLP-76 SH2 domain binds the SLAP-130/FYB adaptor protein, which may exert a negative regulatory effect on TCR activation. Together, SLP adaptors can regulate antigen receptor stimulation in numerous ways and at different levels. The importance of SLP-mediated protein–protein interactions is obvious from the profound and early block of T-cell development in SLP-76 knock out mice (Clements et al. 1998; Pivniouk et al. 1998). The available data from B and T cells suggest that SLP adaptors collect the majority of signals elicited by the antigen receptors on mature and immature lymphocytes. Thus, SLP adaptors would be functionally similar to the insulin receptor substrate 1 (IRS-1). The constitutive association of SLP-65,with central signaling elements, such as Grb2 and Vav, shows that a substantial part of the signaling complex is already preformed in unstimulated cells. This may contribute to a structural organization of signaling pathways, which enables the B cell to respond to BCR aggregation without time delay and without interference or cross talk to other receptors. It allows an incoming signal from the BCR to flow only through BCR-signaling pathways and not through those from other receptors that employ common effector proteins. The early and BCR-dependent phosphorylation of SLP-65, together with its restricted expression pattern, is consistent with SLP-65 being an important element of a preformed BCR transducer complex.

Preassembly of signaling complexes has been directly demonstrated for the photoreceptor system in drosophila and for the osmoadaptation-pheromone responses in yeast (Elion 1995; Posas and Saito 1997; Tsunoda et al. 1997; Craven and Bredt 1998). In both cases, specific adaptor proteins, the PDZ domain-containing InaD and the multi-domain protein Ste5, were found to organize the physical and functional coupling of intracellular signaling proteins with the receptor. Caveolines are another type of protein adaptor that are able to direct the assembly of signaling proteins at the plasma membrane (Okamoto et al. 1998). They interact with a number of proteins, including Src-family PTKs, ras and trimeric G proteins. The multiple and receptor-proximal defects in lymphocytes from Vav-deficient mice have led to the speculation that also cytoskeleton components are required for a functional architecture of antigen-receptor signaling cascades in the absence and presence of antigen. In support of this, a stimulation-dependent and -independent association of the BCR and TCR with the cytoskeleton has been reported (Caplan et al. 1995; Rozdzial et al. 1995; Park and Jongstra-Bilen 1997). Antigen-mediated reorganization of the microtubule-organizing center (MTOC) in T cells and the F-actin assembly in B cells is ITAM-dependent (Cox et al. 1996; Lowin-Kropf et al. 1998). Finally, drugs that affect the integrity of the cytoskeleton can interfere with antigen-receptor signaling (Huby et al. 1998). To further elucidate the BCR transducer complex, it will be important to analyze exactly how SLP-65 is coupled to the receptor.

## 5.6 SLP-65-Independent Signaling Pathways

Except for SLP-76 itself, the TCR-induced protein tyrosine phosphorylation pattern was not drastically altered in the SLP-76-negative T-cell variant. It is thus likely that SLP-independent pathways exist. One of these may be the HS1 (hematopoietic-specific protein 1) pathway. HS1 phosphorylation by Lyn-activated Syk is essential for BCR-induced apoptosis in an immature B-cell line, and impaired lymphocyte development is observed in HS1-negative mouse mutants (Fukuda et al. 1995; Taniuchi et al. 1995; Yamanashi et al. 1997).

Upon BCR aggregation, tyrosine phosphorylation of the adaptor proteins Cbl (Casitas B lineage lymphoma protein), Crk (CT10-regulator of kinase) and Cas (Crk-associated substrate) induces their association with C3G, a GDP/GTP exchange factor for the Rap1 family of small G proteins (Cory et al. 1995; Ingham et al. 1996; Panchamoorthy et al. 1996; Smit et al. 1996). Activated Rap1A competes with ras for binding to Raf. Accordingly, complex formation of either Cbl·Crk·C3G or Cas·Crk·C3G, and their translocation to the membrane fraction in activated B cells may interfere with the effects of ras activation.

Cbl may also play a role in the regulation of PI3K activity, together with or independent of Src-like PTKs and the BCR-co-receptor protein CD19 (Tuveson et al. 1993; Pleiman et al. 1994; Weng et al. 1994; Kim et al. 1995; O'Rourke et al. 1998). PI3K consists of the p110 catalytic and p85 (Grb1) regulatory subunit. After BCR occupancy, p85 binds to the SH3 domain of Lyn or Fyn and to a tyrosine-phosphorylated YxxM motif in CD19 via its SH2 domain. Both interactions lead to increased PI3K activity. The YxxM consensus-binding site, found in IRS-1 for example (Myers et al. 1992), is not present in SLP-65 and, in contrast to other receptor systems, p85 does not become phosphorylated in activated B cells. The different phospholipid products of activated PI3K affect a number of PH/PTB-containing signaling molecules, like Btk (see above), PKB (Akt) and isoforms of PKC (for review, see Toker and Cantley 1997). Inhibition of p110 by wortmannin leads to multiple BCR signaling defects, including $Ca^{2+}$ mobilization. Together the results provide compelling evidence for the critical role of PI3K in B-cell activation.

Other BCR-regulated signaling proteins are known and the formation of additional protein complexes has been documented. They are likely to be important for B-cell activation although many of them could not yet be functionally confined. Their comprehensive review is beyond the scope of this article, and the reader is referred to recent reviews published elsewhere (Birkeland and Monroe 1997; DeFranco 1997).

## 6 Modulation of the BCR Signal: Co-Receptors and Tyrosine Phosphatases

Proliferation and/or differentiation of any cell type has to be tightly controlled. Some of the mechanisms that operate to diminish or eventually terminate BCR activation have been recently clarified. They involve two classes of signaling molecules; transmembrane proteins that act as BCR co-receptors and cytoplasmic phosphatases, specific either for phosphotyrosine or for the 5′-phosphate of phosphoinositide phospholipids (Fig. 2d).

CD22 is a B-cell-specific type-I transmembrane glycoprotein that can be co-purified with the BCR under mild conditions (Leprince et al. 1993; Peaker and Neuberger 1993). When CD22 is co-ligated with the BCR, activation of BCR signaling cascades is inhibited; but when sequestered from the BCR, the threshold at which antigen activates the B cell is lowered by a factor of 100 (Doody et al. 1995 and references therein). Together with the enhanced BCR-response in CD22-deficient mice, the data demonstrate a negative regulatory role for CD22 on B-cell activation. Following BCR engagement, CD22 becomes phosphorylated, most likely by Lyn (Cornall et al. 1998; Smith et al. 1998), within peptide motifs that are also found in other inhibitory receptors and that are known as the immunoreceptor tyrosine-based inhibitory motifs (ITIMs). The ITIM consensus sequence is V/Ix-YxxL (Muta et al. 1994). ITIM phosphorylation of CD22 recruits the SH2 domain-containing protein tyrosine phosphatase SHP-1 (Doody et al. 1995; Smith et al. 1998). In turn, SHP-1 may counteract BCR signaling by dephosphorylation of PTKs and/or their substrates, including Ig-α/Ig-β itself (Pani et al. 1995). This leads to the inhibition of IP3 generation and hence to the loss of $Ca^{2+}$ release from intracellular stores. The pivotal role of SHP-1 for normal B-cell function is demonstrated by the two *motheaten* mouse mutants, *me* and *me*$^{v}$, which produce no or mutant SHP-1, respectively. The homozygous loss-of-function in *me* mice is lethal and B cells from the viable *me*$^{v}$ mice are hyperactive and develop spontaneous autoimmune diseases. Thus, SHP-1 is critical in determining the BCR signaling threshold. In addition to SHP-1, Syk has also been co-purified with tyrosine-phosphorylated CD22 (Leprince et al. 1993; Peaker and Neuberger 1993; Wienands et al. 1995). An attractive model is, therefore, that following phosphorylation by Lyn, CD22 binds both SHP-1 and activated Syk via their SH2 domains, facilitating dephosphorylation and inhibition of Syk by SHP-1. In this scenario, CD22 acts independently of occupancy by an extracellular ligand as a "membrane adaptor molecule". Binding of external ligands, such as CD72, may provide additional levels of regulation. The results indicate that antigen binding to the BCR not only activates positive signaling elements but simultaneously induces feedback inhibition.

A similar but distinct mechanism mediates inhibition of BCR signaling by FcγRIIb1, which is the only FcR for IgG on B cells. It has been known since the early 1970s that IgG-antigen immune complexes are potent inhibitors of the antibody response. This results from co-ligation of the BCR with FcγRIIb1, which leads to recruitment of the SH2-containing phosphoinositol 5′-phosphatase SHIP to the phosphorylated ITIM of the FcγRIIb1 (Ono et al. 1996; Kiener et al. 1997; Ono et al. 1997). Activated SHIP hydrolyses the 5′-phosphate of PtdIns 3,4,5-P3, hence interfering with PH domain-mediated attachment of Btk (Bolland et al. 1998; Scharenberg et al. 1998) and presumably of other signaling molecules. Following co-ligation of the FcγRIIb1 with the BCR, tyrosine phosphorylation of CD19 is significantly decreased compared with BCR stimulation alone (Hippen et al. 1997). Whether CD19 is a specific substrate for Btk has not been investigated so far, but decreased CD19 phosphorylation leads to reduced association with and activation of PI3K, thereby preventing the (re-) synthesis of polyphosphate phospholipids. Note, that PI3K specifically phosphorylates the 3′-OH of PtdIns-phosphates, while SHIP removes the 5′-phosphate. Altogether, FcγRIIb1-mediated signaling inhibits the uptake of extracellular $Ca^{2+}$ ions and terminates the proliferative signal from the BCR. The system provides a powerful tool to measure the IgG titer of a particular antibody species and, further, to distinguish between free antigen and IgG-antigen complexes. This helps the immune system to decide whether or not further activation of B cells is required to combat the pathogen. In support of this, FcγRIIb1-deficient mice display enhanced antibody titers especially in the late phase of the humoral response.

In summary, SH2-containing phosphatases counteract either early or late BCR signaling events by being recruited to the membrane via phosphorylated ITIMs of BCR co-receptors. Based on these findings, one simple model that could account for the diverse biological outcomes of BCR engagement is that "negative signaling" is the result of activated effector elements that shift the equilibrium of protein tyrosine phosphorylation in the direction of dephosphorylation, hence preventing a continuous signal flow from the BCR. As described before, this can delete B cells at all stages of development by a default pathway that results in apoptosis. The activation of specific death-delivering signaling pathways by the activated BCR is not needed in this model and has not been found so far.

## 7 Concluding Remarks

The history of antibodies or anti-toxins since their discovery by von Behring and Kitasato (1890) is a long story, and it is hardly possible to make it short. It will become even more complex in the future because a number of important questions are still unanswered. For example, do the multiple ITAMs in the BCR, TCR and FcRs perform a redundant or specific function? It is noteworthy that a mouse mutant that lacks the cytoplasmic part of Ig-α displays an impaired generation of peripheral B cells and antibody response to T-independent antigens (Torres et al. 1997). Other questions are, what is the structural nature of the BCR transducer complex and how exactly does BCR feedback inhibition work? Conclusive answers will help to understand how a signal from the BCR and its co-receptor(s) induces altered gene transcription in a specific manner. The study of BCR signaling is reminiscent of a chess game, where playing at the level of a grand master requires the ability to understand higher order structures and their influence on each other. It can be hoped that, soon, our ability to play a good signaling game will no longer be hampered by not knowing all the pieces and the rules for their movement.

*Acknowledgements.* I thank Drs. Roberta Pelanda and Peter J. Nielsen for critically reading this manuscript over the weekend. I am particularly grateful to Dr. Michael Reth for the generous support and for the many discussions. Work from our laboratory was supported by the *Deutsche Forschungsgemeinschaft* through SFB 388 and the Leibniz program.

## References

Afar DEH, Park H, Howell BW, Rawlings DJ, Cooper J, Witte ON (1996) Regulation of Btk by Src family tyrosine kinases. Mol Cell Biol 16:3465–3471
Bae YS, Cantley LG, Chen CS, Kim SR, Kwon KS, Rhee SG (1998) Activation of phospholipase C-gamma by phosphatidylinositol 3,4,5-triphosphate. J Biol Chem 273:4465–4469
Baumann G, Maier D, Freuler F, Tschopp C, Baudisch K, Wienands J (1994) In vitro characterization of major ligands for Src homology 2 domains derived from tyrosine kinases, from the adaptor protein SHC and from GTPase-activating protein in Ramos B cells. Eur J Immunol 24:1799–1807
Benatar T, Carsetti R, Furlonger C, Kamalia N, Mak T, Paige CJ (1996) Immunoglobulin-mediated Signal Transduction in B Cells from CD45-deficient Mice. J Exp Med 183:329–334
Birkeland ML, Monroe JG (1997) Biochemistry of antigen receptor signaling in mature and developing B lymphocytes. Critical Rev Immunol 17:353–385
Bolland S, Pearse RN, Kurosaki T, Ravetch JV (1998) SHIP modulates immune receptor responses by regulating membrane association of Btk. Immunity 8:509–516
Burkhardt AL, Brunswick M, Bolen JB, Mond JJ (1991) Anti-immunoglobulin stimulation of B lymphocytes activates src-related protein-tyrosine kinases. Proc Natl Acad Sci USA 88:7410–7414
Campbell KS, Cambier JC (1990) B lymphocyte antigen receptors (mIg) are noncovalently associated with a disulphide-linked, inducibly phosphorylated glycoprotein complex. EMBO J 9:441–448
Campbell KS, Hager EJ, Friedrich RJ, Cambier JC (1991) IgM antigen receptor complex contains phosphoprotein products of B29 and mb-1 genes. Proc Natl Acad Sci USA 88:3982
Campbell MA, Sefton BM (1990) Protein tyrosine phosphorylation is induced in murine B lymphocytes in response to stimulation with anti-immunoglobulin. EMBO J 9:2125–2132
Campbell MA, Sefton BM (1992) Association between B-lymphocyte membrane immunoglobulin and multiple members of the Src family of protein tyrosine kinases. Mol Cell Biol 12:2315–2321
Caplan S, Zeliger S, Wang L, Baniyash M (1995) Cell-surface-expressed T-cell antigen-receptor ζ chain is associated with the cytoskeleton. Proc Natl Acad Sci USA 92:4768–4772
Chan VWF, Meng F, Soriano P, DeFranco AL, Lowell CA (1997) Characterization of the B lymphocyte populations in Lyn-deficient mice and the role of Lyn in signal initiation and down regulation. Immunity 7:69–81
Carter RH, Park, DJ, Rhe SG, Fearon DT (1991) Tyrsoine phosphorylation of phospholipase C induced by membrane immunoglobulin in B lymphocytes. Proc Natl Acad Sci USA 88:2745–2749
Cheng AM, Rowley B, Pao W, Hayday A, Bolen JB, Pawson T (1995) Syk tyrosine kinase required for mouse viability and B-cell development. Nature 378:303–306
Clark MR, Campbell KS, Kazlauskas A, Johnson SA, Hertz M, Potter TA, Pleiman C, Cambier JC (1992) The B cell antigen receptor complex: association of Ig-alpha and Ig-beta with distinct cytoplasmic effectors. Science 258:123–126

Clark MR, Johnson SA, Cambier JC (1994) Analysis of Ig-alpha-tyrosine kinase interaction reveals two levels of binding specificity and tyrosine phosphorylated Ig-alpha stimulation of Fyn activity. EMBO J 13:1911–1919

Clements JL, Yang B, Ross-Barta SE, Eliason SL, Hrstka RF, Williamson RA, Koretzky GA (1998) Requirement for the leukocyte-specific adapter protein SLP-76 for normal T cell development. Science 281:416–419

Cornall RJ, Cyster JG, Hibbs ML, Dunn AR, Otipoby KL, Clark EA, Goodnow CC (1998) Polygenic autoimmune traits: Lyn, CD22 and SHP-1 are limiting elements of a biochemical pathway regulating BCR signaling and selection. Immunity 8:497–508

Cory GOC, Lovering RC, Hinshelwood S, MacCarthy-Morrogh L, Levinsky RJ, Kinnon C (1995) The protein product of the c-cbl Protooncogene is phosphorylated of B cell receptor stimulation and binds the SH3 domain of Bruton's tyrosine kinase. J Exp Med 182:611–615

Cox D, Chang P, Kurosaki T, Greenberg S (1996) Syk tyrosine kinase is required for immunoreceptor tyrosine activation motif-dependent actin assembly. J Biol Chem 271:16597–16602

Craven SE, Bredt DS (1998) PDZ proteins organize synaptic signaling pathways. Cell 93:495–498

Crespo P, Schuebel KE, Ostrom AA, Gutkind JS, Bustelo XR (1997) Phosphotyrosine-dependent activation of Rac-1 GDP/GTP exchange by the vav proto-oncogene product. Nature 385:169–172

Cyster JG, Healy JI, Kishihara K, Mak TW, Thomas ML, Goodnow CC (1996) Regulation of B-lymphocyte negative and positive selection by tyrosine phosphatase CD45. Nature 381:325–328

Daeron M (1997) Fc receptor biology. Ann Rev Immunol 15:203–234

DeFranco AL (1997) The complexity of signaling pathways activated by the BCR. Curr Opi Immunol 9:296–308

Doody GM, Justement LB, Delibrias CC, Matthews RJ, Lin J, Thomas ML, Fearon DT (1995) A role in B cell activation for CD22 and the protein tyrosine phosphatase SHP. Science 269:242–244

Elion EA (1995) Ste5: A meeting place for MAP kinases and their associates. Trends Cell Biol 5:322–327

Falasca M, Logan SK, Letho VP, Baccante G, Lemmon MA, Schlessinger J (1998) Activation of phospholipase C$\gamma$ by PI 3-kinase-induced PH domain-mediated membrane targeting. Embo J 17: 414–422

Fischer K-D, Kong Y-Y, Tedford K, Marengère LEM, Kozieradzki I, Sasaki T, Starr M, Chan G, Gardener S, Nghiem MP, Bouchard D, Barbacid M, Bernstein A, Penninger JM (1998) Vav is a regulator of the cytoskeletal reorganization mediated by the T-cell receptor. Curr Biol 8:554–562

Flaswinkel H, Reth M (1994) Dual role of the tyrosine activation motif of the Ig-alpha protein during signal transduction via the B cell antigen receptor. Embo J 13:83–89

Fluckiger A-C, Li Z, Kato RM, Wahl MI, Ochs HD, Longnecker R, Kinet JP, Witte ON, Scharenberg AM, Rawlings DJ (1998) Btk/Tec kinases regulate sustained increases in intracellular $Ca^{2+}$ following B-cell receptor activation. Embo J 17:1973–1985

Fu C, Turck CW, Kurosaki T, Chan AC (1998) BLNK: A central linker protein in B cell activation. Immunity 9:93–103

Fukuda T, Kitamura D, Taniuchi I, Maekawa Y, Benhamou LE, Sarthou P, Watanabe T (1995) Restoration of surface IgM-mediated apoptosis in an anti-IgM-resistent variant of WEHI-231 lymphoma cells by HS1, a protein-tyrosine kinase substrate. Proc Natl Acad Sci USA 92:7302–7306

Gold MR, Chan VW, Turck CW, DeFranco AL (1992) Membrane Ig cross-linking regulates phosphatidylinositol 3-kinase in B lymphocytes. J Immunol 148:2012–2022

Gold MR, Crowley MT, Martin GA, McCormick F, DeFranco AL (1993) Targets of B lymphocyte antigen receptor signal transduction include the p21ras GTPase-activating protein (GAP) and two GAP-associated proteins. J Immunol 150:377–386

Gold MR, Law DA, DeFranco AL (1990) Stimulation of protein tyrosine phosphorylation by the B-lymphocyte antigen receptor. Nature 345:810–813

Gold MR, Matsuuchi L, Kelly RB, DeFranco AL (1991) Tyrosine phosphorylation of components of the B-cell antigen receptors following receptor crosslinking. Proc Natl Acad Sci USA 88:3436–3440

Gong S, Nussenzweig MC (1996) Regulation of an early developmental checkpoint in the B cell pathway by Ig beta. Science 272:41141–4

Harnett MM (1994) An Atlas of B cell signalling. Immunol Today 15:422–423

Hata A, Sabe H, Kurosaki T, Takata M, Hanafusa H (1994) Functional analysis of Csk in signal transduction through the B-cell antigen receptor. Mol Cell Biol 14:7306–7313

Hibbs ML, Tarlinton DM, Armes J, Grail D, Hodgson G, Maglitto R, Stacker SA, Dunn AR (1995) Multiple defects in the immune system of Lyn-deficient mice, culminating in autoimmune disease. Cell 83:301–311

Hippen KL, Buhl AM, D'Ambrosio DD, Nakamura K, Persin C, Cabier JC (1997) FcgRIIB1 inhibition of BCR-mediated phosphoinositide hydrolysis and $Ca^{2+}$ mobilization is integrated by CD19 dephosphorylation. Immunity 7:49–58

Holsinger LJ, Graef IA, Swat W, Chi T, Bautista DM, Davidson L, Lewis RS, Alt FW, Crabtree GR (1998) Defects in actin-cap formation in Vav-deficient mice implicate an actin requirement for lymphocyte signal transduction. Curr Biol 8:563–572

Hombach J, Lottspeich F, Reth M (1990a) Identification of the genes encoding the Ig-alpha and Ig-beta components of the IgM antigen receptor complex. Eur J Immunol 20:2795–2799

Hombach J, Tsubata T, Leclercq L, Stappert H, Reth M (1990b) Molecular components of the B-cell antigen receptor complex of the IgM class. Nature 343:760

Huby RDJ, Weiss A, Ley SC (1998) Nocodazole inhibits signal transduction by the T cell antigen receptor. J Biol Chem 273:12024–12031

Ingham RJ, Krebs DL, Barbazuk SM, Turck CW, Hirai H, Matsuda M, Gold MR (1996) B cell antigen receptor signaling induces the formation of complexes containing the Crk adapter proteins. J Biol Chem 271:32306–32314

Jackman JK, Motto DG, Sun Q, Tanemoto M, Turck CW, Peltz GA, Koretzky GA, Findell PR (1995) Molecular cloning of SLP-76, a 76-kDa tyrosine phosphoprotein associated with Grb2 in T cells. J Biol Chem 270:7029–3702

Justement LB, Campbell KS, Chien NC, Cambier JC (1991) Regulation of B Cell Antigen Receptor Signal Transduction and Phosphorylation by CD45. Science 252:1839–1842

Justement LB, Wienands J, Hombach J, Reth M, Cambier JC (1990) Membrane IgM and IgD molecules fail to transduce $Ca^{2+}$ mobilizing signals when expressed on differentiated B lineage cells. J Immunol 144:3272–3280

Kerner JD, Appleby MW, Mohr RN, Chien S, Rawlings DJ, Maliszewski CR, Witte ON, Perlmutter RM (1995) Impaired expansion of mouse B cell progenitors lacking Btk. Immunity 3:301–312

Khan WN, Alt FW, Gerstein RM, Malynn BA, Larsson I, Rathbun G, Davidson L, Müller S, Kantor AB, Herzenberg LA (1995) Defective B cell development and function in Btk-deficient mice. Immunity 3:283–299

Kiener PA, Lioubin MN, Rohrschneider LR, Ledbetter JA, Nadler SG, Diegel ML (1997) Co-ligation of the antigen and Fc receptors gives rise to the selective modulation of intracellular signaling in B cells. J Biol Chem 272:3838–3844

Kim KM, Alber G, Weiser P, Reth M (1993) Differential signaling through the Ig-alpha and Ig-beta components of the B cell antigen receptor. Eur J Immunol 23:911–916

Kim TJ, Kim YT, Pillai S (1995) Association of Activated Phosphatidylinositol 3-Kinase with p120cbl in Antigen Receptor-ligated B Cells. J Biol Chem 270:27504–27509

Kitamura D, Kudo A, Schaal S, Müller W, Melchers F, Rajewsky K (1992) A critical role of lambda 5 protein in B cell development. Cell 69:823–831

Kitamura D, Roes J, Kühn R, Rajewsky K (1991) A B cell-deficient mouse by targeted disruption of the membrane exon of the immunoglobulin mu gene. Nature 350:423–426

Kolanus W, Romeo C, Seed B (1993) T cell activation by clustered tyrosine kinases. Cell 74:171–183

Kurosaki T, Johnson SA, Pao L, Sada K, Yamamura H, Cambier JC (1995) Role of the Syk Autophosphorylation Site and SH2 Domains in B Cell Antigen Receptor Signaling. J Exp Med 182:1815–1823

Kurosaki T, Takata M, Yamanashi Y, Inazu T, Taniguchi T, Yamamoto T, Yamamura H (1994) Syk activation by the Src-family tyrosine kinase in the B cell receptor signaling. J Exp Med 179:1725–1729

Lam KP, Kühn R, Rajewsky K (1997) In Vivo Ablation of surface immunoglobulin on mature B cells by inducible gene targeting results in rapid cell death. Cell 90:1073–1083

Lane PJL, Ledbetter JA, McConnell FM, Draves K, Deans J, Schieven GL, Clark EA (1991) The role of tyrosine phosphorylation in signal tranduction through surface Ig in human B cells. J Immunol 146:715–722

Langdon WY, Hartley JW, Klinken SP, Ruscetti SK, Morse HCd (1989) v-cbl, an oncogene from a dual-recombinant murine retrovirus that induces early B-lineage lymphomas. Proc Natl Acad Sci USA 86:1168–1172

Lankester AC, van Schijndel GMW, Rood PML, Verhoeven AJ, van Lier RAW (1994) B cell antigen receptor cross-linking induces tyrosine phosphorylation and membrane translocation of a multimeric Shc complex that is augmented by CD19 co-ligation. Eur J Immunol 24:2818–2825

Larbolette O, Wollscheid B, Schweikert J, Nielsen PJ, Wienands J (1998) SH3P7 is a cytoskeleton adapter protein and is coupled to signal transduction from lymhocyte antigen receptors. Mol Cell Biol 19:1539–1406

Lemmon MA, Ferguson KM (1998) Pleckstrin homology domains. Curr Topics Microbiol Immunol 228:39–74
Leprince C, Draves KE, Geahlen RL, Ledbetter JA, Clark EA (1993) CD22 associates with the human surface IgM-B-cell antigen receptor complex. Proc Natl Acad Sci USA. 90:3236–3240
Lin J, Justement LB (1992) The MB-1/B29 heterodimer couples the B cell antigen receptor to multiple src family protein tyrosine kinases. J Immunol 149:1548–1555
Liu K-Q, Bunnell SC, Gurniak CB, Berg LJ (1998) T cell receptor-initiated calcium release is uncoupled from capacitative calcium entry in Itk-deficient T cells. J Exp Med 187:1721–1727
Lowin-Kropf B, Smith Shapiro V, Weiss A (1998) Cytoskeletal polarization of T cells is regulated by an immunoceptor tyrosine-based activation motif-dependent mechanism. J Cell Biol 140:861–971
Mahajan S, Fargnoli J, Burkhardt AL, Kut SA (1995) Src family protein tyrosine kinases induce autoactivation of Bruton's tyrosine kinase. Mol Cell Biol 5:5304–5311
Mayer BJ, Gupta G (1998) Functions of SH2 and SH3 domains. Curr Topics Microbiol Immunol 228: 1–22
Melchers F, Rolink A, Grawunder U, Winkler TH, Karasuyama H, Ghia P, Anderson J (1995) Positive and negative selection events during B lymphopoiesis. Curr Opin Immunol 7:214–227
Muta T, Kurosaki T, Misulovin Z, Sanchez M, Nussenzweig MC, Ravetch JV (1994) A 13-amino-acid motif in the cytoplasmic domain of FcgammaRIIB modulates B-cell receptor signaling. Nature 368:70–74
Myers MGj, Backer JM, Sun XJ, Shoelson S, Hu P, Schlessinger J, Yoakim M, Schaffhausen B, White MF (1992) IRS-1 activates phosphatidylinositol 3′kinase by associating with Src homology 2 domains of p85. Proc Natl Acad Sci USA 89:10350–10354
Nagai K, Takata M, Yamamura H, Kurosaki T (1995) Tyrosine phosphorylation of Shc is mediated through Lyn and Syk in B cell receptor signaling. J Biol Chem 270:6824–6829
Nishizumi H, Taniuchi I, Yamanashi Y, Kitamura D, Ilic D, Mori S, Watanabe T, Yamamoto T (1995) Impaired proliferation of peripheral B cells and indication of autoimmune disease in lyn-deficient mice. Immunity 3:549–560
O'Rourke LM, Tooze R, Turner M, Sandoval DM, Carter RH, Tybulewicz VLJ, Fearon DT (1998) CD19 as a membrane-anchored adaptor protein of B lymphocytes: Costimulation of lipid and protein kinases by recruitment of Vav. Immunity 8:635–645
Okamoto T, Schlegel A, Scherer P, Lisanti MP (1998) Caveolins, a family of scaffolding proteins for organizing "preassembled signaling complexes" at the plasma membrane. J Biol Chem 273:5419–5422
Ono M, Bolland S, Tempst P, Ravetch JV (1996) Role of the inositol phosphatase SHIP in negative regulation of the immune system by the receptor FcgammaRIIB. Nature 383:263–266
Ono M, Okada H, Bolland S, Yanagi S, Kurosaki T, Ravetch JV (1997) Deletion of SHIP or SHP-1 reveals two distinct pathways for inhibitory signaling. Cell 90:293–301
Panchamoorthy G, Fukazawa T, Miyake S, Soltoff S, Reedquist K, Druker B, Shoelson S, Cantley L, Band H (1996) p120cbl is a major sSubstrate of tyrosine phosphorylation upon B cell antigen receptor stimulation and interacts in vivo with Fyn and Syk tyrosine kinases, Grb-2 and Shc adaptors, and the p85 subunit of phosphytidylinositol 3-kinase. J Biol Chem 271:3187–3194
Pani G, Kozlowski M, Cambier JC, Mills GB, Siminovitch KA (1995) Identification of the tyrosine phosphatase PTP1 C as a B cell antigen receptor-associated protein involved in the regulation of B cell signaling. J Exp Med 181:2077–2084
Park JY, Jongstra-Bilen J (1997) Interactions between membrane IgM and the cytoskeleton involve the cytoplasmic domain of the immunoglobulin receptor. Eur J Immunol 27:3001–3009
Peaker CJG, Neuberger MS (1993) Association of CD22 with the B cell antigen receptor. Eur J Immunol 23:1358–1363
Pivniouk V, Tsitsikov E, Swinton P, Rathbun G, Alt FW, Geha RS (1998) Impaired viability and profound block in thymocyte development in mice lacking the adaptor protein SLP-76. Cell 94: 229–238
Pleiman CM, Hertz WM, Cambier JC (1994) Activation of Phosphatidylinositol-3′ Kinase by Src-Family Kinase SH3 Binding to the p85 Subunit. Science 263:1609–1612
Posas F, Saito H (1997) Osmotic activation of the HOG MAPK pathway via Ste11p MAPKKK: Scaffold role of PBS2p MAPKK. Science 276:1702–1705
Rawlings DJ, Scharenberg AM, Park H, Wahl MI, Lin S, Kato RM, Fluckiger A-C, Witte ON, Kinet J-P (1996) Activation of BTK by a phosphorylation mechanism initiated by SRC family kinases. Science 271:822–825
Rawlings DJ, Witte ON (1994) Bruton's tyrosine kinase is a key regulator in B-cell development. Immunol Rev 138:105–119

Reth M (1989) Antigen receptor tail clue. Nature 338:383

Reth M (1992) Antigen receptors on B lymphocytes. Ann Rev Immunol 10:97–121

Rhee SG, Bae YS (1997) Regulation of phosphinositide-specific phospholipase C isozymes. J Biol Chem 272:15045–15048

Rigley KP, Harnett MM, Phillips RJ, Klaus GGB (1989) Analysis of signaling via surface immunoglobulin receptors on B cells from CBA/N mice. Eur J Immunol 19:2081–2086

Rivera VM, Brugge JS (1995) Clustering of Syk is sufficient to induce tyrosine phosphorylation and release of allergic mediators from rat basophilic leukemia cells. Mol Cell Biol 15:1582–1590

Rowley RB, Burkhardt AL, Chao HG, Matsueda GR, Bolen JB (1995) Syk protein-tyrosine kinase is regulated by tyrosine-phosphorylated Ig alpha/Ig beta immunoreceptor tyrosine activation motif binding and autophosphorylation. J Biol Chem 270:11590–11594

Rozdzial MM, Malissen B, Finkel TH (1995) Tyrosine-phosphorylated T cell receptor ζ chain associates with the actin cytoskeleton upon activation of mature T lymphocytes. Immunity 3:623–633

Sakai R, Iwamatsu A, Hirano N, Ogawa S, Tanaka T, Mano H, Yazaki Y, Hirai H (1994) A novel signaling molecule, p130, forms stable complexes in vivo with v-Crk and v-Src in a tyrosine phosphorylation-dependent manner. EMBO J 13:3748–3756

Salim K, Bottomley MJ, Querfurth E, Zvelebil MJ, Gout I, Scaife R, Margolis RL, Gigg R, Smith CIE, Driscoll PC, Waterfield MD, Panayotou G (1996) Distinct specificity in the recognition of phosphoinositides by the pleckstrin homology domains of dynamin and Bruton's tyrosine kinase. Embo J 15:6241–6250

Sanchez M, Misulovin Z, Burkhardt AL, Mahajan S, Costa T, Bolen JB, Nussenzweig M (1993) Signal transduction by immunoglobulin is mediated through Ig-α and Ig-β. J Exp Med 178:1049–1055

Saouaf SJ, Mahajan S, Rowley RB, Kut SA, Fargnoli J, Burkhardt AL, Tsukada S, Witte ON, Bolen JB (1994) Temporal differences in the activation of three classes of non-transmembrane protein tyrosine kinases following B-cell antigen receptor surface engagement. Proc Natl Acad Sci USA 91:9524–9528

Saxton TM, van Oostveen I, Bowtell D, Aebersold R, Gold MR (1994) B Cell Antigen Receptor Cross-Linking Induces Phosphorylation of the p21ras Oncoprotein Activators SHC and mSOS1 As Well As Assembly of Complexes Containing SHC, GRB-2, mSOS1, and a 145-kDa Tyrosine-Phosphorylated Protein. J Immunol 153:623–636

Scharenberg AM, El-Hillal O, Fruman DA, Beitz LO, Li Z, Lin S, Gout I, Cantley LC, Rawlings DJ, Kinet J-P (1998) Phosphatidylinositol-3,4,5-trisphosphate (PtdIns-3,4,5-P3)/Tec kinase-dependent calcium signaling pathway: a target for SHIP-mediated inhibitory signals. Embo J 17:1961–1972

Sicheri F, Moarefi I, Kuriyan J (1997) Crystal structure of the Src family tyrosine kinase Hck. Nature 385:602–609

Smit L, de Vries-Smits MM, Bos JL, Borst J (1994) B Cell Antigen Receptor Stimulation Induces Formation of a Shc-Grb2 Complex Containing Multiple Tyrosine-phosphorylated Proteins. J Biol Chem 269:20209–20212

Smit L, van der Horst G, Borst J (1996) Sos, Vav, and C3G Participate in B Cell Receptor-induced Signaling Pathways and Differentially Associate with Shc-Grb2, and Crk-L Adaptors. J Biol Chem 271:8564–8569

Smith KGC, Tarlinton DM, Doody GM, Hibbs ML, Fearon DT (1998) Inhibition of the B cell by CD22: A requirement for Lyn. J Exp Med 187:807–811

Sparks AB, Hoffmann NG, McConnell SJ, Fowlkes DM, Kay BK (1996) Cloning of ligand targets: Systematic isolation of SH3 domain-containing proteins. Nature Biotech 14:741–744

Taddie JA, Hurley TR, Hardwick BS, Sefton BM (1994) Activation of B- and T-cells by the cytoplasmic domains of the B-cell antigen receptor proteins Ig-alpha and Ig-beta. J Biol Chem 269:13529–13535

Takata M, Kurosaki T (1996) A role for Bruton's tyrosine kinase in B cell antigen receptor-mediated activation of phospholipase C-g2. J Exp Med 184:31

Takata M, Sabe H, Hata A, Inazu T, Homma Y, Nukada T, Yamamura H, Kurosaki T (1994) Tyrosine kinases Lyn and Syk regulate B cell receptor-coupled Ca2+ mobilization through distinct pathways. Embo J 13:1341–1349

Taniuchi I, Kitamura D, Maekawa Y, Fukuda T, Kish H, Watanabe T (1995) Antigen-receptor induced clonal expansion and deletion of lymphocytes are impaired in mice lacking HS1 protein, a substrate of the antigen-receptor-coupled tyrosine kinases. EMBO J 14:3664–3678

Tarakhovsky A, Turner M, Schaal S, Mee PJ, Duddy LP, Rajewsky K, Tybulewicz VL (1995) Defective antigen receptor-mediated proliferation of B and T cells in the absence of Vav. Nature 374:467–470

Teh YM, Neuberger M (1997) The immunoglobulin (Ig) alpha and Ig beta cytoplasmic domains are independently sufficient to signal B cell maturation and activation in transgenic mice. J Exp Med 185:1753–1758

Toker A, Cantley LC (1997) Signalling through the lipid products of phosphoinositide-3-OH kinase. Nature 387:673–676

Torres RM, Flaswinkel H, Reth M, Rajewsky K (1997) Aberrant B cell development and immune response in mice with a compromised BCR complex. Science 272:1804–1808

Tsunoda S, Sierralta J, Sun Y, Bodner R, Suzuki E, Becker A, Socolich M, Zuker CS (1997) A multivalent PDZ-domain protein assembles signalling complexes in a G-protein-coupled cascade. Nature 388:243–249

Turner M, Mee PJ, Costello PS, Williams O, Price AA, Duddy LP, Furlong MT, Geahlen RL, Tybulewicz VL (1995) Perinatal lethality and blocked B-cell development in mice lacking the tyrosine kinase Syk. Nature 378:298–302

Tuveson DA, Carter RH, Soltoff SP, Fearon DT (1993) CD19 of B cells as a surrogate kinase insert region to bind phosphatidylinositol 3-kinase. Science 260:986–989

Venkitaraman AR, Williams GT, Dariavach P, Neuberger MS (1991) The B-cell antigen receptor of the five immunoglobulin classes. Nature 352:777

von Behring E, Kitasato S (1890) Über das Zustandekommen der Diphterie-Immunität und der Tetanus-Immunität bei Tieren. Deutsche Medizinische Wochenschrift 49:1113

Weng WK, Jarvis L, LeBien TW (1994) Signaling through CD19 activates Vav/mitogen-activated protein kinase pathway and induces formation of a CD19/Vav/phosphatidylinositol 3-kinase complex in human B cell precursors. J Biol Chem 269:32514–32521

Wienands J, Freuler F, Baumann G (1995) Tyrosine-phosphorylated forms of Ig-beta, CD22, TCR-zeta and HOSS are major ligands for tandem SH2 domains of Syk. Intern Immunol 7:1701–1708

Wienands J, Hombach J, Radbruch A, Riesterer C, Reth M (1990) Molecular components of the B-cell antigen receptor complex of class IgD differ partly from those of IgM. EMBO J 9:449–455

Wienands J, Larbolette O, Reth M (1996) Evidence for a preformed transducer complex organized by the B cell antigen receptor. Proc Natl Acad Sci USA 93:7865–7870

Wienands J, Schweikert J, Wollscheid B, Jumaa H, Nielsen PJ, Reth M (1998) SLP-65: A new signaling component in B lymphocytes which requires expression of the antigen receptor for phosphorylation. J Exp Med 188:791–794

Williams GT, Peaker CJ, Patel KJ, Neuberger MS (1994) The alpha/beta sheath and its cytoplasmic tyrosines are required for signaling by the B-cell antigen receptor but not for capping or for serine/threonine-kinase recruitment. Proc Natl Acad Sci USA 91:474–478

Xu W, Harrison SC, Eck MJ (1997) Three-dimensional structure of the tyrosine kinase c-Src. Nature 385:595–602

Yablonski D, Kuhne MR, Kadlecek T, Weiss A (1998) Uncoupling of nonreceptor tyrosine kinases from PLC-gamma1 in an SLP-76-deficient T cell. Science 281:413–416

Yamanashi Y, Baltimore D (1997) Identification of the Abl- and rasGAP-associated 62kDa protein as a docking protein, Dok. Cell 88:205–211

Yamanashi Y, Fukuda T, Nishizumi H, Inazu T, Higashi K, Kitamura D, Ishida T, Yamamura H, Watanabe T, Yamamoto T (1997) Role of tyrosine phosphorylation of HS1 in B cell antigen receptor-mediated apoptosis. J Exp Med 185:1387–1392

Yamanashi Y, Kakiuchi T, Mizuguchi J, Yamamoto T, Toyoshima K (1991) Association of B cell antigen receptor with protein tyrosine kinase Lyn. Science 251:192–194

Yanagi S, Sugawara H, Kurosaki M, Sabe H, Yamamura H, Kurosaki T (1996) CD45 modulates phosphorylation of both autophosphorylation and negative regulatory tyrosines of Lyn in B cells. J Biol Chem 271:30487–30492

Zhang R, Alt FW, Davidson L, Orkin SH, Swat W (1995) Defective signalling through the T- and B-cell antigen receptors in lymphoid cells lacking the vav proto-oncogene. Nature 374:470–473

# Intermediary Signaling Effectors Coupling the B-Cell Receptor to the Nucleus

M.R. Gold



Department of Microbiology and Immunology, University of British Columbia, 6174 University Blvd., Vancouver, British Columbia V6T 1Z3, Canada

# 1 B-Cell Receptor Signaling: Complexity and Flexibility

As described in the previous chapters, the B-cell receptor (BCR) activates multiple tyrosine kinases which, in turn, phosphorylate a number of proteins. These tyrosine kinase substrates include proteins involved in the activation of key signal-transduction pathways. The role of these signaling pathways is to convey the signal from the BCR to cellular targets. These targets include: (1) transcription factors that regulate the expression of genes involved in B-cell activation, proliferation, and differentiation; (2) cell-cycle regulatory proteins; and (3) the cytoskeleton. The goal of this chapter is to provide a current review of the cytoplasmic transducers of BCR signaling, with an emphasis on signaling to the nucleus. Since this subject has been reviewed frequently over the last several years (Bireland and Monroe 1997; DeFranco 1997; Kurosaki 1997), this chapter will focus on recent findings as well as the challenges that lie ahead.

Engaging the BCR with a multivalent antigen or with anti-immunoglobulin (Ig) antibodies initiates a large number of signaling reactions. These include activation of the Ras GTPase, the production of lipid second messengers by phosphatidylinositol 3-kinase (PI 3-kinase), and hydrolysis of phosphatidylinositol 4,5-bisphosphate ($PIP_2$) by phospholipase C-$\gamma$ (PLC-$\gamma$). In addition to these three well-studied signaling pathways that are activated by many different receptors, the BCR also signals via the HS-1 protein, the Vav protein, the c-Jun N-terminal kinase (JNK) and p38 mitogen-activated protein (MAP) kinases, the Crk/C3G pathway, and the Rap1 GTPase. This is certainly just a partial list.

Why does the BCR need to activate so many different signaling pathways? Do they all have different targets or are they parallel pathways leading to the same targets? Perhaps such complexity of signaling pathways reflects the fact that the response of B cells to BCR engagement is highly contextual. Depending on (1) the maturation state of the B cell and whether it is a B-1 or B-2 B cell; (2) the physical nature of the antigen and whether the antigen is coated with complement fragments or with antibodies; (3) the duration of the B cell's exposure to antigen (i.e., chronic exposure to a self antigen versus acute exposure to a foreign antigen); and (4) the presence of co-stimulatory signals derived from either T cells, e.g., CD40 ligand, interleukin (IL)-4, or bacteria, e.g., lipopolysaccharide (LPS), BCR engagement can result in activation, proliferation, survival, apoptosis, anergy, or maturational arrest. A simple on/off switch mode of signaling would not allow such a diversity of responses. B cells know that combinatorial processes allow one to obtain the greatest diversity of outcomes from a small number of pre-programmed elements. In this case, the pre-programmed elements are the different signaling pathways that the BCR can access.

Different combinations of signaling pathways may specify different responses. As will be discussed later, there is now evidence that the BCR activates different sets of signaling pathways in mature versus immature B cells and in tolerant versus naive. Signals from co-receptors may selectively enhance the activation of some signaling pathways and not others and thereby change the "quality" of the net

BCR signal. Similarly, signals from co-stimulatory receptors, such as CD40, may change the amplitude of one or more of the BCR signaling pathways. Alternatively, a target molecule may require inputs from both a BCR signaling pathway and a co-stimulatory receptor pathway in order to be effectively activated. Thus, once we have identified the major BCR signaling pathways, the important task is to determine to what extent each pathway is activated in the various situations in which BCR signaling can lead to different outcomes. The more difficult task after that will be to provide a molecular explanation for these correlations, e.g., why do B cells undergo apoptosis if BCR signaling pathways A, B, and C are activated but BCR signaling pathway D is not?

## 2 The Role of Docking Proteins and Adaptor Proteins in BCR Signaling

A common theme that we will encounter repeatedly when describing how the BCR activates signaling pathways is the concept of receptor-induced changes in the subcellular localization and organization of signaling proteins. Almost all of the signaling pathways we will discuss involve cytoplasmic signaling proteins that must be recruited to cell membranes where either their substrates or activators are located. This seems like a good way to design a signaling pathway since keeping the components of the pathway physically separated until a receptor is engaged ensures a low basal level of signaling and a good signal-to-noise ratio. The key event in activating the pathway is therefore recruiting the signaling proteins to the correct membrane locations where they will be in close proximity to their activators, their substrates, and other signaling proteins that function in the same pathway.

For BCR signaling, a number of cytoplasmic enzymes must be recruited to the cellular membranes where their substrates are located. PLC-$\gamma$ and PI 3-kinase are cytoplasmic enzymes whose substrates are the plasma membrane lipid $PIP_2$. Similarly, the SOS (the mammalian homologue of the Drosophila son of sevenless protein) and C3G exchange factors are cytoplasmic enzymes that act on GTPases (Ras and Rap1, respectively) that are tethered to the cytoplasmic faces of cellular membranes.

How do the BCR and other receptors recruit signaling proteins to cellular membranes? Many signaling enzymes such as PLC-$\gamma$, PI 3-kinase, and Vav contain Src homology 2 (SH2) domains, which bind to specific phosphotyrosine-containing sequences in other proteins. Other signaling enzymes, such as the nucleotide exchange factors SOS and C3G, do not have their own SH2 domains but instead associate with adaptor proteins such as Grb2 and Crk, which contain SH2 domains. In either case, these proteins or protein complexes can be recruited to membrane locations by binding to transmembrane proteins or peripheral membrane-associated proteins that are phosphorylated on the appropriate tyrosine residues after receptor engagement. Since the SH2 domains of different proteins recognize different phosphotyrosine-containing amino acid sequences (Songyang et al. 1994), a tyrosine kinase-linked receptor, such as the BCR, "selects" which signaling pathways it will access by the set of phosphotyrosine-containing sequences it creates in membrane-associated proteins. The platelet-derived growth factor (PDGF) receptor, which has intrinsic tyrosine-kinase activity, provides an excellent illustration of this concept. In response to binding PDGF, PDGF receptors dimerize and phosphorylate each other on a number of tyrosine residues in their cytoplasmic domains. Different phosphorylated tyrosine residues act as binding sites for the SH2 domains of PLC-γ, PI 3-kinase, and Grb2 (Fantl et al. 1992). Substituting a phenylalanine for the tyrosine that one of these signaling proteins binds to prevents that signaling protein from binding to the PDGF receptor and abrogates the ability of the PDGF receptor to activate that signaling pathway (Arvidsson et al. 1994).

Unlike the PDGF receptor, the BCR uses a number of different membrane-associated proteins as docking sites for SH2-domain-containing signaling proteins (Table 1). In response to BCR ligation, the Ig-α and Ig-β components of the BCR are phosphorylated on tyrosine residues that are part of the immunoreceptor tyrosine-based activation motifs (ITAMs) present in these polypeptides. This is essential for the BCR to recruit both Src family tyrosine kinases and the Syk tyrosine kinase, which prefer to bind to the pYxxL/I sequences (single letter code; x is any amino acid) of the ITAMs. However, this limited diversity of phosphotyrosine-containing sequences does not allow the BCR to recruit the vast array of signaling proteins and tyrosine-kinase substrates that are involved in BCR signaling. The BCR solves this problem by directing the phosphorylation of other membrane-associated proteins on tyrosine residues, thus creating binding sites that can recruit a variety of signaling proteins.

One of the most important of these "docking" proteins for BCR signaling is CD19, a transmembrane protein that is strongly phosphorylated on tyrosine residues after BCR ligation (Tuveson et al. 1993). The phosphotyrosine-containing sequences created in the CD19 cytoplasmic domain are able to recruit both PI 3-kinase and Vav to the plasma membrane (Tuveson et al. 1993; Weng et al. 1994; Li et al. 1997b). CD22, another transmembrane protein, is also tyrosine phosphorylated after

**Table 1.** Docking proteins used by the B-cell receptor (BCR) to recruit SH2 domain-containing signaling proteins to cellular membranes

| Docking protein | Proteins recruited to membrane |
|---|---|
| Transmembrane docking proteins | |
| Ig-α/β | Src kinases, Syk tyrosine kinase |
| CD19 | PI 3-kinase, Vav |
| CD22 | SHP1, PLC-γ1 |
| **Docking protein** | **Proteins (or protein complexes) recruited to membrane** |
| Membrane-associated docking proteins | |
| Cbl | PI 3-kinase, Crk (Crk/C3G) |
| Gab1 | PI 3-kinase, Shc (Shc/Grb2/SOS), SHP2 (SHP2/Grb2/SOS) |
| Cas | Crk (Crk/C3G) |

BCR ligation. The phosphotyrosine-containing sequences in CD22 may be able to recruit PLC-$\gamma$ but probably play a more important role in negative regulation of BCR signaling by recruiting the SHP1 tyrosine phosphatase (LAW et al. 1996b). In addition to using these transmembrane proteins as docking sites for signaling proteins, the BCR also uses the p120 Cbl, p120 Gab1, and p130 Cas proteins to form signaling complexes and to recruit signaling proteins to membrane locations.

Cbl, Gab1, and Cas belong to a family of docking proteins that include IRS-1, which is important for insulin-receptor signaling, and IRS-2, which is important for IL-4-receptor signaling (WATERS and PESSIN 1996). In response to receptor engagement, these docking proteins are phosphorylated on multiple tyrosine residues which can then act as binding sites for SH2-domain-containing proteins. BCR ligation leads to tyrosine phosphorylation of Cbl, Gab1, and Cas, creating binding sites for a number of important signaling proteins (CORY et al. 1995; INGHAM et al. 1996, 1998; PANCHAMOORTHY et al. 1996). In anti-Ig-stimulated B cells, Cbl binds the SH2 domain of PI 3-kinase and the Crk adaptor protein (KIM et al. 1995; INGHAM et al. 1996; PANCHAMOORTHY et al. 1996; SMIT et al. 1996b), Gab1 binds the SH2 domains of Shc, PI 3-kinase, and the SHP-2 tyrosine phosphatase (INGHAM et al. 1998), and Cas binds the SH2 domain of Crk (INGHAM et al. 1996). Although Cbl, Gab1, and Cas are cytoplasmic proteins, subcellular fractionation experiments have shown that a significant portion of each of these proteins is present in the membrane-enriched particulate fraction of B cells (INGHAM et al. 1996, 1998). Complexes of signaling proteins with Cas, Gab1, and Cbl are also found in this particulate fraction of the cells. This suggests that Cas, Gab1, and Cbl could be peripheral membrane proteins and that their interactions with signaling proteins could recruit these signaling proteins to cell membranes. Although an attractive model, the membrane localization of the Cbl, Gab1, and Cas docking proteins needs to be confirmed by microscopy. Nevertheless, these results suggest that the BCR uses a variety of transmembrane and membrane-associated docking proteins to recruit signaling proteins to cell membranes.

Recently, another mechanism for recruiting signaling proteins to the plasma membrane has received a great deal of attention. Phosphatidylinositol 3,4,5-trisphosphate ($PIP_3$), a membrane lipid produced by PI 3-kinase, has been shown to bind to pleckstrin homology (PH) domains (TOKER and CANTLEY 1997), which are found in a number of proteins involved in BCR signaling. The $PIP_3$/PH domain interaction may be sufficient to recruit some of these proteins to the plasma membrane. For example, the BCR recruits the Bruton's tyrosine kinase (Btk) to the plasma membrane in this way (see Sect. 5.1). The production of $PIP_3$ by PI 3-kinase may therefore play a key role in initiating multiple BCR signaling pathways.

## 3 The Ras-Signaling Pathway

The p21$^{ras}$ oncoprotein (Ras) acts as a binary molecular switch that is active when it has GTP bound to it and inactive when GDP is bound. BCR ligation causes a rapid (maximal after 2–3 min) increase in the fraction of cellular Ras that is in the active GTP-bound state (HARWOOD and CAMBIER 1993; LAZARUS et al. 1993; TORDAI et al. 1994). This response is likely to be important for B-cell activation since Ras is a potent inducer of cell growth and differentiation in many cell types. Consistent with this idea, transgenic mice expressing a dominant-negative form of Ras only in the B-cell lineage show greatly impaired B-cell development and the B cells that do develop show decreased anti-Ig-induced proliferation (IRITANI et al. 1997).

Ras controls a kinase cascade that culminates in the activation of the ERK MAP kinases (Fig. 1). GTP-bound Ras binds to and activates the Raf-1 kinase (AVRUCH et al. 1994). Raf-1 then phosphorylates and activates a kinase called MEK which, in turn, phosphorylates and activates the ERK1 and ERK2 MAP kinases. BCR ligation has been shown to activate Raf-1 and MEK (TORDAI et al. 1994) and to cause a large transient activation of ERK2 (GOLD et al. 1992; SUTHERLAND et al. 1996). Interestingly, BCR ligation causes very little activation of the ERK1 kinase (SUTHERLAND et al. 1996) even though it is highly expressed in B cells. The basis for the selective activation of ERK2 as opposed to ERK1 is not known.

The ERK kinases provide a direct signaling link to the nucleus. Upon activation, they can migrate to the nucleus where they phosphorylate and activate Ets domain-containing transcription factors including Ets-1, Ets-2, and members of the ternary complex factor (TCF) family, such as Elk-1 and Sap1a (SU and KARIN 1996; TREISMAN 1996; JANKNECHT and HUNTER 1997). The Elk-1 and Sap1a TCFs form complexes with serum response factor (SRF) and stimulate the transcription of genes containing serum response elements (SREs) (SU and KARIN 1996; TREISMAN 1996). Several key early-response genes whose expression is induced by BCR ligation contain SREs. These include the Fos and Egr-1 transcription factors. Fos is a key component of the AP-1 transcription factor complex. AP-1 stimulates transcription both on its own and in conjunction with the cytoplasmic component of nuclear factor of activated T cells (NF-AT) (see Sect. 5.3). Egr-1 mediates the ability of the BCR to induce expression of several genes, including those encoding CD44 and ICAM-1 (MALTZMAN and MONROE 1996a,b). Work by MCMAHON and MONROE (1995) experimentally confirmed that the induction of Egr-1 expression by the BCR is dependent on Ras activation. They found that BCR-induced Egr-1 expression was blocked by expressing in B cells a dominant-negative form of Ras that prevents activation of the endogenous Ras. In addition to directly activating transcription factors, the ERK kinases phosphorylate and activate the p90$^{rsk}$ kinase (DALBY et al. 1998) which can phosphorylate and activate the CREB (cAMP-responsive element binding protein) transcription factor (XING et al. 1996). BCR ligation has been shown to activate p90$^{rsk}$ (TORDAI et al. 1994; LI et al. 1997a) and to stimulate phosphorylation of CREB on serine 133 (XIE and ROTHSTEIN 1995), a modification that allows it to recruit the transcriptional machinery. Thus, the Ras/Raf-1/MEK/ERK pathway links the BCR to a number of pre-existing latent transcription factors. Moreover, once these pre-existing transcription factors are activated, they stimulate the transcription of immediate early genes, such as c-*fos* and *egr-1*, which encode additional transcription factors.

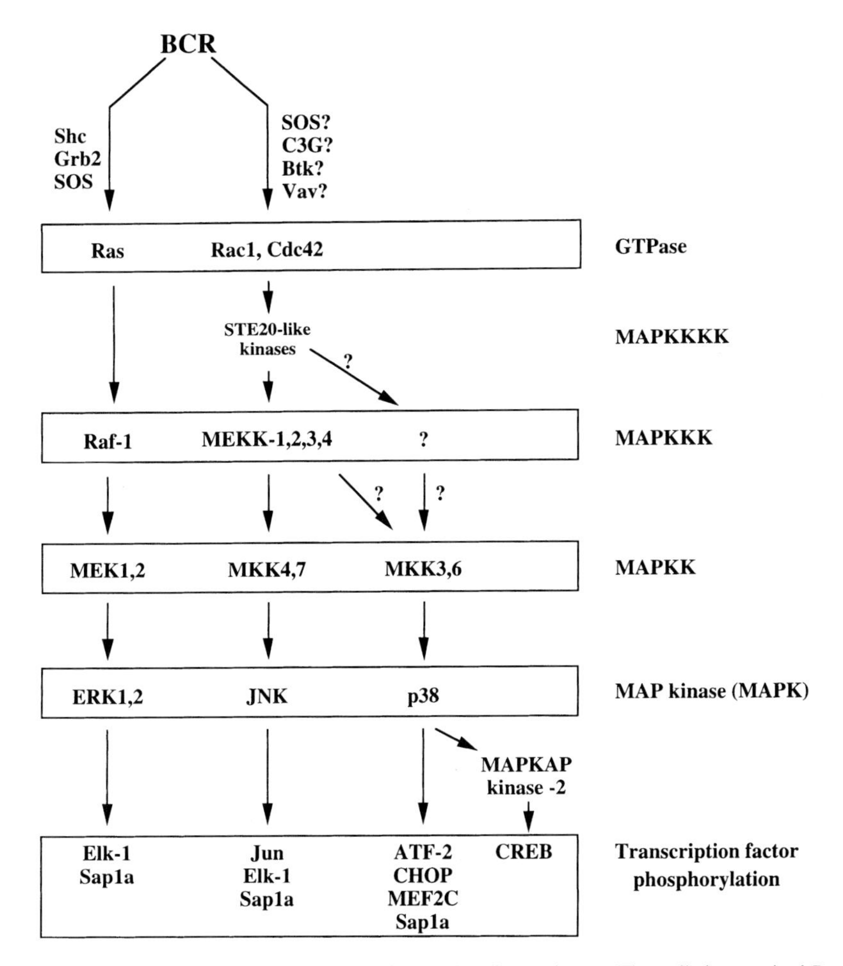


**Fig. 1.** Mitogen-activated protein (MAP) kinase signaling pathways. The well-characterized Ras/Raf-1/MEK/ERK pathway is described in this section. The JNK and p38 MAP kinase pathways are described in Sect. 8. Note that there are still many unanswered questions as to how the JNK and p38 MAP kinase pathways are activated. While the B-cell receptor (BCR) activates Ras, presumably via the formation of Shc/Grb2/SOS complexes, BCR-induced activation of Rac1 or Cdc42 has not been demonstrated. The possible links between the BCR and Rac1 include SOS and Vav, which are Rac1 exchange factors, as well as C3G and Btk, which activate JNK by unknown mechanisms. The connections between the GTPases and the MEKKs that regulate the JNK and p38 MAP kinase pathways are also unclear. Based on work in yeast, the STE20-like kinases are thought to link Rac1 and Cdc42 to the MEKKs. A large number of mammalian STE20-like kinases have been cloned but it is not known which ones regulate the JNK and p38 MAP kinases in B cells or other cells. There is some evidence that the mammalian STE20-like kinases may not phosphorylate the MEKKs directly, but instead activate yet another family of kinases that phosphorylate the MEKKs. Finally, it is not known whether different MEKKs selectively activate JNK versus the p38 MAP kinase

The Ras/Raf-1/MEK/ERK pathway is also directly involved in cell-cycle progression, presumably by controlling the transcription of key cell-cycle regulators. Overexpression of activated forms of Raf-1 in fibroblasts increases the expression of cyclins D and E1, which are required for the G1 to S transition (Woods et al. 1997). It also causes a corresponding decrease in the expression of $p27^{Kip1}$, a protein that inhibits the G1 to S transition. Expression of activated forms of Ras and MEK can also induce cyclin D1 expression (FILMUS et al. 1994; LAVOIE et al. 1996). Consistent with the idea that the Ras/Raf-1/MEK/ERK pathway promotes cell-cycle progression and proliferation, mutant forms of Ras that are constitutively active can transform many types of cells, including B cells (SIRANIAN et al. 1993). Conversely, PD98059, a compound that prevents activation of ERK, strongly inhibits the ability of anti-IgM antibodies to stimulate the proliferation of resting splenic B cells (DEFRANCO 1997).

The Raf-1/MEK/ERK kinase cascade is probably not the only target for activated Ras. Ras-GTP can bind to and activate PI 3-kinase (RODRIGUEZ-VICIANA et al. 1994) and there is some older evidence that the p120 Ras GTPase-activating protein (Ras-GAP) may transmit signals from Ras (MARTIN et al. 1992; DUCHESNE et al. 1993). A number of other potential Ras effectors, whose functions are unknown, have also been identified recently (KURIYAMA et al. 1996; MARSHALL 1996; PONTING and BENJAMIN 1996; WOLTHIUS et al. 1996; VAVVAS et al. 1998). These novel Ras effectors include the Nore1, Rin1, RalGDS, Rlf, AF-6, and Canoe proteins.

## 3.1 Activation of Ras by the BCR

Ras-dependent signaling is controlled by regulating the amount of active GTP-bound Ras in the cell. Two types of proteins are involved in this process (BOGUSKI and MCCORMICK 1993), proteins that stimulate the binding of GTP to Ras (guanine-nucleotide exchange factors or GNEFs) and proteins that stimulate the intrinsic GTPase activity of Ras to convert the bound GTP to GDP (GTPase-activating proteins or GAPs). Cellular Ras activity reflects a balance between GNEF activity and GAP activity. Receptor signaling can increase the amount of GTP-bound Ras either by stimulating GNEF activity and/or by inhibiting GAP activity. There are two distinct Ras-GAPs that favor the inactive GDP-bound form of Ras, p120 Ras-GAP and the *NF-1* gene product, neurofibromin. In permeabilized B cells, BCR ligation decreases total cellular Ras-GAP activity (LAZARUS et al. 1993) by an unknown mechanism. The regulation of Ras-GAP activity by the BCR has not been confirmed or further elucidated since this initial report. Although it may be quite important, attention has focused instead on regulation of Ras GNEFs by the BCR. The best-characterized Ras GNEFs that are expressed in B cells are the mammalian homologues of the *Drosophila son of sevenless* gene products, SOS1 and SOS2 (BOWTELL et al. 1992). There is no evidence that receptor signaling alters the specific activity of SOS. Instead, as described below, the ability of SOS to activate Ras is regulated by controlling the subcellular distribution of SOS.

The 170-kDa SOS proteins are cytosolic enzymes whose substrate, Ras, is tethered to the inner face of the plasma membrane by lipid modifications. Thus, the key event in Ras activation is the recruitment of SOS to the plasma membrane. Consistent with this idea, artificially targeting SOS to the plasma membrane by adding a lipid modification signal is sufficient to activate Ras in T cells (HOLSINGER et al. 1995). While SOS does not contain an SH2 domain that would allow it to bind to membrane proteins that are tyrosine phosphorylated after receptor engagement, it binds constitutively to a 23-kDa adaptor protein called Grb2, which does have an SH2 domain. This interaction is mediated by proline-rich sequences in SOS binding to the SH3 domains of Grb2 (ROZAKIS-ADCOCK et al. 1993). Grb2 mediates the recruitment of SOS to the plasma membrane in response to receptor signaling. In epidermal growth factor (EGF)-stimulated fibroblasts, Grb2 binds via its SH2 domain to phosphotyrosine residues on activated EGF receptors (BUDAY and DOWNWARD 1993; ROZAKIS-ADCOCK et al. 1993). This results in translocation of SOS1 from the cytosol to the plasma membrane, where Ras is located (BUDAY and DOWNWARD 1993).

In anti-Ig-stimulated B cells, the major target of the Grb2 SH2 domain is not a membrane-associated protein but another cytosolic protein called Shc (SAXTON et al. 1994). BCR ligation induces strong tyrosine phosphorylation of both the 46-kDa and 52-kDa forms of Shc and this leads to the formation of Shc/Grb2/SOS complexes (LANKESTER et al. 1994; SAXTON et al. 1994; SMIT et al. 1994). Shc has an SH2 domain and may therefore act as an additional adaptor protein that couples Grb2/SOS complexes to membrane-associated proteins that are tyrosine phosphorylated after BCR ligation. Consistent with this idea, BCR engagement increases the amount of Shc in the membrane-enriched particulate fraction of B cells, and Shc/Grb2/SOS complexes can be found in this fraction after BCR ligation (LANKESTER et al. 1994; SAXTON et al. 1994).

Until recently, the membrane target for the Shc SH2 domain has remained elusive and our understanding of how the BCR recruits SOS to the membrane has been incomplete. Although the Shc SH2 domain binds strongly to the tyrosine-phosphorylated ITAM motifs of Ig-$\alpha$ and Ig-$\beta$ in vitro (BAUMANN et al. 1994; D'AMBROSIO et al. 1996), this interaction has been difficult to demonstrate in vivo by co-precipitation of Shc with Ig-$\alpha$/$\beta$ from B cells. This suggested that another membrane-associated protein might be the main target for the Shc SH2 domain in activated B cells. Recently, INGHAM et al. (1998) found that BCR ligation causes a substantial amount of Shc to bind via its SH2 domain to the Gab1 docking protein in the RAMOS human B cell line. Gab1 is a 120-kDa docking protein that is strongly tyrosine phosphorylated in response to BCR ligation. Although it is not a transmembrane protein, cell fractionation studies showed that a significant portion of Gab1 is in the membrane-enriched particulate fraction of RAMOS cells before BCR ligation. Moreover, Gab1/Shc/Grb2 complexes are found in the particulate fraction of anti-IgM-stimulated RAMOS cells.

Gab1 appears to be a multifunctional docking protein that is involved in multiple BCR signaling pathways. BCR-induced tyrosine phosphorylation of Gab1 creates binding sites not only for the Shc SH2 domain but also for the SH2 domains of PI 3-kinase and the SHP2 tyrosine phosphatase (INGHAM et al. 1998). Moreover, Far western blots [blots probed with a glutathione *S*-transferase (GST) fusion protein containing the Grb2 SH2 domain] showed that the Grb2 SH2 domain could bind not only to the tyrosine-phosphorylated Shc that was associated with Gab1 but also to the tyrosine-phosphorylated SHP2 that was associated with Gab1. Thus, Gab1 may be able to recruit Grb2/SOS complexes to the membrane using both Shc and SHP2 as adaptor proteins.

Although Gab1 can bind significant amounts of Shc and Grb2 after BCR ligation, the role of Gab1 in BCR-induced Ras activation has not been tested. One important caveat is that Gab1 appears to be expressed at significantly higher levels in RAMOS cells than in normal human B cells or in other mouse and human B cell lines (Ingham et al. 1998). Thus, the generality of this model in which Gab1 recruits Shc/Grb2/SOS complexes needs to be confirmed. In addition, the role of the Shc/Grb2/SOS pathway in BCR-induced Ras activation has not been tested experimentally by expressing dominant-negative forms of these proteins in B cells.

One alternative model is that the BCR activates Ras primarily by inhibiting Ras-GAP (Lazarus et al. 1993), which would cause Ras-GTP to accumulate. In T cells, phorbol esters that activate protein kinase C (PKC) have been shown to activate Ras by decreasing Ras-GAP activity (Downward et al. 1990). Phorbol esters can also activate Ras in B cells (Harwood and Cambier 1993; Lazarus et al. 1993; Tordai et al. 1994), but it is not known whether this is due to inhibition of Ras-GAP. Since BCR-induced $PIP_2$ breakdown leads to activation of PKC (see Sect. 6), the BCR could activate Ras via a PKC/Ras-GAP pathway. Although Harwood and Cambier (1993) showed that calphostin, a PKC inhibitor, does not block BCR-induced Ras activation, BCR-induced activation of ERK is partially blocked by a different PKC inhibitor (Gold et al. 1992). Since activation of ERK by PKC (at least in fibroblasts) is dependent on the presence of Ras-GTP/Raf-1 complexes (Marais et al. 1998), the latter result suggests that PKC could contribute to the activation of Ras by the BCR. Thus, the role of PKC in BCR-induced Ras activation needs to be reevaluated and the relative contributions of this putative PKC/Ras-GAP pathway and the Shc/Grb2/SOS pathway to BCR-induced Ras activation need to be determined.

## 4 The PI 3-Kinase Signaling Pathway

Inositol phospholipids play a major role in signal transduction. As shown in Fig. 2, both PLC and PI 3-kinase use these lipids as substrates to generate intracellular second messengers. PLC preferentially cleaves phosphatidylinositol 4,5-bisphosphate [PI(4,5)$P_2$] to produce inositol 1,4,5-trisphosphate ($IP_3$), which causes the release of $Ca^{2+}$ from intracellular stores, and diacylglycerol (DAG), which activates PKC enzymes (Berridge 1993). In contrast, the PI 3-kinase family of enzymes phosphorylate phosphatidylinositol (PI), PI(4)P and PI(4,5)$P_2$ on the 3-position of the inositol ring, yielding PI(3)P, PI(3,4)$P_2$, and PI(3,4,5)$P_3$. PI(3)P is produced by PI 3-kinase family members that are homologous to yeast Vps34p. It is constitutively present in significant amounts in cells and is thought to be involved in the trafficking of intracellular vesicles (Shepherd et al. 1996; Jones and Howell

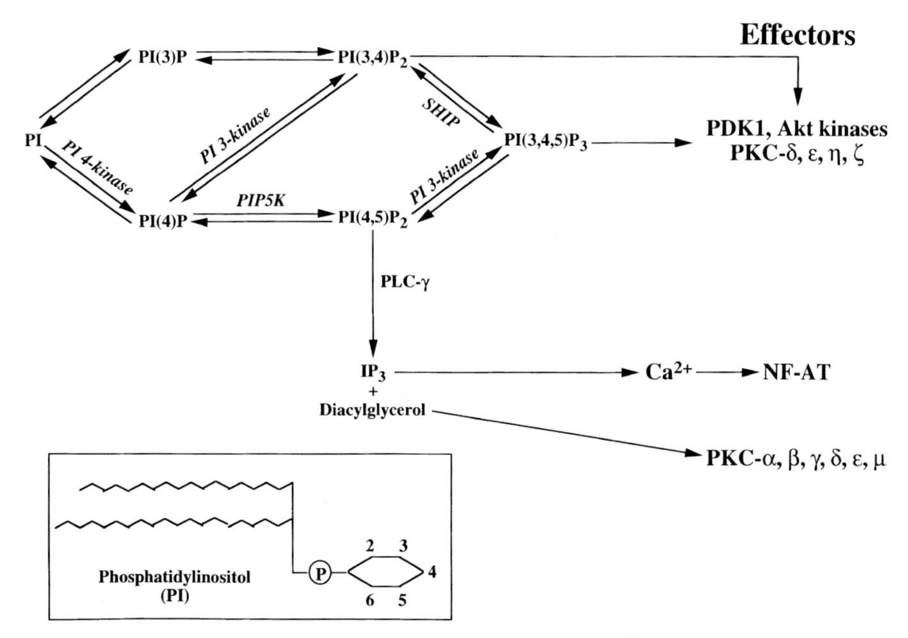


**Fig. 2.** Inositol lipid signaling pathways. The *inset* at the bottom shows the structure of phosphatidylinositol (PI) with the positions on the inositol ring numbered. The various PI kinases phosphorylate either the 3, 4, or 5 position of the inositol ring. $PI(3,4)P_2$ and $PIP_3$ are produced by class-I PI 3-kinases. These lipids activate the PDK1 and Akt protein kinases as well as several of the $Ca^{2+}$-independent isoforms of protein kinase C (PKC). $PIP_3$ also plays a key role in recruiting pleckstrin homology (PH) domain-containing proteins to the plasma membrane. As described in Sect. 5.1 and Fig. 3, $PIP_3$-dependent recruitment of the Bruton's tyrosine kinase to the plasma membrane is important for activation of phospholipase C (PLC)-γ. The 5′-inositol phosphatase SHIP converts $PIP_3$ to $PI(3,4)P_2$. While this might prevent further membrane recruitment of PH domain-containing proteins, $PI(3,4)P_2$ may still be able to act as a second messenger. PLC-γ cleaves $PI(4,5)P_2$ into inositol trisphosphate ($IP_3$) and diacylglycerol. $IP_3$ causes the release of $Ca^{2+}$ from intracellular stores while diacylglycerol activates both the "conventional" and "novel" PKC isoforms which are discussed in Sect. 6. Sustained $PIP_2$ breakdown by PLC-γ requires that the $PIP_2$ pools be replenished by the sequential phosphorylation of PI by PI 4-kinase and PI(4)P 5-kinase (PIP5K). Increased synthesis of PI(4)P by PI 4-kinase requires the phosphatidylinositol transfer protein (PITP) as a cofactor. PITP may present PI to PI 4-kinase so that it can be phosphorylated. As described in Sect. 5.2 and Sect. 10.4, PIP5K is regulated by the Rac1 GTPase. Figure 4 shows that the BCR may activate Rac1 via a pathway that involves PI 3-kinase and Vav

1997). $PI(3,4)P_2$ and $PIP_3$, however, are likely to be involved in signal transduction since they are normally present at very low levels in cells but are produced rapidly in response to engagement of many different receptors (Auger and Cantley 1991). Since $PI(3,4)P_2$ and $PIP_3$ are not substrates for PLC or other known phospholipases (Serunian et al. 1989), they are thought to act directly as second messengers. As described in Sect. 4.1, these lipids can bind to and activate several protein kinases. In addition, $PIP_3$ can recruit cytoplasmic signaling proteins to the plasma membrane by binding to their PH domains.

$PI(3,4)P_2$ and $PIP_3$ are produced by the class-I PI 3-kinases, which all have a 110-kDa catalytic subunit (reviewed by Vanhaesbroeck et al. 1997). Four related p110 proteins have been identified. p110γ associates with a 101-kDa adaptor protein and is involved in signaling by seven transmembrane G-protein-coupled receptors. In contrast, p110α, β, and δ associate with an 85-kDa adaptor subunit that has two SH2 domains and participates in tyrosine kinase-based signal transduction.

BCR ligation increases the specific activity of PI 3-kinase species that bind the p85 adaptor subunit (AAGAARD-TILLERY and JELINEK 1996) and increases the levels of PI 3-kinase products in B cells (GOLD and AEBERSOLD 1994). In the murine B cell lines BAL17 and WEHI-231, BCR ligation causes a rapid but transient increase in the levels of $PIP_3$, which peak after 15–30s. In contrast, a much larger increase in $PI(3,4)P_2$ levels is seen 1–2min after BCR engagement, and $PI(3,4)P_2$ levels remain elevated for at least 15min. The sequential appearance of $PIP_3$ and then $PI(3,4)P_2$ may reflect phosphorylation of $PI(4,5)P_2$ by PI 3-kinase to yield to $PIP_3$, followed by dephosphorylation of $PIP_3$ by p145 SHIP, an inositol phosphatase that specifically removes the phosphate group from the 5-position of the inositol head group to produce $PI(3,4)P_2$ (DAMEN et al. 1996) (Fig. 2). SHIP is recruited to sites of BCR signaling by binding to the Shc adaptor protein and to Syk (CROWLEY et al. 1996; KAVANAUGH et al. 1996; TRIDANDAPANI et al. 1997). Thus, $PIP_3$ is the initial PI3-kinase product, while $PI(3,4)P_2$ is produced later in the response. This shift from $PIP_3$ to $PI(3,4)P_2$ may have interesting consequences. PH domains bind preferentially to $PIP_3$, suggesting that membrane recruitment of PH-domain-containing signaling proteins is the earliest PI 3-kinase-mediated event. In contrast, both $PI(3,4)P_2$ and $PIP_3$ can activate downstream kinases including PKC-ε, PKC-λ, and Akt (TOKER and CANTLEY 1997). $PI(3,4)P_2$ may therefore be responsible for sustained PI 3-kinase-mediated signaling in B cells.

Activation of the PI 3-kinase signaling pathway involves recruiting PI 3-kinase to the plasma membrane, where its substrates are located, as well as increasing the specific activity of the enzyme. This can be mediated by a number of different protein–protein interactions. The first involves the two SH2 domains of the p85 regulatory subunit of PI 3-kinase. Both SH2 domains bind with high affinity to the motif phosphoYxxM (single letter code; x is any amino acid) in other proteins. SH2-mediated binding of PI 3-kinase to membrane proteins containing this motif can bring PI 3-kinase to the plasma membrane. In anti-Ig-stimulated B cells, PI 3-kinase binds to the CD19 co-receptor (TUVESON et al. 1993). CD19 contains two YxxM motifs in its cytoplasmic domain and is tyrosine phosphorylated after BCR ligation (CHALUPNY et al. 1993). Changing the tyrosine residues in these two motifs of CD19 to phenylalanines prevents the association of PI 3-kinase with CD19 (TUVESON et al. 1993). In anti-IgM-stimulated B cells, PI 3-kinase also binds via its SH2 domains to Cbl and Gab1 (KIM et al. 1995; PANCHAMOORTHY et al. 1996; INGHAM et al. 1998), two membrane-associated docking proteins that are tyrosine phosphorylated after BCR ligation. Not only does the binding of the p85 SH2 domains to phosphoYxxM motifs in membrane-associated proteins bring PI 3-kinase close to its substrates, it may also increase PI 3-kinase activity. In vitro, the binding of peptides containing phosphoYxxM motifs to the p85 SH2 domains increases the specific activity of PI 3-kinase (BACKER et al. 1992; CARPENTER et al. 1993). Peptides with two such motifs are more potent activators of PI 3-kinase, suggesting that both SH2 domains of p85 must be engaged for maximal activation of PI 3-kinase (Rordorf-Nikolic et al. 1995).

PI 3-kinase activity and subcellular localization may also be regulated by the binding of a proline-rich region in the p85 subunit of PI 3-kinase to the SH3 domains of some Src family tyrosine kinases. The SH3 domains of Lyn, Fyn, and Lck, but not Blk, can bind to the p85 subunit in vitro (Prasad et al. 1993; Pleiman et al. 1994). Since these tyrosine kinases are located at the inner face of the plasma membrane, this interaction could also attract PI 3-kinase to the plasma membrane where its substrates are located. PI 3-kinase has been shown to associate with Lyn in anti-Ig-stimulated B cells (Yamanishi et al. 1992). In B cells, Lyn is tightly associated with CD19 (van Noesel et al. 1993), suggesting that PI 3-kinase associates both directly and indirectly with CD19. In addition to recruiting PI 3-kinase to the membrane, these interactions may also activate PI 3-kinase. Isolated SH3 domains from Src kinases can activate PI 3-kinase in vitro (Pleiman et al. 1994). The SH3 domain of Src kinases may be available to bind to and activate PI 3-kinase only after the Src kinase has been activated by a receptor. Crystallographic data suggests that activation of a Src kinase induces a conformational change that disrupts intramolecular associations involving the kinase's SH3 domain (Moarefi et al. 1997; Sicheri et al. 1997).

Finally, the active GTP-bound form of Ras, which is located at the plasma membrane, can bind to and activate PI 3-kinase (Rodriguez-Viciana et al. 1994). The mechanism of PI 3-kinase activation by Ras is not well understood but is thought to involve direct binding of Ras to the p110 catalytic subunit of PI 3-kinase. Thus, there are multiple mechanisms by which PI 3-kinase can be recruited to the plasma membrane and activated. It is not known which of these mechanisms is most important for the BCR to activate PI 3-kinase.

## 4.1 Downstream Targets of the PI 3-Kinase-Derived Second Messengers

Since no known phospholipase cleaves $PIP_3$ or $PI(3,4)P_2$, it has long been thought that these lipid products of PI 3-kinase (as opposed to hydrolysis products such as DAG, phosphatidic acid, or IPs) act directly as second messengers. In 1993 and 1994, it was reported that several isoforms of PKC, in particular the $\delta$, $\varepsilon$, $\eta$, and $\zeta$ isoforms, could be activated in vitro by $PIP_3$ and $PI(3,4)P_2$ (Nakanishi et al. 1993; Toker et al. 1994). Although treating cells with PI 3-kinase inhibitors such as wortmannin can block receptor activation of PKC-$\zeta$ and PKC-$\lambda$ in some cells (Akimoto et al. 1996; Herrera-Velit et al. 1997), the connection between PI 3-kinase and PKC enzymes remains somewhat controversial. In 1995, the first bona fide downstream target of the PI 3-kinase pathway, the Akt kinase (also called PKB; described in Sect. 4.1.1), was identified (Burgering and Coffer 1995; Franke et al. 1995; Kohn et al. 1995). Detailed studies on Akt showed that the binding of $PIP_3$ to the PH domain of Akt was important for recruiting Akt to the plasma membrane where it could be phosphorylated and activated by upstream kinases.

PH domains are found in more than 90 proteins, including a number of cytosolic signaling enzymes involved in BCR signaling such as PLC-$\gamma$, the Ras activator SOS, the Ras GAP, the Btk, the PDK1 kinase, PKC-$\mu$, and Vav (LEMMON et al. 1997; TOKER and CANTLEY 1997). Recent work has shown that the PH domains in many of these proteins can bind $PIP_3$ and that this interaction plays a role in the recruitment of these proteins to the plasma membrane where they can perform their signaling functions or interact with other proteins that regulate their activity (TOKER and CANTLEY 1997). In some cases, the binding of $PIP_3$ to the PH domains of these proteins can influence their enzymatic activity directly. Since PH-domain-containing proteins are involved in pathways leading to the activation of PLC-$\gamma$ and Ras, as well as the JNK and p38 MAP kinases, the production of $PIP_3$ by PI 3-kinase is therefore an important initial event in the activation of multiple signaling pathways. Thus, in addition to having its own signaling pathway, PI 3-kinase plays a central role in most, if not all, of the signaling pathways activated by the BCR. The role of $PIP_3$/PH domain interactions in other signaling pathways will be discussed in the individual sections relating to those signaling pathways, while the activation of PKC enzymes by PI 3-kinase-derived lipids will be discussed in Sect. 6. Sect. 4.1.1 will focus on Akt as a major downstream effector of PI 3-kinase.

#### 4.1.1 Akt (Protein Kinase B)

Akt is a 60-kDa serine/threonine protein kinase, which is the cellular homologue of the v-*akt* oncogene. Its kinase domain has a high degree of sequence similarity to those of protein kinase A and PKC-$\varepsilon$, leading some workers to refer to it as PKB. A wide variety of receptors that activate PI 3-kinase, including the BCR (M. Gold, unpublished observations), activate Akt. It is now well documented that activation of PI 3-kinase is both necessary and sufficient for Akt to be activated. Receptor-induced activation of Akt is blocked by PI 3-kinase inhibitors and by expression of dominant-negative forms of PI 3-kinase (BURGERING and COFFER 1995; FRANKE et al. 1995). Conversely, expression of constitutively-active forms of PI 3-kinase activates Akt (KLIPPEL et al. 1996).

As reviewed recently (DOWNWARD 1998), considerable progress has been made towards understanding how PI 3-kinase activates Akt. Akt has a PH domain which binds the PI 3-kinase-derived lipids $PIP_3$ and $PI(3,4)P_2$ (FRECH et al. 1997). This interaction recruits Akt to the plasma membrane where it can be phosphorylated and activated by upstream kinases (ANDJELKOVIC et al. 1997). In response to receptor engagement, human Akt is phosphorylated on threonine 308 and serine 473. Phosphorylation of threonine 308 in the activation loop of Akt is both necessary and sufficient to activate Akt (ALESSI et al. 1996), while phosphorylation of serine 473 is required for maximal activation (WELCH et al. 1998). A kinase called PDK1 has been isolated which phosphorylates threonine 308, but not serine 473 (ALESSI et al. 1997). The kinase that phosphorylates serine 473 has yet to be identified. Interestingly, the PDK1 kinase also contains a PH domain and production of $PIP_3$

by PI 3-kinase is required to recruit PDK1 to the plasma membrane where it can phosphorylate Akt (Anderson et al. 1998).

Activated Akt is likely to have several important functions in BCR signaling. First, Akt may regulate the activity of the NF-AT transcription factor by promoting its accumulation in the nucleus (described more fully in Sect. 5.3). NF-AT consists of a nuclear component as well as a cytoplasmic component that is called NF–AT$_c$ (Rao et al. 1997). BCR-induced increases in intracellular $Ca^{2+}$ activate the phosphatase calcineurin which dephosphorylates NF-AT$_c$. This reveals a nuclear localization signal that allows NF-AT$_c$ to translocate into the nucleus, bind to the nuclear component, and stimulate transcription. This process is opposed by glycogen synthase kinase-3 (GSK3), a constitutively-active kinase that phosphorylates NF-AT$_c$ and changes its conformation so that a nuclear export signal is revealed. Once phosphorylated by GSK3, NF-AT$_c$ is rapidly exported from the nucleus (Beals et al. 1997; Klemm et al. 1997). Thus, unless GSK3 is inhibited, NF-AT$_c$ cannot accumulate in the nucleus and stimulate transcription. Akt has been shown to directly phosphorylate GSK3 and inhibit its activity (Cross et al. 1995). Akt may therefore have an important role in allowing the BCR to induce the transcription of genes that are regulated by NF-AT.

Another important function for the PI 3-kinase/Akt pathway is the prevention of apoptosis. In an number of cell types, the ability of growth factors to prevent apoptosis is blocked by PI 3-kinase inhibitors, while expression of constitutively-active forms of PI 3-kinase prevents apoptosis (Scheid et al. 1995; Yao and Cooper 1995; Philpott et al. 1997). Akt appears to mediate this protection from apoptosis. Expression of constitutively-active forms of Akt is sufficient to protect cells from apoptosis caused by growth-factor withdrawal, while dominant-negative forms of Akt can cause apoptosis (Dudek et al. 1997; Philpott et al. 1997). One way in which Akt may prevent apoptosis is by phosphorylating Bad, a death-promoting member of the Bcl-2 family (del Peso et al. 1997; Datta et al. 1997). Phosphorylation of Bad by Akt creates a binding site for a member of the 14-3-3 protein family (Zha et al. 1996). Once bound to 14-3-3, Bad cannot dimerize with Bcl-2 or Bcl-$x_L$ and neutralize their anti-apoptotic activities.

Although it is not known whether BCR signaling causes phosphorylation of Bad, the diverse signals that control B-cell activation versus apoptosis could be integrated by the PI 3-kinase/Akt pathway. For example, T-cell-dependent B-cell activation involves signaling by the BCR, CD40 and the IL-4 receptor, all of which activate PI 3-kinase (Gold et al. 1994; Gold and Aebersold 1994; Aagaard-Tillery and Jelinek 1996). A testable hypothesis is that simultaneous signaling by these receptors has additive or perhaps synergistic effects on the activation of Akt and the phosphorylation of Bad. This would yield a strong anti-apoptotic signal and promote B-cell activation.

The PI 3-kinase/PDK1/Akt pathway may also regulate the activity of the $p70^{S6K}$ kinase. $p70^{S6K}$ is regulated in a complicated manner by multiple upstream kinases. PDK1, the kinase that phosphorylates and activates Akt, can also phosphorylate and activate $p70^{S6K}$ (Alessi et al. 1998). Akt may also phosphorylate and activate $p70^{S6K}$ since expressing the constitutively active viral form of Akt in fibroblasts stimulates $p70^{S6K}$ phosphorylation (BURGERING and COFFER 1995). Finally, the mammalian target of rapamycin (mTOR), another kinase that is regulated by PI 3-kinase, may also play a role in activating $p70^{S6K}$ (ABRAHAM 1998). Thus, $p70^{S6K}$ is a target of multiple PI 3-kinase-regulated protein kinases, PDK1, Akt, and mTOR. $p70^{S6K}$ phosphorylates the ribosomal S6 protein and this preferentially increases the translation of mRNAs containing 5′-terminal oligopolypyrimidine tracts (ABRAHAM 1998). Since the mRNAs for several components of the protein synthesis machinery contain these polypyrimidine tracts, the net result of $p70^{S6K}$ activation may be to increase the overall rate of protein synthesis when cells are activated.

The PDK1/Akt pathway appears to be a multifunctional mediator of PI 3-kinase-dependent signaling. In addition to opposing apoptosis, inhibiting GSK3, and activating $p70^{S6K}$, this pathway may also signal directly to the nucleus. There are several reports that Akt translocates to the nucleus after it is activated at the plasma membrane (MEIER et al. 1997). Nuclear targets of Akt remain to be identified.

## 5 The Phospholipase C-γ Signaling Pathway

The BCR activates both PLC-γ1 and PLC-γ2 (BIJSTERBOSCH et al. 1985; FAHEY and DEFRANCO 1987; CARTER et al. 1991; HEMPEL et al. 1992), enzymes that cleave the plasma membrane lipid $PI(4,5)P_2$ into two second messengers, $IP_3$ and DAG (BERRIDGE 1993) (Fig. 2). $IP_3$ increases cytosolic free calcium concentrations by stimulating the release of $Ca^{2+}$ from the endoplasmic reticulum. This release of calcium into the cytosol is mediated by an $IP_3$-gated $Ca^{2+}$ channel, whose properties have been studied in detail (DAWSON 1997). $IP_3$ is rapidly converted into a large number of other inositol phosphate species by the action of kinases and phosphatases (BERRIDGE 1993). It is not known whether these other inositol phosphate isomers also have signaling functions, although a specific binding protein for inositol 1,3,4,5-tetrakisphosphate ($IP_4$) has been cloned and shown to be a Ras-GAP whose activity is stimulated by $IP_4$ (CULLEN et al. 1996). In B cells, increased levels of $IP_3$, $IP_4$, and other inositol phosphates as well as elevated levels of intracellular calcium can be detected within seconds of antigen receptor ligation (BIJSTERBOSCH et al. 1985; FAHEY and DEFRANCO 1987). DAG activates multiple isoforms of the PKC family of serine/threonine kinases (NISHIZUKA 1995) and this is accompanied by translocation of these enzymes from the cytosol to the membrane. BCR cross-linking has been shown to increase DAG levels in B cells (BIJSTERBOSCH et al. 1985) and to cause a transient increase in the amount of PKC activity associated with the membrane fraction of the cells (CHEN et al. 1986; NEL et al. 1986). When the subcellular localization of various PKC isoforms was examined in the A20 B cell line, BRAS et al. (1997) found that BCR ligation caused an increase in the amount of PKC-α, $\beta_{II}$, δ, ε, and ζ in the membrane fraction of these cells. The role of $Ca^{2+}$ fluxes and PKC activation in BCR signaling will be discussed in Sect. 5.2 and Sect. 6, respectively.

## 5.1 Activation of PLC-$\gamma$ by the BCR

Activation of PLC-$\gamma$ by the BCR is a multi-step process that requires both membrane recruitment and tyrosine phosphorylation of PLC-$\gamma$. PLC-$\gamma$1 and PLC-$\gamma$2 are cytosolic enzymes and must therefore be recruited to the inner face of the plasma membrane where $PIP_2$ is located. PLC-$\gamma$1 and PLC-$\gamma$2 both have two SH2 domains and could therefore bind to membrane-associated docking proteins that are tyrosine phosphorylated in response to BCR ligation. Until recently, however, the ligands for the PLC-$\gamma$ SH2 domains in B cells remained elusive. In T cells, T-cell receptor (TCR) signaling causes PLC-$\gamma$ to bind via its SH2 domains to a 76-kDa cytosolic adaptor protein called SLP-76 (Motto et al. 1996) and to a 36-kDa transmembrane protein called LAT (Zhang et al. 1998), both of which are tyrosine phosphorylated after TCR engagement. Although PLC-$\gamma$1 binds to tyrosine phosphorylated proteins of 68–70kDa in anti-Ig-stimulated B cells (Fu and Chan 1997), SLP-76 is not expressed in B cells. Fu et al. (1998) have recently purified and cloned this B cell-specific 70-kDa PLC-$\gamma$1-associated protein which they call BLNK for B-cell linker protein. BLNK is tyrosine phosphorylated in response to BCR ligation and binds PLC-$\gamma$ only after BCR engagement. The C-terminal SH2 domain of both PLC-$\gamma$1 and PLC-$\gamma$2 can bind to tyrosine-phosphorylated BLNK. Thus, BLNK may fulfill the same role in B cells as SLP-76 does in T cells, a cytosolic adaptor protein whose tyrosine phosphorylation creates binding sites for the SH2 domains of PLC-$\gamma$.

Fu et al. (1998) have now performed a number of functional assays suggesting that BLNK has a role in BCR-induced $PIP_2$ breakdown. Overexpression of wild-type BLNK in B-cell lines potentiates the BCR-induced increase in intracellular $Ca^{2+}$. In contrast, overexpression of a mutant form of BLNK in which four potential sites of tyrosine phosphorylation are changed to phenylalanine residues causes a slight decrease in BCR-induced $Ca^{2+}$ fluxes. Although this suggests that BLNK may not be the major target for the PLC-$\gamma$ SH2 domains, it is known that maximal $Ca^{2+}$ fluxes can be achieved at submaximal levels of $IP_3$. It is possible that if $IP_3$ production were examined, the effect of this mutant BLNK protein might be more significant. The relative contribution of BLNK to BCR-induced $PIP_2$ breakdown will be better defined once the genes encoding BLNK are disrupted either in mice or in the DT40 B-cell line.

While the identification of BLNK as a target for the PLC-$\gamma$ SH2 domains in B cells is a major advance, this only partially answers the question of how PLC-$\gamma$ is recruited to the plasma membrane after BCR engagement. Although BLNK is a cytosolic protein, it can be found in the membrane-enriched particulate fraction of B cells after BCR ligation (Fu et al. 1998). The question then is how is BLNK recruited to the membrane. BLNK has an SH2 domain which it may use to bind a membrane-associated protein that becomes tyrosine phosphorylated after BCR ligation. The identity of this protein is not known at this time. Nevertheless, a current working model for the recruitment of PLC-$\gamma$ to the plasma membrane in B cells involves (1) creation of a membrane binding site for the BLNK SH2 domain by phosphorylation of a membrane-associated protein on a suitable tyrosine residue, (2) binding of cytosolic BLNK to this membrane-associated protein, (3) tyrosine phosphorylation of BLNK by the Syk tyrosine kinase, and (4) binding of PLC-$\gamma$1 or PLC-$\gamma$2 to phosphorylated BLNK via their SH2 domains. There may be additional ways in which PLC-$\gamma$ can be recruited to the plasma membrane after BCR ligation. PLC-$\gamma$1 has been reported to bind via its SH2 domain to both Syk and CD22 after BCR engagement (Sillman and Monroe 1995; Law et al. 1996a).

Recruitment of PLC-$\gamma$ to the plasma membrane is essential but not sufficient to cause increased hydrolysis of $PIP_2$ to $IP_3$ and DAG. Goldschmidt-Clermont et al. (1991) showed that, in the presence of profilin, an abundant cytosolic $PIP_2$-binding protein, tyrosine-phosphorylated PLC-$\gamma$ isolated from growth factor-treated cells could cleave $PIP_2$ while the unphosphorylated PLC-$\gamma$ from resting cells could not. Thus, tyrosine phosphorylation of PLC-$\gamma$ may be required for receptor-induced $PIP_2$ breakdown in cells. Tyrosine phosphorylation of both PLC-$\gamma$1 and PLC-$\gamma$2 has been observed after BCR ligation (Carter et al. 1991; Hempel et al. 1992). The critical tyrosine residues that must be phosphorylated have not been mapped nor is it known how this modification overcomes inhibition of $PIP_2$ breakdown by profilin.

Several lines of evidence suggest that Btk (or a kinase activated by Btk) phosphorylates and activates PLC-$\gamma$ once the BCR is engaged. Btk-deficient DT40 B cells fail to mobilize $Ca^{2+}$ in response to BCR cross-linking (Takata and Kurosaki 1996). Similarly, B cells from *xid* mice or X-linked agammaglobulinemia (XLA) patients, both of which lack functional Btk, also exhibit decreased BCR-induced $IP_3$ production and $Ca^{2+}$ mobilization (Rigley et al. 1989; Fluckiger et al. 1998). Corresponding gain-of-function experiments performed by Fluckiger et al. (1998) support the idea that Btk is an upstream activator of PLC-$\gamma$ in B cells. They showed that overexpression of Btk in B cells greatly increases BCR-induced $IP_3$ production and $Ca^{2+}$ fluxes. Moreover, co-expression of Btk and PLC-$\gamma$2 in fibroblasts results in tyrosine phosphorylation of PLC-$\gamma$2.

The requirements for activation of Btk then become key factors in the ability of the BCR to activate PLC-$\gamma$. Work from several groups strongly suggests that production of $PIP_3$ by PI 3-kinase is important for recruiting Btk to the plasma membrane and that this, in turn, is essential for Btk to activate PLC-$\gamma$. Both Bolland et al. (1998) and Buhl et al. (1997) showed that treating B cells with the PI 3-kinase inhibitor wortmannin decreases the magnitude of BCR-induced $Ca^{2+}$ fluxes. Conversely, BCR-induced $Ca^{2+}$ increases are much larger in DT40 cells, in which the genes encoding the 5′-inositol phosphatase SHIP have been disrupted (Bolland et al. 1998). The absence of SHIP, which converts $PIP_3$ to $PI(3,4)P_2$, presumably allows greater accumulation of $PIP_3$ after BCR engagement. Targeting SHIP to the membrane by fusing it to a membrane protein has the opposite effect, allowing it to efficiently degrade $PIP_3$, and this mimics the effect of wortmannin and decreases BCR-induced $Ca^{2+}$ fluxes (Bolland et al. 1998). Thus, $PIP_3$ production correlates with the ability of the BCR to activate PLC-$\gamma$. The effects of manipulating $PIP_3$ levels on PLC-$\gamma$ activation appear to involve Btk. Btk has a PH domain which binds $PIP_3$ with high affinity (Salim et al. 1996), suggesting that $PIP_3$ can recruit Btk to the plasma membrane where it can phosphorylate PLC-$\gamma$ which has also been recruited to the plasma membrane (Fig. 3). Several pieces of evidence support this model. Bolland et al. (1998) showed that BCR ligation increases the amount of Btk in the membrane-enriched particulate fraction of B cells and that membrane translocation of Btk is enhanced in SHIP-deficient DT40 cells which presumably have elevated levels of $PIP_3$. Furthermore, expressing a membrane-targeted form of Btk in B cells overcomes the effects of wortmannin and membrane-targeted SHIP (both of which reduce $PIP_3$ levels) and restores BCR-induced PLC-$\gamma$ activation (Bolland et al. 1998). This argues that the major role for $PIP_3$ in BCR-induced PLC-$\gamma$ activation is to recruit Btk to the plasma membrane by providing a ligand for its PH domain. Consistent with this idea, BCR-induced $Ca^{2+}$ signaling is greatly reduced in *xid* mice, which have a point mutation in the Btk PH domain that prevents it from binding $PIP_3$ (Salim et al. 1996).

The role of PI 3-kinase and Btk in activating PLC-$\gamma$ may also explain, at least in part, the role of CD19 in BCR-induced $Ca^{2+}$ signaling. In response to BCR ligation, CD19 is tyrosine phosphorylated and binds the PI 3-kinase SH2 domains (Tuveson et al. 1993). In B cells from CD19 −/− mice, the ability of the BCR to activate both PI 3-kinase and PLC-$\gamma$ is greatly reduced (Bolland et al. 1998). The defect in PLC-$\gamma$ activation may be due to an inability to activate PI 3-kinase and recruit Btk to the plasma membrane. In support of this idea, expressing CD19 in the CD19-negative B cell line J558 greatly increases BCR-induced $Ca^{2+}$ signaling, whereas a mutant form of CD19 that cannot bind PI 3-kinase does not potentiate BCR-induced $Ca^{2+}$ signaling (Buhl et al. 1998).

Figure 3 synthesizes the data described in the previous paragraphs. The key elements of this model are: (1) tyrosine phosphorylation of CD19 (and perhaps other membrane-associated proteins) recruits PI 3-kinase to the plasma membrane, resulting in the production of $PIP_3$, (2) Btk is recruited to the plasma membrane using its PH domain to bind to $PIP_3$, (3) membrane-associated Btk phosphorylates and activates the PLC-$\gamma$ that has been recruited to the plasma membrane by binding to tyrosine-phosphorylated BLNK.

## 5.2 Other Components Required for Sustained $PIP_2$ breakdown by PLC-$\gamma$

The $PI(4,5)P_2$ pool in most cells is very small and it has been estimated that activation of PLC would deplete this supply of $PIP_2$ within seconds if the pool were not refilled with newly synthesized $PIP_2$ (Stephens et al. 1993). Thus, sustained receptor-stimulated $PIP_2$ breakdown requires that receptor signaling also stimulate $PIP_2$ biosynthetic pathways. $PI(4,5)P_2$ is produced by the sequential phosphorylation of PI on the 4 and 5 positions of the inositol head group. Recent work has identified a pathway by which BCR signaling stimulates the activity of PI(4)P 5-kinase (PIP5K), a kinase that phosphorylates PI(4)P on the 5 position of the inositol ring (Fig. 2). As described in more detail in Sect. 10.4 and Fig. 4, the Rac1 GTPase binds to and activates PIP5K (Chong et al. 1994; Tolias et al. 1995; Ren et al. 1996). As for other GTPases, activation of Rac1 involves the release of GDP

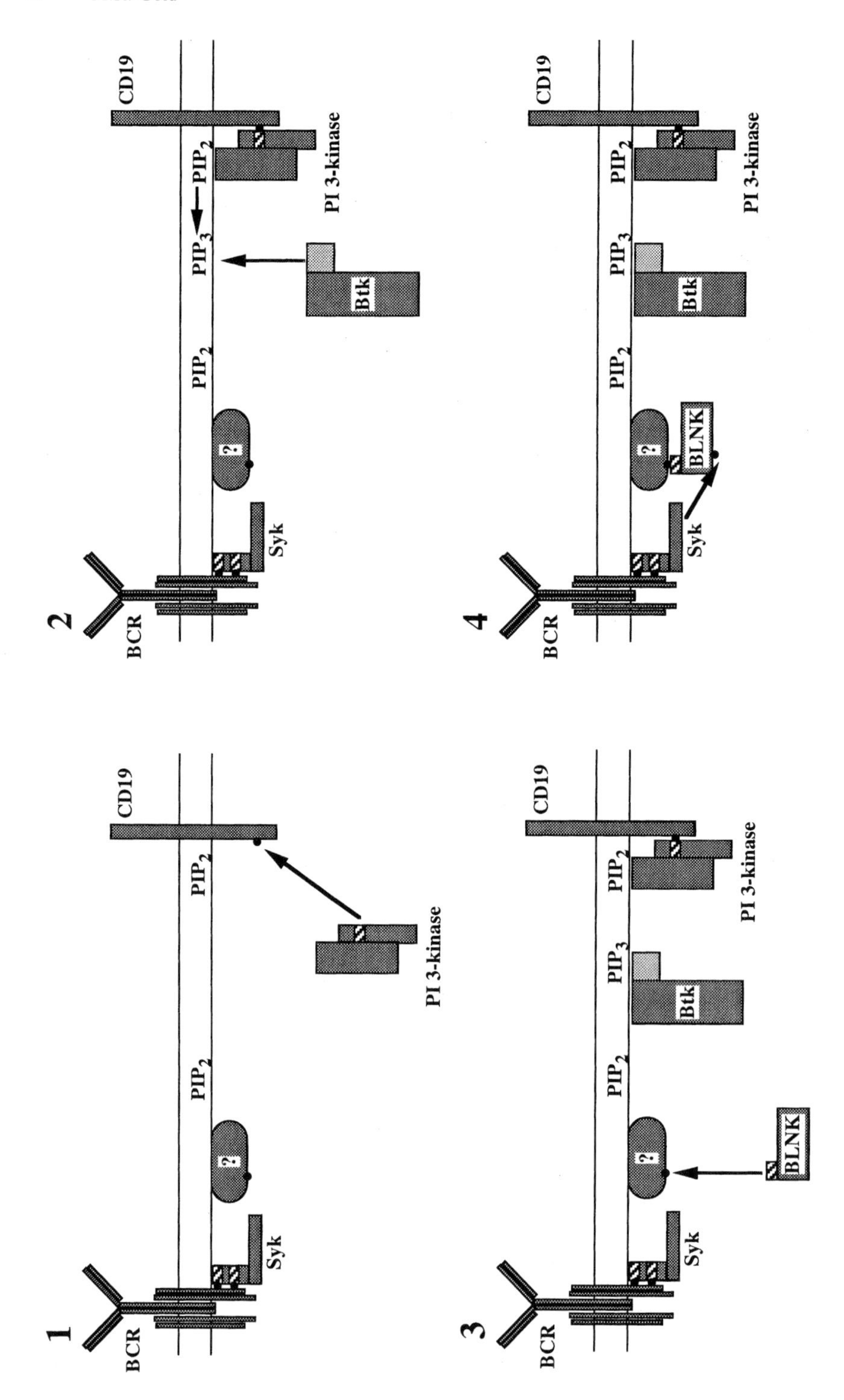
1
BCR
Syk
?
PIP2
PIP2
CD19
PI 3-kinase
2
BCR
Syk
?
PIP2
Btk
PIP3
PIP2
PI 3-kinase
CD19
3
BCR
Syk
?
BLNK
PIP2
Btk
PIP3
PIP2
PI 3-kinase
CD19
4
BCR
Syk
?
BLNK
PIP2
Btk
PIP3
PIP2
PI 3-kinase
CD19

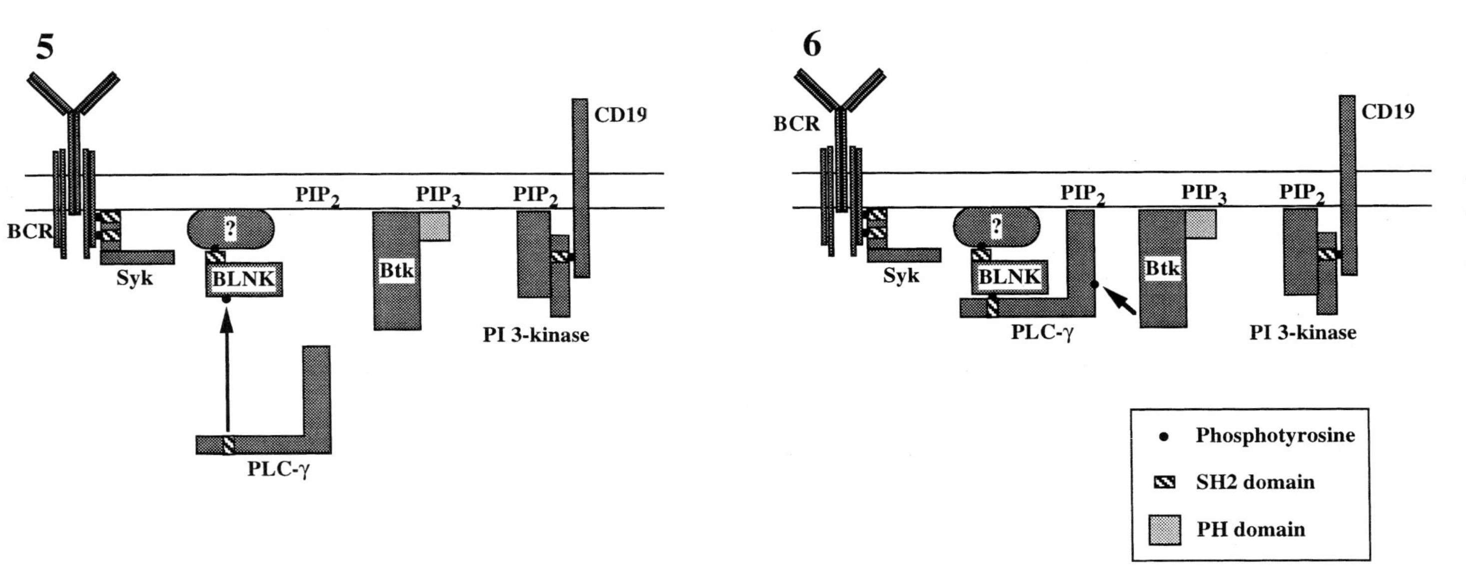


**Fig. 3.** Activation of phospholipase C (PLC)-γ by the B-cell receptor (BCR). The key elements of this model are: (1) tyrosine phosphorylation of CD19 (and perhaps other membrane-associated proteins such as Cbl and Gab1) recruits PI 3-kinase to the plasma membrane, (2) PI 3-kinase produces $PIP_3$ (Btk is recruited to the plasma membrane using its PH domain to bind to $PIP_3$), (3) BLNK uses its SH2 domain to bind to a tyrosine-phosphorylated membrane-associated protein that has not been identified, (4) BLNK is phosphorylated by Syk, (5) PLC-γ uses its SH2 domain to bind tyrosine-phosphorylated BLNK and (6) membrane-associated Btk phosphorylates and activates the PLC-γ that has been recruited to the plasma membrane by binding to tyrosine-phosphorylated BLNK

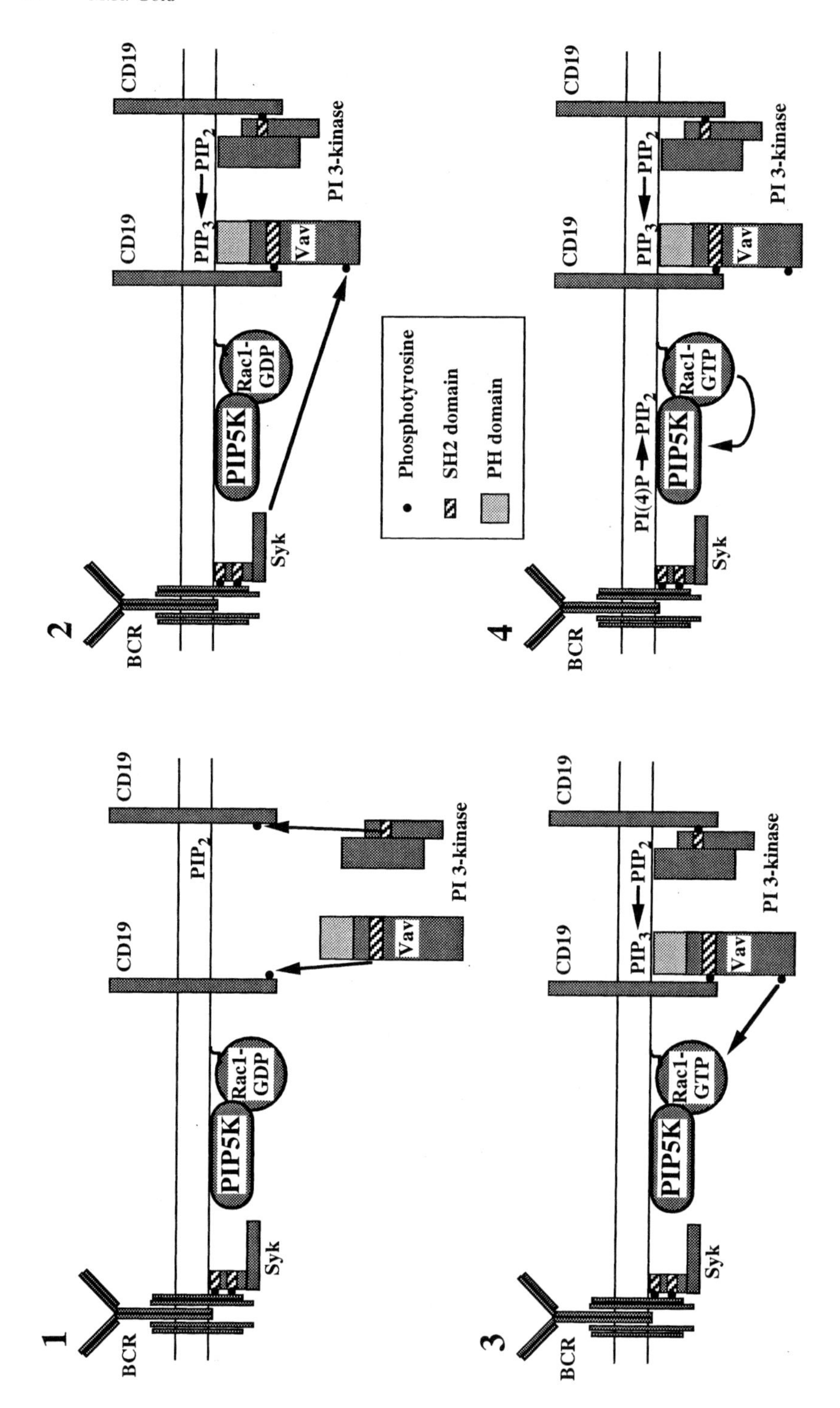
1
BCR
Syk
PIP5K
Rac1-GDP
CD19
Vav
PIP2
PI 3-kinase
CD19
2
BCR
Syk
PIP5K
Rac1-GDP
CD19
PIP3
PIP2
Vav
PI 3-kinase
CD19
Phosphotyrosine
SH2 domain
PH domain
3
BCR
Syk
PIP5K
Rac1-GTP
CD19
PIP3
PIP2
Vav
PI 3-kinase
CD19
4
BCR
Syk
PI(4)P
PIP2
PIP5K
Rac1-GTP
CD19
PIP3
PIP2
Vav
PI 3-kinase
CD19

from Rac1 and the subsequent binding of GTP which allows Rac1 to assume its activated conformation. The 95-kDa Vav protein has been identified as an exchange factor that can activate Rac1 (CRESPO et al. 1997). BCR ligation causes strong tyrosine phosphorylation of Vav (BUSTELO and BARBACID 1992b), a modification that stimulates Vav's Rac1 exchange factor activity. This presumably leads to activation of Rac1 and PIP5K, resulting in increased synthesis of $PIP_2$ from PI(4)P. The involvement of Rac1 in providing a continuous supply of $PIP_2$ for sustained production of $IP_3$ and DAG may explain earlier data which indicated that a GTPase is involved in BCR-induced $PIP_2$ breakdown (GOLD et al. 1987; HARNETT and KLAUS 1988; MONROE and HALDAR 1989).

While the Vav/Rac1/PIP5K pathway produces $PI(4,5)P_2$ from PI(4)P, the supply of PI(4)P is also limiting, indicating that a PI 4-kinase is also important for sustained $PIP_2$ breakdown (Fig. 2). There is a large pool of PI in most cells, suggesting that only PI 4-kinase and PIP5K need to be activated to provide a continuous supply of $PIP_2$. However, some intriguing work by Cockcroft and colleagues suggests that only a small portion of the PI in the cells may be available for conversion to $PIP_2$ by the PI kinases. Using a permeabilized cell system, they found that sustained $PIP_2$ breakdown in response to epidermal growth factor (EGF) required that a soluble component that leaked out of the cells be added back to the permeabilized cells (KAUFFMANN-ZEH et al. 1995). Purification of this component revealed that it was the phosphatidylinositol transfer protein (PITP) whose function is to carry newly-synthesized PI from the endoplasmic reticulum to the plasma membrane. Moreover, they found that EGF treatment induced the formation of a complex that contained PITP, PI 4-kinase, and PLC-$\gamma$.

Further kinetic studies have suggested a speculative model in which PI 4-kinase and PIP5K are a coupled enzyme system and that PITP is a cofactor required for these kinases to phosphorylate PI (CUNNINGHAM et al. 1995). Thus, only the PI bound to PITP would be phosphorylated to yield $PIP_2$, and this newly synthesized $PIP_2$ would be preferentially cleaved by PLC-$\gamma$ which is co-localized at the site of $PIP_2$ synthesis. This is consistent with earlier data suggesting that cells have a small pool of "hormone-responsive" phosphoinositides in addition to a much larger pool of PI. While the role of PITP and PI 4-kinase in BCR signaling has not been examined directly, it has been shown that BCR ligation causes a rapid increase in

**Fig. 4.** Proposed PI 3-kinase/Vav/Rac1/PIP5K pathway for B-cell receptor (BCR)-induced $PIP_2$ synthesis. The key elements of this model are discussed. (1) PI 3-kinase and Vav are recruited to the plasma membrane. The SH2 domains of PI 3-kinase and Vav bind to CD19, which is tyrosine phosphorylated after BCR ligation. PI 3-kinase may also be recruited to the plasma membrane using its SH2 domain to bind to Cbl and Gab1, two membrane-associated docking proteins that are tyrosine phosphorylated after BCR ligation. (2) $PIP_3$ made by PI 3-kinase binds to the pleckstrin homology (PH) domain of Vav and promotes the phosphorylation of Vav by tyrosine kinases. (3) Tyrosine-phosphorylated Vav acts as an exchange factor that promotes the binding of GTP to Rac1. (4) The active GTP-bound Rac1 stimulates the activity of PI(4)P 5-kinase (PIP5K). PIP5K produces $PIP_2$ by phosphorylating PI 4-phosphate. This ensures a continuous supply of $PIP_2$ despite the hydrolysis of $PIP_2$ by PLC-$\gamma$. The newly synthesized $PIP_2$ can be used as a substrate by both PLC-$\gamma$ and PI 3-kinase, allowing for sustained production of $IP_3$, diacylglycerol, and $PIP_3$

the production of PI(4)P from PI (Gold and Aebersold 1994). Whether this reflects increased delivery of PI to the PI 4-kinase by PITP remains to be determined. It has recently been proposed that PITP not only delivers PI to the plasma membrane but may also act as a "liftase" that lifts PI out of the plasma membrane in order to present it to PI 4-kinase (Kearns et al. 1998). Another interesting recent finding is that PI 4-kinase associates with CD81 (also known as TAPA-1) which is part of the CD19 complex (Berditchevski et al. 1997). Thus, CD19 may have an additional role in BCR-induced $PIP_2$ breakdown besides recruiting PI 3-kinase.

## 5.3 NF-AT

BCR-induced $PIP_2$ breakdown results in the production of $IP_3$ which causes the release of $Ca^{2+}$ from the endoplasmic reticulum. After the initial spike of $Ca^{2+}$ release into the cytosol, the intracellular stores are refilled via plasma membrane $Ca^{2+}$ channels which may be directly linked to the endoplasmic reticulum. This allows for sustained increases in cytoplasmic $Ca^{2+}$ concentrations that are dependent on extracellular $Ca^{2+}$ to refill the intracellular stores. One of the main targets of $Ca^{2+}$-mediated signaling in T- and B cells is NF-$AT_c$, the cytoplasmic component of the NF-AT transcription factor, which was briefly described in Sect. 4.1.1. Increases in intracellular $Ca^{2+}$ activate a serine/threonine phosphatase called calcineurin which dephosphorylates NF-$AT_c$ (Rao et al. 1997). This reveals a nuclear localization signal that allows NF-$AT_c$ to translocate into the nucleus where it can bind to the nuclear component of NF-AT and stimulate transcription. BCR ligation has been shown to cause NF-$AT_c$ to translocate from the cytosol to the nucleus (Healy et al. 1997).

As described in Sect. 4.1.1., nuclear translocation of NF-$AT_c$ is opposed by GSK3, a constitutively-active kinase that phosphorylates NF-$AT_c$ and changes its conformation so that a nuclear export signal is revealed. Once phosphorylated by GSK3, NF-$AT_c$ is rapidly exported from the nucleus (Beals et al. 1997; Klemm et al. 1997). However, the BCR (M. Gold, unpublished results) and other receptors that signal via PI 3-kinase activate Akt, a kinase that phosphorylates GSK3 and inhibit its activity (Cross et al. 1995). Thus, the accumulation of NF-$AT_c$ in the nucleus after BCR signaling requires both increases in cytoplasmic $Ca^{2+}$ concentrations as well as activation of the PI 3-kinase/Akt pathway.

The nuclear component of NF-AT is an AP-1 transcription factor complex that consists of Fos and Jun proteins (Rao et al. 1997). Although NF-$AT_c$ can bind weakly to DNA by itself, its affinity for DNA is increased about 20-fold when it interacts with AP-1 that has bound to an adjacent site on the DNA (Rao et al. 1997). Thus, NF-$AT_c$ and AP-1 cooperate to stimulate transcription from a composite NF-AT/AP-1 promoter element. Fos expression is very low in resting cells but is rapidly induced by receptor signaling. The BCR induces Fos expression by activating the Ras/Raf-1/MEK/ERK pathway described in Sect. 3. The ERK MAP kinase phosphorylates and activates Elk-1, Sap1a and other members of the TCF family of transcription factors. Phosphorylated Elk-1 and Sap1a can cooperate with SRF to stimulate transcription of the c-*fos* gene (KARIN et al. 1997). Jun expression is also very low in resting cells. Receptor-induced increases in Jun expression are mediated by Jun/ATF-2 transcription factor complexes, which are regulated by the JNK and p38 MAP kinases (SU and KARIN 1996; TREISMAN 1996; KARIN et al. 1997). The p38 MAP kinase phosphorylates ATF-2 while JNK phosphorylates the pre-existing Jun. The complex of phosphorylated ATF-2 and phosphorylated Jun can recruit the transcriptional machinery and induce transcription of the c-*jun* gene. As described in Sect. 8, the BCR activates both the JNK and p38 MAP kinases to a modest extent (SUTHERLAND et al. 1996; HEALY et al. 1997; GRAVES et al. 1996; SALMON et al. 1997). While BCR signaling alone may be able to induce the formation of active NF-AT complexes that can stimulate the transcription of NF-AT responsive genes, CD40 is a much stronger activator of JNK and p38 MAP kinase than the BCR (SUTHERLAND et al. 1996). Thus, full induction of NF-AT-mediated transcription may require signaling by both the BCR and CD40. In any case, the regulation of NF-AT nicely illustrates the various ways in which receptors regulate transcription factors: translocation of latent transcription factors from the cytosol to the nucleus, phosphorylation of pre-existing transcription factors, and de novo synthesis of transcription factors.

# 6 Protein Kinase C Enzymes

Members of the PKC family of serine/threonine kinases are downstream effectors of multiple lipid-dependent signaling pathways including the PLC pathway and the PI 3-kinase pathway (Fig. 2). The eleven PKC isoforms can be grouped into three families according to which second messengers are required for their activation (NISHIZUKA 1995). The conventional PKCs (PKC-$\alpha$, $\beta$, $\beta_{II}$, $\gamma$) require both DAG and $Ca^{2+}$ to be activated. Activation of these PKC isoforms depends primarily on PLC activation which yields DAG as well as $IP_3$ which causes increases in cytoplasmic $Ca^{2+}$ concentrations (BERRIDGE 1993). The novel PKCs (PKC-$\delta$, $\varepsilon$, $\eta$, $\theta$, $\mu$) require DAG for their activation but do not require $Ca^{2+}$. These PKC isoforms can be activated by the classical PLC pathway and presumably by other reactions that generate DAG, such as phosphatidylcholine hydrolysis. The novel PKC isoforms PKC-$\delta$, $\varepsilon$, and $\eta$ can also be activated in vitro by $PIP_3$ and $PI(3,4)P_2$ which are produced by PI 3-kinase (TOKER et al. 1994). The atypical PKCs (PKC-$\zeta$, $\lambda/\iota$; iota is the human homologue of the murine PKC-$\lambda$) are not activated by either DAG or $Ca^{2+}$. There is some evidence that the PI 3-kinase-derived lipid $PIP_3$ can activate PKC-$\zeta$ in vitro (NAKANISHI et al. 1993), although this has been disputed (TOKER et al. 1994). Nevertheless, the PI 3-kinase inhibitor wortmannin can block receptor activation of PKC-$\zeta$ and PKC-$\lambda$ in some cells (AKIMOTO et al. 1996; HERRERA-VELIT et al. 1997), suggesting that the atypical PKCs are either direct or indirect targets of PI 3-kinase signaling. PKC-$\zeta$ can be also be activated by ceramide, a second messenger produced by hydrolysis of sphingomyelin (MULLER et al. 1995). Since the BCR activates both PLC-$\gamma$ and PI 3-kinase, it is likely that some or all of the PKC isoforms present in B cells are activated following BCR ligation.

The various PKC isoforms have different tissue distributions and are likely to have different substrates as well as unique functions (DEKKER and PARKER 1994; NISHIZUKA 1995). PKC-$\alpha$, $\beta$, $\gamma$, $\delta$, $\varepsilon$, $\zeta$, $\eta$, and $\mu$ have been detected in B cells (MISCHAK et al. 1991; SIDORENKO et al. 1996). Translocation of PKC enzymes from the cytosol to the membrane allows them to bind their lipid activators and this is usually used as an indirect measure of PKC activation. BCR engagement has been shown to increase the amount of total PKC enzyme activity in the membrane fraction of B cells (CHEN et al. 1986; NEL et al. 1986). BRAS et al. (1997) have investigated the effect of BCR signaling on individual PKC isoforms and found that anti-IgM treatment increases the amount of PKC-$\alpha$, $\beta_{II}$, $\delta$, $\varepsilon$, and $\zeta$ in the membrane fraction of the A20 $IgG^+$ B-cell line. BCR engagement also causes very strong tyrosine phosphorylation of PKC-$\delta$, although the significance of this modification is not known (BARBAZUK and GOLD 1999). Finally, PKC-$\mu$ can associate with the BCR and its specific activity is increased upon BCR ligation (SIDORENKO et al. 1996). It is not known whether the BCR activates other PKC isoforms or whether the spectrum of PKC isoforms activated by the BCR varies depending on the maturation state or activation state of the B cell. Although the role of individual PKC isoforms in BCR signaling has not been fully delineated, PKC-$\beta$ clearly has an important role in BCR signaling. Mice in which the genes encoding PKC-$\beta$/$\beta_{II}$ have been knocked out have defects in B-cell development as well as defects in BCR-induced activation of mature splenic B cells (LEITGES et al. 1996).

Activation of PKC enzymes is likely to make a significant contribution to the ability of the BCR to activate a variety of transcription factors. One way in which PKC links the BCR to transcription factors is by activating the ERK2 MAP kinase. Treating B cells with phorbol esters that activate PKC leads to rapid activation of ERK2 (CASILLAS et al. 1991; GOLD et al. 1992). Moreover, PKC inhibitors partially block BCR-induced activation of ERK2 (GOLD et al. 1992). The mechanism by which PKC activates ERK2 is not clear, but phorbol ester treatment can activate Ras in B cells (HARWOOD and CAMBIER 1993; TORDAI et al. 1994). Consistent with the idea that PKC activates ERK2 via the classical Ras/Raf-1/MEK/ERK2 pathway, recent work has shown that the ability of PKC to activate ERK in COS cells depends upon the formation of Ras-GTP/Raf-1 complexes (MARAIS et al. 1998). Alternatively, PKC may bypass Ras and activate either Raf-1 or MEK. In either case, as described in Sect. 3, activated ERK2 migrates to the nucleus where it phosphorylates and activates a number of key transcription factors including Elk-1 and other members of the Ets family.

PKC is also involved in activation of other transcription factors by the BCR. For example, PKC can phosphorylate and activate the CREB transcription factor which is known to be a target of BCR signaling (XIE and ROTHSTEIN 1995). In addition, the BCR causes transient activation of the NF-$\kappa$B transcription factor via a PKC-dependent mechanism (LIU et al. 1991). NF-$\kappa$B regulates the expression of a number of genes including c-*myc* (DUYAO et al. 1990), a transcription factor that plays an important role in B-cell activation. In resting B cells, NF-$\kappa$B is found mostly in the cytosol complexed with an inhibitory subunit, IκB, which prevents translocation of NF-κB to the nucleus. Receptor signaling results in phosphorylation of IκB which targets it for degradation by the proteasome. This allows NF-κB to migrate to the nucleus and activate transcription (LENARDO and BALTIMORE 1989). The mechanism by which PKC activates NF-κB is not clear. Recent work has identified a kinase complex that phosphorylates IκB and there is some evidence that phorbol esters can activate this complex (DIDONATO et al. 1997). However, it is not known whether PKC directly phosphorylates a component of the IκB kinase complex. NF-κB is a dimer composed of p50 and Rel family proteins. There are five p50 and Rel proteins which can form a variety of homodimers and heterodimers that may have different transactivating functions (BALDWIN Jr. 1996). In B cells, some NF-κB is constitutively active and plays a role in maintaining expression of the Ig κ chain gene. However, BCR cross-linking can activate the latent cytosolic NF-κB, which may contain different subunits than the constitutively-active NF-κB. Of the Rel family members, c-Rel appears to be important for BCR signaling since splenic B cells lacking this protein show impaired proliferation in response to anti-IgM antibodies (GRUMONT et al. 1998).

By promoting the activation of pre-existing transcription factors, such as Elk-1, CREB, and NF-κB, PKC plays a major role in the induction of early response genes such as c-*myc*, c-*fos*, and *egr*-1 by the BCR (KLEMSZ et al. 1989; SEYFERT et al. 1990; MITTELSTADT and DEFRANCO 1993). These early response genes encode transcription factors that are likely to be important regulators of B-cell survival, activation, and proliferation.

In addition to playing a central role in BCR signaling, PKC may act as a feedback inhibitor that limits the magnitude or duration of BCR signaling. PKC activation downregulates BCR-stimulated PLC activation (GOLD and DEFRANCO 1987; MIZUGUCHI et al. 1987) and may also downregulate BCR-induced Ras activation by promoting phosphorylation of the Ras activator SOS (SAXTON et al. 1994; WATERS et al. 1995). Recent work by SIDORENKO et al. (1996) suggests a mechanism by which PKC could inhibit BCR-induced activation of PLC-γ. They showed that PKC-μ, which is activated by the BCR, can phosphorylate PLC-γ1 as well as the Btk and Syk tyrosine kinases. Phosphorylation of Syk by PKC-μ decreased its enzymatic activity. Since Syk is essential for the activation of multiple BCR signaling pathways, this could be an important mechanism for downregulating BCR signaling. The effects of PKC-μ-mediated phosphorylation of PLC-γ1 and Btk have not been determined but could represent other sites at which PKC-μ inhibits BCR signaling.

## 7 The Crk-C3G Pathway and the Rap1 GTPase

Like Grb2 (the adaptor protein that controls the subcellular localization of the Ras activator SOS), the Crk proteins are adaptor proteins that consist almost entirely of SH2 and SH3 domains. Three different Crk proteins, termed Crk II, Crk I, and

Crk-L have been identified. The 40-kDa Crk II protein has an N-terminal SH2 domain and two SH3 domains. The 28-kDa Crk I protein is an alternatively spliced form of the *crk II* gene that lacks the C-terminal SH3 domain (MATSUDA et al. 1992). The 38-kDa Crk-L protein, which is the most abundant form of Crk in hematopoietic cells, is encoded by a separate gene (TEN HOEVE et al. 1993). While the amino acid similarity between Crk II and Crk-L is only 60% overall, the SH2 and SH3 domains are highly conserved. This suggests that Crk II and Crk-L could interact with the same proteins and be functionally redundant.

Much of the initial interest in the Crk proteins stemmed from observations suggesting that they might participate in Ras activation. MATSUDA et al. (1994) showed that overexpression of Crk I or Crk II in the neuronal PC12 cell line potentiated nerve-growth-factor-induced activation of Ras. Moreover, they found that the Ras exchange factor SOS bound constitutively to the endogenous Crk proteins in these cells. SOS binds to the N-terminal SH3 domain of Crk protein much the way that it binds to Grb2. This suggested that receptors could activate Ras by recruiting either Crk/SOS complexes or Grb2/SOS complexes to the plasma membrane. Finally, Hanafusa and colleagues showed that expressing a dominant negative form of Ras (RasN17) in chicken embryo fibroblasts could reverse transformation caused by the v-Crk oncogene (GREULICH and HANAFUSA 1996). In addition to binding SOS, the SH3 domains of Crk proteins can bind the Abl tyrosine kinase (FELLER et al. 1994), the DOCK 180 and Eps15 proteins (SCHUMACHER et al. 1995; HASEGAWA et al. 1996), and C3G (MATSUDA et al. 1994; TANAKA et al. 1994), a nucleotide exchange factor for the Ras-related GTPase Rap1 (GOTOH et al. 1995). Thus, Crk proteins could mediate a number of important signaling interactions. This prompted several groups to examine whether the BCR used Crk proteins to form signaling complexes.

INGHAM et al. (1996) showed that BCR ligation causes both Crk-L and Crk II to bind via their SH2 domains to p120 Cbl and to p130 Cas, two docking proteins that are tyrosine phosphorylated when the BCR is engaged. The interaction of Crk with Cbl was also demonstrated by Borst and colleagues (SMIT et al. 1996b). INGHAM et al. (1996) went on to show that a significant fraction of Cbl and Cas is constitutively present in the membrane-enriched particulate fraction of the RAMOS human B-cell line. Moreover, they showed that BCR ligation increased the amount of Crk proteins in this particulate fraction and also induced the formation of Cbl/Crk and Cas/Crk complexes in the particulate fraction of the RAMOS cells. This suggests that BCR-induced tyrosine phosphorylation of membrane-associated Cbl and Cas creates binding sites for the Crk SH2 domain and recruits Crk proteins to cellular membranes. Confocal microscopy studies are needed to confirm that Cbl/Crk and Cas/Crk complexes are indeed associated with cellular membranes in activated B cells and to determine whether these complexes are located at the plasma membrane. Nevertheless, a working model is that proteins that bind to the Crk SH3 domains can be recruited to cell membranes after BCR engagement. This may be relevant for SOS and C3G, nucleotide exchange factors that activate small GTPases which are tethered to the cytoplasmic faces of cell membranes by lipid modifications.

Although in vitro analysis had shown that the Crk SH3 domains could bind both SOS and C3G, INGHAM et al. (1996) found that, in RAMOS B cells, both Crk-L and Crk II bind substantial amounts of C3G but very little SOS. In contrast, Grb2 binds SOS, but not C3G in these cells (INGHAM et al. 1996; SMIT et al. 1996a). This indicated that Grb2 and Crk are likely to have distinct functions in B cells. SOS-induced activation of Ras in B cells is likely to be mediated primarily by Grb2 and not by Crk. In contrast, the Crk proteins appear to be the major adaptor proteins responsible for directing the subcellular localization of C3G in B cells. Since MATSUDA and colleagues had shown that C3G is a nucleotide exchange factor that can activate the Rap1 GTPases (GOTOH et al. 1995), an obvious question was whether the BCR activates Rap1.

## 7.1 The Rap1 GTPases

The Rap1 GTPases may be negative regulators of Ras-mediated signaling. The two mammalian Rap1 proteins, Rap1A and Rap1B, are 97% identical at the protein level and are closely related to Ras. Rap1A (also known as Krev-1) was first identified by its ability to reverse the transformation of NIH 3T3 cells by an activated form of Ki-Ras (KITAYAMA et al. 1989). When expressed in fibroblasts, a constitutively-active form of Rap1 can inhibit Ras-dependent activation of the ERK kinases (COOK et al. 1993). Similarly, activated Rap1 also opposes the actions of Ras in *Drosophila* eye development and in the maturation of *Xenopus* oocytes (CAMPA et al. 1991; HARIHARAN et al. 1991).

Since the effector-binding domain of Rap1 is identical to that of Ras, activated Rap1 may block Ras-mediated signaling by binding to and sequestering Ras effectors. In vitro experiments have shown that Rap1-GTP binds to Raf-1 and other downstream targets of Ras, but does not activate them (FRECH et al. 1990; URANO et al. 1996; HU et al. 1997). In cells, Ras-GTP and Rap1-GTP may compete for these Ras effectors. Recent work has shown that Rap1-GTP has higher affinity than Ras-GTP for the cysteine-rich region of Raf-1 and that binding of Rap1-GTP to this region of Raf-1 interferes with the ability of Ras-GTP to activate it (HU et al. 1997). Rap1 may also keep Ras effectors physically separated from Ras. Rap1 is located primarily on the cytoplasmic face of the Golgi apparatus (BERANGER et al. 1991), while Ras is tethered to the inner face of the plasma membrane. Thus, activation of Rap1 may be an important mechanism by which the magnitude or duration of Ras-mediated signaling is limited.

Although transfection experiments have indicated that ectopic expression of activated Rap1 can block Ras-mediated signaling, there is no evidence in any system that the endogenous Rap1 proteins normally perform this function. If Rap1 does in fact limit Ras-mediated signaling, one would expect that expressing a dominant-negative form of Rap1 would relieve this inhibition and potentiate Ras-mediated signaling. However, this has not been successfully tested. In *Drosophila*, loss-of-function mutations in Rap1 are lethal (HARIHARAN et al. 1991), suggesting that Rap1 has an important function and that any test of its function would require inducible expression of a dominant-negative form of Rap1.

In addition to opposing Ras-mediated signaling and Ras-induced proliferation, Rap1 may also activate a distinct signaling pathway that inhibits proliferation. Genetic analysis in *Drosophila* has shown that the *dacapo* gene product is a target of Rap1 (DE NOOIJ et al. 1996). *Dacapo* encodes a cyclin-dependent kinase (cdk) inhibitor similar to $p27^{Kip1}$ which blocks cell-cycle progression during *Drosophila* eye development.

As a potential antagonist of Ras-mediated proliferation and a mediator of cell-cycle arrest, Rap1 may be an important tumor suppressor. One hypothesis is that Ras and Rap1 are like yin and yang and that simultaneous activation of both these GTPases by a receptor would ensure that Ras-mediated signaling is transient and leads to limited cell activation and proliferation, as opposed to oncogenic transformation. This model predicts that every receptor that activates Ras would also activate Rap1. As receptor-induced activation of Rap1 has only been reported for a small number of receptors, the correlation between Ras activation and Rap1 activation remains to be established.

## 7.2 Mechanism of Rap1 Activation by the BCR

Recent work by MCLEOD et al. (1998) showed that BCR ligation causes a rapid increase in the amount of activated GTP-bound Rap1 in a variety of B-cell lines and in murine splenic B cells. Rap1 activation was measured using a novel assay (FRANKE et al. 1997) that is based on the observation that the RalGDS protein binds with high affinity to active GTP-bound Rap1 but not to the inactive GDP-bound form of Rap1. Thus, a glutathione S-transferase (GST) fusion protein containing the Rap1-binding domain of RalGDS can be used to selectively precipitate activated Rap1, which can then be detected by immunoblotting with an antibody to Rap1.

MCLEOD et al. (1998) went on to investigate the mechanism by which the BCR activates Rap1. The initial assumption was that it involved the binding of Crk/C3G complexes to tyrosine-phosphorylated Cbl and Cas (INGHAM et al. 1996). Since Cbl and Cas can associate with cellular membranes, this interaction would recruit cytosolic Crk/C3G complexes to membranes where Rap1 is located. Subsequently, FRANKE et al. (1997) showed that increases in intracellular $Ca^{2+}$ were both necessary and sufficient for the thrombin receptor to activate Rap1 in platelets. Thus, there were at least two possible pathways for Rap1 activation, a Crk/C3G-dependent pathway and a $Ca^{2+}$-dependent pathway. However, MCLEOD et al. (1998) found that treating B cells with $Ca^{2+}$ ionophores caused little or no activation of Rap1. They went on to show that increases in intracellular $Ca^{2+}$ concentrations are not necessary for the BCR to activate Rap1 The BCR can activate Rap1 in a variant of the DT40 chicken B-cell line, in which BCR ligation does not induce $Ca^{2+}$ fluxes because the genes encoding all three $IP_3$ receptors have been disrupted. While the $Ca^{2+}$-dependent pathway for Rap1 activation appears not to be present in B cells, McLeod et al. found that DAG, as well as phorbol esters that mimic the actions of DAG, cause strong activation of Rap1 in B cells. Thus, B cells have a novel DAG-dependent pathway for Rap1 activation.

To determine whether BCR-induced Rap1 activation was mediated by this DAG-dependent pathway as opposed to the Cbl/Crk/C3G pathway, McLeod et al. (1998) used a PLC-γ-deficient variant of the DT40 cell line, in which BCR ligation does not cause $PIP_2$ breakdown or generate DAG. They found that BCR-induced Rap1 activation was dramatically reduced in the PLC-γ-deficient DT40 cells even though Crk associated normally with tyrosine-phosphorylated proteins including Cbl. Phorbol esters were still able to cause strong activation of Rap1 in these PLC-γ-deficient DT40 cells, but did not cause Crk to associate with Cbl, Cas, of other tyrosine-phosphorylated proteins. Thus, the association of Crk complexes with Cbl, Cas, or other tyrosine-phosphorylated proteins is neither necessary nor sufficient for the majority of BCR-induced Rap1 activation which appears to proceed via the PLC-γ- and DAG-dependent pathway. These findings do not, a priori, rule out a role for Crk and C3G in activation of Rap1 by the BCR. It is possible that the DAG-dependent pathway recruits C3G or Crk/C3G complexes to Rap1 in some way that does not involve the binding of Crk to tyrosine-phosphorylated docking proteins such as Cbl or Cas.

At this point, the components of this novel signaling pathway that link DAG to Rap1 activation in B cells are not known. An obvious question is whether activation of Rap1 by DAG and phorbol esters is mediated by PKC enzymes. Downregulating PKC expression by treating B cells overnight with phorbol esters resulted in a higher basal level of activated Rap1 in the cells but abolished any further increase in Rap1 activation due to BCR ligation (S. McLeod and M. Gold, unpublished observations). While this supports the idea that PKC enzymes are involved in BCR-induced activation of Rap1, additional studies are necessary. It is also not known whether the ultimate target of this DAG-dependent Rap1 activation pathway is a Rap1 GNEF, such as C3G, or a Rap1 GAP. An increase in the amount of activated Rap1 in the cell could be due to an increase in the rate at which Rap1 exchange factors activate Rap1 or to a decrease in the rate at which Rap1-GAPs stimulate hydrolysis of the GTP bound to Rap1. There are likely to be multiple Rap1 exchange factors in addition to C3G, and three Rap1 GAPs have been identified so far, SPA-1, RapGAP1, and tuberin (Rubinfeld et al. 1991; Wienecke et al. 1995; Kurachi et al. 1997). SPA-1 is expressed primarily in lymphoid cells (Kurachi et al. 1997) and it will be interesting to determine whether DAG regulates its activity or subcellular localization. In any case, much remains to be learned about how the BCR activates Rap1.

## 7.3 Roles for Rap1 in B Cells?

The functions of the Rap1 proteins in B cells remain to be elucidated. In particular, an important question is whether activation of the endogenous Rap1 by the BCR has a negative effect on Ras-mediated signaling. If Rap1 does function as an antagonist of Ras-mediated signaling in B cells, then Rap1 may play a key role in regulating B-cell activation. Depending on whether or not a B cell receives co-stimulatory signals from T cells, BCR ligation can result in either activation and proliferation, functional inactivation (anergy), or apoptosis. Since activated Ras can transform B cells (Siranian et al. 1993), activation of Ras by the BCR most likely promotes B-cell proliferation. The concomitant activation of Rap1 may oppose this by limiting Ras-mediated signaling and perhaps by activating other signaling pathways that induce the expression of cell-cycle inhibitors. Thus, the balance between the activation of Ras and Rap1 may be a key parameter that determines the outcome of BCR ligation. T-cell-derived co-stimulatory signals in the form of CD40 ligand and IL-4 may shift the balance in favor of Ras and proliferation, while inhibitory signals delivered by Fc receptors may shift the balance in favor of Rap1, leading to cell-cycle arrest and either anergy or apoptosis. It will be of interest to see whether signaling by CD40, the IL-4 receptor, or Fc receptors modulates activation of Rap1 by the BCR.

The function of Rap1 as a modulator of Ras-mediated signaling may be cell-type-specific, depending on whether or not the cell expresses B-Raf, a relative of Raf-1 that can also activate the MEK/ERK signaling module. Rap1-GTP binds to both Raf-1 and B-Raf but has opposite effects on them. While Rap1-GTP sequesters Raf-1 in inactive complexes and prevents Ras from activating it, Rap1-GTP activates B-Raf (Ohtsuka et al. 1996; Vossler et al. 1997). Stork and colleagues have shown that, in cells expressing B-Raf, expression of activated Rap1 causes sustained activation of the B-Raf/MEK/ERK signaling pathway (York et al. 1998). This is in contrast to B-Raf-negative cells, in which Rap1-GTP is thought to compete with Ras for Raf-1 and limit the activation of ERK. B cells appear not to express B-Raf (or at least the isoform of B-Raf that activates MEK) and would therefore belong to the class of cells in which Rap1-GTP is a negative regulator of Ras-mediated signaling. This is consistent with observations that BCR ligation causes only transient activation of ERK (Sutherland et al. 1996).

Rap1A and Rap1B may also have other functions in B cells that are not related to Ras-mediated signaling. Maly et al. (1994) have shown that the ability of the BCR to stimulate the oxidative burst in a Burkitt's lymphoma cell line is dependent on Rap1A. In neutrophils, Rap1A associates with cytochrome b558 and this may be important for assembly of the NADPH oxidase complex, which produces reactive oxygen species (Gabig et al. 1995). The production of these reactive oxygen species may allow B cells to kill bacteria that they bind via their BCR.

## 7.4 Roles for the Cas/Cbl/Crk/C3G Pathway in B cells?

As described in Sect. 7.1, tyrosine phosphorylation of the Cas and Cbl docking proteins may recruit Crk/C3G complexes to cell membranes where C3G can act on membrane-associated targets such as Rap1. However, as discussed in Sect. 7.2, this does not appear to be the major mechanism by which the BCR activates Rap1. The Cas–Cbl/Crk/C3G pathway may play a minor role in BCR-induced Rap1 activation and could account for the small amount of Rap1 activation seen when PLC-γ-deficient DT40 cells are stimulated via their BCR (see above). Alternatively, the main function of the Cas–Cbl/Crk/C3G pathway may be to initiate other signaling events. Recently, TANAKA et al. (1997) showed that overexpression of C3G in NIH 3T3 cells activates the JNK, one of the MAP kinases. Although the mechanism by which C3G activates JNK is not known, as described in the next section, activated JNK migrates to the nucleus where it phosphorylates and activates several transcription factors including c-Jun, which is a component of the AP-1 transcription factor complex. BCR ligation activates JNK (SUTHERLAND et al. 1996; HEALY et al. 1997) but it is not known whether C3G is involved in this process.

## 8 The JNK and p38 MAP Kinases

In addition to the ERKs, mammalian cells express two other families of MAP kinases, the JNKs and the p38 MAP kinases. There are multiple isoforms of both these MAP kinases and, like the ERKs, they are activated by phosphorylation of the threonine and tyrosine residues in a threonine-X-tyrosine motif. BCR engagement activates both the JNK and p38 MAP kinases, although to a lesser extent than it activates ERK2 (GRAVES et al. 1996; SUTHERLAND et al. 1996; HEALY et al. 1997; SALMON et al. 1997). In addition, CD40 is a much stronger activator of JNK and p38 MAP kinase than the BCR (SUTHERLAND et al. 1996).

The mechanism by which the BCR activates these MAP kinases is not completely understood. Analogous to the Ras/Raf-1/MEK/ERK signaling module, both JNK and the p38 MAP kinase are regulated by multi-layered kinase cascades that are controlled by a GTPase (Fig. 1) (SU and KARIN 1996; ROBINSON and COBB 1997). The dual specificity kinases MKK4 and MKK7 phosphorylate and activate JNK, while MKK3 and MKK6 selectively activate p38 MAP kinase (ROBINSON and COBB 1997; FOLTZ et al. 1998). Although this suggests that JNK and p38 MAP kinase can be regulated independently, most stimuli that activate JNK also activate p38 MAP kinase and vice versa. Thus, the JNK and p38 pathways may be controlled by the same upstream activator, although it is not clear at what point the pathway branches. The MKKs that activate JNK and p38 are, in turn, phosphorylated and activated by kinases called MEKKs, which occupy the position analogous to Raf-1 in the Ras/Raf-1/MEK/ERK pathway. It is not known whether some MEKKs activate the JNK pathway while others activate the p38 MAP kinase pathway. Unlike Raf-1, which is directly activated by Ras, the MEKKs are not direct targets of GTPases. Instead, kinases related to the yeast STE20 kinase link the Rac1 and Cdc42 GTPases to the MEKKs. A large number of mammalian STE20-like kinases have been cloned recently (reviewed by DIENER et al. 1997). These kinases include the mixed lineage kinases (MLKs), the germinal center kinases (GCKs), the PAK kinases, HPK1, KHS, SOK, and NIK. It is not known which, if any of these kinases, do in fact regulate the JNK and p38 MAP kinase pathways in B cells or other cells. In addition, it is not known whether the MEKKs are direct targets of these kinases or whether there is yet another layer of kinases between the STE20-like kinases and the MEKKs.

Despite the confusion about which kinases lie upstream of the MEKKs, there is good evidence that the Rac1 and Cdc42 GTPases regulate the kinase cascades leading to JNK and p38 MAP kinase. In fibroblasts, expression of activated Rac1 or Cdc42 can activate JNK while dominant-negative forms of Rac1 and, to a lesser extent, Cdc42, block the ability of several receptors to activate JNK (Coso et al. 1995; Minden et al. 1995). It is not known whether the BCR activates Rac1 or Cdc42 and the role of these GTPases in activation of JNK and p38 MAP kinase by the BCR has not been tested (by expressing a dominant negative form of Rac1 or Cdc42 in B cells).

If BCR-induced activation of JNK and p38 MAP kinase is mediated by Rac1, there are a number of possible mechanisms by which the BCR might activate Rac1. One potential mechanism involves the 95-kDa Vav protein (which is discussed in detail in Sect. 10.3), a protein that is strongly tyrosine phosphorylated in response to BCR ligation (Bustelo and Barbacid 1992a). Vav has a Dbl homology (DH) domain which is characteristic of exchange factors for Rac1 and Rho. Bustelo et al. (Crespo et al. 1997) showed that Vav can catalyze GDP/GTP exchange on Rac1 and that tyrosine phosphorylation greatly enhances its exchange factor activity. Although a Vav/Rac1/JNK pathway is an appealing model, a recent paper showed that BCR-induced JNK activation is normal in B cells from Vav knockout (*vav* −/−) mice (Fischer et al. 1998). There are three other potential mechanisms by which the BCR could activate JNK. First, the Ras exchange factor SOS also has a DH domain (Cerione and Zheng 1996) and can activate Rac1 (Nimnual et al. 1998). Thus, SOS may link the BCR to both Ras and Rac1, mediating both Ras-dependent activation of ERK2 as well as Rac1-dependent activation of JNK and p38 MAP kinase. Another possible connection between the BCR and JNK involves the Rap1 exchange factor C3G. When overexpressed in NIH 3T3 cells, C3G can activate JNK by an unknown mechanism that does not involve Rac1 (Tanaka et al. 1997; Tanaka and Hanafusa 1998). Since BCR ligation mobilizes both Grb2/SOS complexes and Crk/C3G complexes, causing them to bind to docking proteins and perhaps associate with cellular membranes, it is possible that SOS and/or C3G contribute to BCR-induced activation of JNK and p38 MAP kinase. Finally, the Btk, which is activated by the BCR, has been shown to activate JNK by an unknown mechanism (Kawakami et al. 1997).

Although the pathway linking the BCR to JNK and p38 MAP kinase is not completely understood, these MAP kinases are likely to have important roles in linking the BCR to changes in gene expression. Like ERK, the activated forms of JNK and p38 MAP kinase translocate to the nucleus where they phosphorylate and activate different sets of transcription factors. JNK can phosphorylate and activate the Elk-1 and Sap1a transcription factors as well as c-Jun, which is an important component of the AP-1 transcription factor (Gupta et al. 1996; Treisman 1996; Cohen 1997; Whitmarsh et al. 1997). p38 MAP kinase can phosphorylate and activate the ATF-2, Sap1a, CHOP, and MEF2C transcription factors (Cohen 1997; Wang and Ron 1996; Han et al. 1997; Janknecht and Hunter 1997;

WHITMARSH et al. 1997). p38 MAP kinase also phosphorylates and activates a serine/threonine kinase called MAPKAP kinase-2 (CUENDA et al. 1995) which in turn phosphorylates and activates the CREB transcription factor (TAN et al. 1996). The BCR may therefore activate CREB via PKC, $p90^{rsk}$, and the p38 MAP kinase/ MAPKAP kinase-2 pathway. In summary, the JNK and p38 MAP kinases may link the BCR to a number of nuclear transcription factors. Whether all of these transcription factors are in fact activated by the BCR via JNK and p38 MAP kinase remains to be established. The connections between these MAP kinases and the various transcription factors appear to be cell-type dependent (JANKNECHT and HUNTER 1997; WHITMARSH et al. 1997).

# 9 Signals Mediated by the HS-1 protein

HS1 is a 75-kDa protein expressed only in hematopoietic cells (KITAMURA et al. 1989). In 1993, Watanabe and colleagues (YAMANASHI et al. 1993) showed that HS1 is a major substrate of the BCR-activated tyrosine kinases. Subsequent work by this group showed that HS1 plays an important role in the ability of the BCR to induce both proliferation and apoptosis. In mice in which the genes encoding HS1 are disrupted, B-cell development appears to be normal, but the ability of mature splenic B cells to proliferate in response to anti-IgM antibodies is reduced by about 70% compared with cells from wild-type mice (TANIUCHI et al. 1995). The peritoneal B-1 B cells from these mice also show impaired responses to anti-IgM antibodies. Unlike their normal counterparts, they do not undergo apoptosis in response to anti-IgM. The role of HS1 in BCR-induced apoptosis was nicely illustrated by FUKUDA et al. (1995) who showed that a variant of the WEHI-231 cell line that failed to undergo apoptosis in response to anti-IgM had greatly decreased expression of HS1. Moreover, expression of an exogenous HS1 in these mutant WEHI-231 cells gene could restore BCR-induced apoptosis (FUKUDA et al. 1995). Interestingly, BCR-induced apoptosis was not restored by expressing HS1 in another HS1-deficient variant of WEHI-231. This indicates that HS1 is necessary but not sufficient for the BCR to initiate apoptosis in WEHI-231 cells.

Although the function of HS1 is not known, several models have been proposed. First, tyrosine-phosphorylated HS1 can bind with high affinity to the SH2 domains of the Lyn, Fyn, and Blk tyrosine kinases, which are activated by the BCR (YAMANASHI et al. 1993; BAUMANN et al. 1994). Since HS1 has an SH3 domain that can interact with proline-rich sequences in other proteins, HS1 could act as an adaptor protein that co-localizes the Src family tyrosine kinases with some of their substrates. However, proteins that bind to the SH3 domain of HS1 have yet to be identified. A second model suggests that HS1 may function like a STAT protein which, upon receptor signaling, translocates from the cytosol to the nucleus, binds DNA, and activates transcription. HS1 has several motifs that resemble DNA binding and dimerization motifs of basic helix-loop-helix transcription factors (KITAMURA et al. 1989). In resting B cells, HS1 is found in both the cytosol and nucleus, but BCR signaling increases the amount of HS1 in the nucleus, at least in WEHI-231 cells (Yamanashi et al. 1997). Interestingly, a mutated version of HS1 in which tyrosines 378 and 397 are changed to phenylalanine is not tyrosine phosphorylated after BCR ligation, does not translocate to the nucleus, and cannot restore BCR-induced apoptosis in the anti-IgM-resistant variant of WEHI-231 (Yamanashi et al. 1997). Although these results argue that nuclear translocation of HS1 is required for it to mediate BCR-induced apoptosis, there is no evidence yet that it binds DNA or acts as a transcription factor.

A recent paper by Watanabe's group suggests another mechanism by which HS1 may regulate apoptosis. Yeast two-hybrid analysis revealed that HS1 interacts directly with a ubiquitously-expressed 35-kDa protein called HAX-1 (Suzuki et al. 1997). Immunoprecipitation studies confirmed that HS1 and HAX-1 interact in B cells. HAX-1 has a potential membrane-spanning region and confocal microscopy showed that HAX-1 is found in the outer mitochondrial membrane, the endoplasmic reticulum, and the nuclear envelope. When HAX-1 is overexpressed in COS-7 cells, HS-1 co-localizes with HAX-1. The cellular distribution of HAX-1 is intriguing given that the Bcl-2 family members show the same localization pattern and that changes in mitochondrial membrane potential as well as the release of cytochrome c from mitochondria are important early events in the apoptotic process (Kroemer et al. 1997; Reed 1997). Moreover, HAX-1 contains regions with some homology to the BH1 and BH2 domains of Bcl-2 as well as to a Nip3, a protein that binds Bcl-2 and prevents apoptosis. Thus, one can propose a model in which HAX-1 promotes cell survival by interacting with death promoting members of the Bcl-2 family and neutralizing them. The binding of HS1 to HAX-1 may disrupt these interactions and thereby promote apoptosis. The subcellular distribution of HS1 may therefore be an important factor in its function as a mediator of BCR-induced apoptosis in peritoneal B-1 B cells and in WEHI-231 cells.

It is harder to understand how the HS1/HAX-1 interaction might be involved in BCR-induced proliferation in conventional splenic B cells. It is possible that the nuclear localization of HS1 is important for it to promote B-cell proliferation, while its mitochondrial localization is more important for it to promote apoptosis. It will be interesting to see whether HS1 goes to different cellular locations in B cells that undergo proliferation in response to anti-Ig antibodies versus those that undergo apoptosis. Alternatively, HS1 may always go to both the nucleus and mitochondria after BCR engagement, but its action at one site or the other may be enhanced or repressed depending on whether this occurs in a mature conventional B cell or a peritoneal B-1 B cell.

## 10 Signals Mediated by Vav

Few signaling proteins have engendered as much confusion and controversy as Vav. This 95-kDa protein was discovered in 1989 by Shulamit Katzav, who named it after the sixth letter of the Hebrew alphabet (KATZAV et al. 1989). Vav is expressed in hematopoietic and trophoblast cells (KATZAV et al. 1989; ZMUIDZINAS et al. 1995), while the closely-related Vav-2 protein is ubiquitously expressed (SCHUEBEL et al. 1996). Vav clearly has an important role in antigen-receptor signaling. It is strongly tyrosine phosphorylated in response to engagement of the BCR, the TCR, and a number of other receptors on hematopoietic cells (reviewed by BONNEFOY-BERARD et al. 1996). In addition, both B cells and T cells from *vav* –/– mice show greatly reduced proliferative responses to antigen receptor engagement (TARAKHOVSKY et al. 1995; ZHANG et al. 1995).

The Vav protein contains many different interaction domains (reviewed by BUSTELO 1996; COLLINS et al. 1997). Vav has an SH2 domain, two SH3 domains, a PH domain, a leucine-rich region, a cysteine-rich region similar to the lipid-binding domains of PKC enzymes, and a DH domain which is characteristic of proteins with nucleotide exchange activity for Rac and Rho family GTPases. The N-terminal portion of Vav has a region with some similarity to the amphipathic helix-loop-helix motifs found in transcription factors. However, there is no functional evidence that Vav acts as a transcription factor and molecular modeling of this region suggests that it may instead be a calponin homology (CH) domain (CASTRESANA and SARASTE 1995) CH domains are found in a number of cytoskeletal proteins and are thought to be involved in binding F-actin. Given the number of different functional domains it has, the confusion as to the role of Vav in signal transduction may be because it really does do many different things. Below is summarized a number of potential roles for Vav in lymphocyte activation. Some of these proposed models are strongly supported by recent data, while others are much more speculative.

## 10.1 Vav as an Adaptor Protein?

Given the number of protein–protein interaction motifs in Vav, Vav could function as an adaptor protein that co-localizes components of signaling pathways. For example, Vav could facilitate BCR-induced tyrosine phosphorylation by bringing substrates close to the BCR-activated tyrosine kinases. DECKERT et al. (1996) showed that BCR ligation causes Vav to bind via its SH2 domain to phosphotyrosine-containing sequences in Syk and that this facilitates tyrosine phosphorylation of Vav by Syk. Thus, SH2 domain-containing proteins that bind to tyrosine-phosphorylated Vav, as well as proteins that bind constitutively to the Vav SH3 domains or to other portions or Vav, could be recruited to a site where they can be phosphorylated by Syk. Although an appealing model, there is no evidence that overexpression of Vav potentiates BCR-induced protein tyrosine phosphorylation or that loss of Vav inhibits this process. The effects of Vav on BCR-induced tyrosine phosphorylation may be limited to those proteins that bind Vav. Thus, the role of Vav as an adaptor protein that recruits substrates to the BCR-associated tyrosine kinases remains to be established.

There are several other possible scenarios in which Vav could act as an adaptor protein. In Con A-stimulated T cells and steel factor-treated P815 cells, Vav binds via its SH2 domain to the SHP1 tyrosine phosphatase (Kon-Kozlowski et al. 1996). Vav could therefore facilitate the dephosphorylation of proteins that bind to it. This Vav/SHP1 interaction, however, has not been demonstrated in B cells. Vav could also have a role in the formation of signaling complexes. As discussed previously, the Cbl docking protein is tyrosine phosphorylated after BCR ligation and this allows it to bind the SH2 domains of a number of signaling proteins including PI 3-kinase. In T cells, TCR engagement leads to phosphorylation of Cbl on a tyrosine residue that allows the Vav SH2 domain to bind to it (Marengere et al. 1997). Although it has not been shown whether this Vav/Cbl interaction occurs after BCR ligation, it suggests that Vav could bring associated proteins into signaling complexes that are nucleated by Cbl. This might facilitate cooperative interactions among signaling proteins or allow crosstalk between different signaling pathways.

## 10.2 Nuclear Functions for Vav?

Yeast two-hybrid analysis using Vav as the bait revealed that Vav can interact with several nuclear proteins (Bustelo et al. 1996; Hobert et al. 1996; Romero et al. 1996). Immunoprecipitation studies confirmed that these interactions can occur in cells. Although Vav is localized predominantly in the cytoplasm, subcellular fractionation and immunofluorescence studies have shown that Vav can be found in the nucleus of certain cells (Clevenger et al. 1995; Romero et al. 1996). The initial sequence analysis of Vav suggested that it might have a nuclear localization signal (Katzav et al. 1989). However, this has not been confirmed by deletion analysis or by showing that the putative Vav nuclear localization signal can induce nuclear translocation when fused onto a cytoplasmic protein. It is possible that Vav localizes to the nucleus only after associating with a protein that contains a bona fide nuclear localization signal. Although the presence of Vav in the nucleus is controversial, the reported interactions of Vav with nuclear proteins suggest some very intriguing roles for Vav.

Two different groups (Hobert et al. 1994; Bustelo et al. 1996) found that the C-terminal SH3 domain of Vav can bind to a proline-rich region in the human heterogeneous nuclear ribonucleoprotein K (hnRNP K). hnRNP K is a multifunctional nuclear protein that binds RNA, DNA, and a number of other proteins (Bomsztyk et al. 1997). As an RNA-binding protein it may participate in mRNA processing or export. Perhaps more significant is the ability of hnRNP K to bind various transcriptional activators and inhibitors (Denisenko et al. 1996; Michelotti et al. 1996; Miau et al. 1998) and modulate their effects on transcription. In particular, hnRNP K can stimulate transcription of the c-*myc* gene (Lee et al. 1996; Michelotti et al. 1996). hnRNP K expression is greatly increased in transformed cells (Dejgaard et al. 1994), consistent with a role for hnRNP K in cell activation or proliferation.

BUSTELO et al. (1996) showed that Vav may influence the subcellular distribution of hnRNP K. In 3T3 cells that don't express Vav, hnRNP K is normally found in nearly equal amounts in the nuclear and cytoplasmic fractions. However, when Vav is expressed in these cells, almost all of the hnRNP K is found in the cytoplasm. This suggests that cytoplasmic Vav binds to hnRNP K and retains it in the cytoplasm where it would be inactive. This raises the possibility that Vav regulates the subcellular distribution of hnRNP K in lymphocytes and that antigen receptor-induced tyrosine phosphorylation of Vav could influence the nuclear localization of hnRNP K. In this regard, one speculative model is that tyrosine phosphorylation of Vav releases hnRNP K, allowing hnRNP K to migrate from the cytosol to the nucleus. Another purely hypothetical model is that tyrosine phosphorylation of Vav would allow Vav/hnRNP K complexes to migrate into the nucleus. CLEVENGER et al. (1995) have shown that prolactin stimulation of the Nb-2T cell line causes Vav to translocate to the nucleus. Although receptor-induced nuclear import of hnRNP K has not been demonstrated in any system, this could be an important event if hnRNP K does in fact have a role in cell activation and proliferation.

Two other nuclear proteins that bind directly to Vav are Ku-70 (HOBERT et al. 1996) and ENX-1 (ROMERO et al. 1996). Both of these interactions have been detected in nuclear fractions from cells. Ku-70 is the regulatory subunit of the DNA-dependent protein kinase. It is responsible for binding double-stranded DNA breaks and initiating DNA repair. ENX-1 is the human homologue of the *Drosophila Enhancer of zeste* gene, which is a member of the *Polycomb* group of genes. In *Drosophila*, *Polycomb* gene products repress transcription of *homeobox* (*Hox*) genes (SIMON 1995). In mammals, *Hox* genes play an important role in hematopoiesis (VAN DER LUGT et al. 1994). Moreover, T-cell proliferation correlates with expression of *HoxB* genes (CARE et al. 1994) and unregulated *Hox* gene expression can lead to B- and T-cell lymphomas (VAN LOHUIZEN et al. 1991). If ENX-1 prevents lymphocyte proliferation by repressing the expression of *Hox* genes, then the binding of Vav to ENX-1 could be one way in which Vav regulates T- and B-cell proliferation. Since antigen receptor-induced proliferation of T and B cells is impaired in *vav* –/– mice, one could propose that Vav neutralizes the growth-suppressing function of ENX-1. An extension of this hypothesis would be that antigen-receptor-induced tyrosine-phosphorylation of Vav would increase its ability to neutralize ENX-1 and promote proliferation. This interesting hypothesis remains to be tested.

## 10.3 Vav as a Rac1 Exchange Factor: A Link to the Cytoskeleton and to MAP kinases?

The most likely and significant role for Vav may be its ability to activate the Rac1 GTPase. The DH domain in Vav is characteristic of guanine nucleotide exchange factors for the Rac/Rho/Cdc42 family of GTPases (CERIONE and ZHENG 1996).

GNEFs activate GTPases by stimulating the release of GDP. The GTPase can then bind GTP and assume its active conformation. In vitro studies (CRESPO et al. 1997) showed that purified Vav can activate Rac1 and, to a lesser extent, RhoA. Vav had no exchange activity towards Cdc42, Ran, or Ras. Previous reports that Vav was an exchange factor for Ras (GULBINS et al. 1993; PRONK et al. 1993) seemed at odds with Vav having a DH domain and have since been attributed to contaminating or co-precipitating Ras exchange factors. Bustelo and colleagues went on to show that tyrosine phosphorylation of Vav by either Fyn or Lck increases the Rac1 GDP/GTP exchange activity of Vav in in vitro reactions and when Vav is co-expressed with Lck or Fyn in COS cells (CRESPO et al. 1997). Thus, the main function of Vav tyrosine phosphorylation may be to increase its enzymatic activity as a Rac1 exchange factor.

Rac1 is involved in several important processes in mammalian cells. First, it plays a key role in receptor-induced actin polymerization and cytoskeletal reorganization (RIDLEY et al. 1992). Consistent with the idea that Vav activates Rac1, two recent papers showed that TCR-induced actin polymerization is defective in T cells from *vav* –/– mice (FISCHER et al. 1998; HOLSINGER et al. 1998). Moreover, antibody-induced patching and capping of the TCR, a process that depends on actin polymerization, was also found to require Vav (FISCHER et al. 1998; HOLSINGER et al. 1998). Thus, a Vav/Rac1 pathway may initiate actin polymerization in the vicinity of antigen receptors that are signaling in response to antigen binding. BCR engagement is known to cause the membrane Ig as well as other signaling components to associate with the actin cytoskeleton (JUGLOFF and JONGSTRA-BILEN 1997). This may immobilize those signaling proteins, concentrating them in a particular locale and facilitating the formation of signaling complexes that transduce signals. Consistent with this idea, HOLSINGER et al. (1998) found that treating T cells with cytochalasin D, a drug that prevents actin polymerization, results in a similar set of signaling defects as those seen in *vav* –/– T cells. Similarly, cytochalasin D prevents BCR-induced activation of the ERK MAP kinase in B cells (MELAMED et al. 1995).

Another potential role for the Vav/Rac1 pathway may be to link receptors to the JNK and p38 MAP kinases. Expression of activated Rac1 in fibroblasts can initiate the protein-kinase cascades that culminate in activation of these two MAP kinases (COSO et al. 1995; MINDEN et al. 1995). In support of the existence of a Vav/Rac1/JNK pathway, co-expression of Lck and Vav in fibroblasts activates JNK and this response can be blocked by expressing a dominant negative form of Rac1 (CRESPO et al. 1997). While BCR engagement activates both JNK and p38 MAP kinase (SUTHERLAND et al. 1996; HEALY et al. 1997), the role of Rac1 in this process has not been tested (by expressing dominant negative Rac1 in B cells). However, two recent reports have shown that BCR- and TCR-induced JNK activation is normal in cells from *vav* –/– mice (FISCHER et al. 1998; HOLSINGER et al. 1998). Thus, activation of Rac1 by Vav induces actin polymerization but is not a major mechanism by which the antigen receptors activate JNK or p38 MAP kinase. As discussed in Sect. 8, the BCR could instead activate JNK and p38 MAP kinase via the SOS or C3G exchange factors or via the Btk.

## 10.4 Vav and $PIP_2$ Hydrolysis

In lymphocytes from *vav* −/− mice, the ability of the TCR and BCR to increase intracellular $Ca^{2+}$ concentrations is greatly impaired (FISCHER et al. 1995; TURNER et al. 1997; HOLSINGER et al. 1998). Several recent findings have revealed how Vav might be involved in generating $Ca^{2+}$ fluxes. As described above, tyrosine-phosphorylated Vav can activate the Rac1 GTPase and, to a lesser extent, the RhoA GTPase (CRESPO et al. 1997). Both Rac1-GTP and RhoA-GTP can stimulate the activity of PIP5K (CHONG et al. 1994; TOLIAS et al. 1995; REN et al. 1996). PIP5K produces $PIP_2$, the substrate for PLC, by phosphorylating PI 4-phosphate on the 5 position of the inositol ring. Since the $PIP_2$ pool in cells is small, activation of PLC-$\gamma$ would deplete the supply of $PIP_2$ within seconds unless new PIP2 is produced by PIP5K (STEPHENS et al. 1993). Thus, the diminished antigen receptor-induced $Ca^{2+}$ fluxes in *vav* −/− lymphocytes may reflect a failure to sustain $IP_3$ production due to a shortage of $PIP_2$ for the PLC-$\gamma$ to cleave.

A model for how this Vav/Rac1/PIP5K pathway might be activated in B cells is presented in Fig. 4. BCR ligation results in tyrosine phosphorylation of CD19 and this creates a binding site for the SH2 domain of Vav (LI et al. 1997b). Recruitment of Vav to CD19 is important for Vav to become tyrosine phosphorylated since anti-IgM-induced Vav tyrosine phosphorylation is decreased in CD19-deficient B cells (SATA et al. 1997). Vav may be phosphorylated either by the CD19-associated Lyn or by the BCR-associated Syk when the BCR and CD19 are co-ligated. Tyrosine phosphorylation of Vav stimulates its ability to activate Rac1 and RhoA (CRESPO et al. 1997). Activated Rac1 and RhoA then activate PIP5K and increase the synthesis of $PIP_2$, allowing for sustained $PIP_2$ breakdown when PLC-$\gamma$ is activated. In support of this model, it has recently been shown that co-ligation of the BCR to CD19 stimulates $PIP_2$ synthesis and that this does not occur in B cells from *vav* −/− mice (O'ROURKE et al. 1998). Thus, the Vav/Rac1/PIP5K pathway appears to be essential for sustained BCR signaling that can promote B-cell activation.

It had been shown previously that PI 3-kinase activity is also required for Rac1 activation (HAWKINS et al. 1995), but it was not known how PI 3-kinase or $PIP_3$ contributed to activation of Rac1. A recent paper by HAN et al. (1998) showed that $PIP_3$ binds to the PH domain of Vav and that this makes Vav a better substrate for tyrosine kinases and, thus, a better activator of Rac1. The Ras activator SOS also has a DH domain and Bar-Sagi and colleagues have now shown that SOS can indeed activate Rac1 and that this is dependent on $PIP_3$ binding to the PH domain of SOS (HAN et al. 1998). Thus, PI 3-kinase is an upstream activator of two Rac1 exchange factors, Vav and SOS. However, since BCR-induced $PIP_2$ synthesis is completely abrogated in *vav* −/− mice, the SOS/Rac1/PIP5K pathway does not appear to make a significant contribution to $PIP_2$ synthesis in B cells. Nevertheless, recruitment of PI 3-kinase to the plasma membrane and the subsequent production of $PIP_3$ is a key element in the activation of the Rac1/PIP5K pathway.

If sustained BCR-induced $PIP_2$ breakdown depends on the PI 3-kinase/Vav/Rac1/PIP5K pathway, it would explain several puzzling observations. The first is the now 10-year-old observation that BCR-induced $IP_3$ production exhibits the classic characteristics of a GTPase-dependent process. It is inhibited by GDP analogs, which favor the inactive state of GTPase and potentiated by non-hydrolyzable GTP analogs which stabilize the activated states of GTPases (GOLD et al. 1987; HARNETT and KLAUS 1988; MONROE and HALDAR 1989). These data never quite fit with the tyrosine phosphorylation-dependent mode of PLC-$\gamma$ activation. We can now explain these results by proposing that the introduction of these guanine nucleotides into B cells alters the activation state of Rac1 or Rho and thereby influences the ability of PIP5K to refill the $PIP_2$ pools and promote sustained $IP_3$ production. The other puzzling observation is that co-ligating the BCR and CD19 causes synergistic activation of the ERK kinases and that this is dependent on the Vav binding site in CD19 (LI et al. 1997b). As opposed to having an effect on Ras-dependent activation of ERK, this can now be explained by Vav potentiating BCR-induced $PIP_2$ hydrolysis and DAG production, which stimulates PKC-dependent activation of ERK (see Sect. 6).

Finally, if the Vav/Rac1/PIP5K pathway is important for maintaining the supply of $PIP_2$, an obvious question is whether this is important not only for sustained $PIP_2$ breakdown by PLC-$\gamma$ but also for sustained production of PI(3,4,5)$P_3$ by PI 3-kinase. It would be interesting to see whether BCR-induced production of the PI 3-kinase-derived inositol lipids is also decreased in cells from *vav* −/− mice due to a shortage of $PIP_2$ for PI 3-kinase to phosphorylate.

# 11 Transcription Factor Targets of BCR Signaling

As described in the previous sections, the signaling pathways activated by the BCR target a number of pre-existing transcription factors, in some cases activating them via phosphorylation, e.g., Elk-1, in other cases inducing their translocation from the cytosol to the nucleus, e.g., NF-κB and NF-$AT_c$. Once activated, these pre-existing transcription factors stimulate the transcription of immediate early genes, a number of which encode additional transcription factors. Transcription factors that are rapidly synthesized in response to BCR ligation include members of the Fos and Jun families, Egr-1, and c-Myc (MONROE 1988; KLEMSZ et al. 1989; SEYFERT et al. 1989; TILZEY et al. 1991; MITTELSTADT and DEFRANCO 1993; Huo and ROTHSTEIN 1995). Figure 5 summarizes the connections between BCR signaling pathways and transcription factors.

Transcription factors must often form heterodimers or higher order complexes in order to bind to promoter elements and stimulate transcription. Examples of transcription-factor complexes that are formed during BCR signaling include AP-1 complexes which are heterodimers of Fos and Jun proteins. AP-1 complexes can bind to TPA-responsive elements (TREs) and stimulate transcription, whereas Fos and Jun monomers or homodimers cannot (CHIU et al. 1988). BCR ligation has been shown to induce the formation of complexes that bind to TREs (CHILES et al.

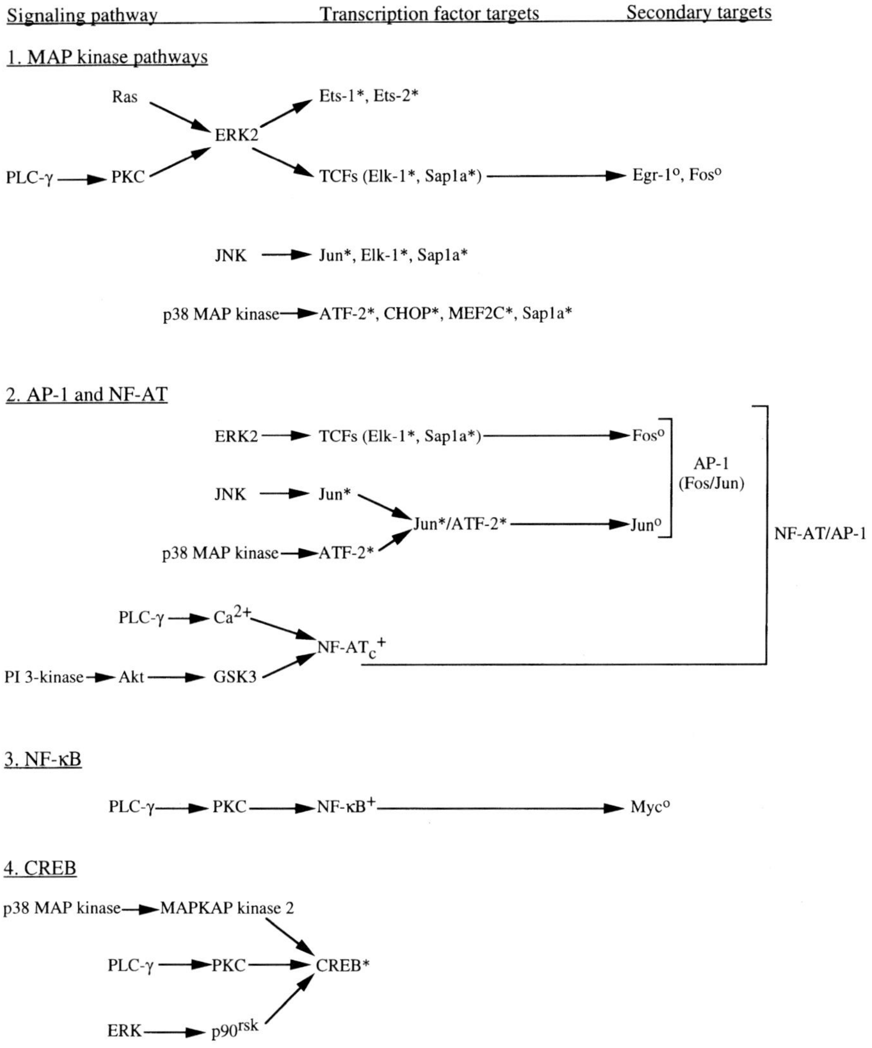

| Signaling pathway | Transcription factor targets | Secondary targets |
|---|---|---|

1. MAP kinase pathways

Ras → ERK2

PLC-γ → PKC → ERK2

ERK2 → Ets-1*, Ets-2*

ERK2 → TCFs (Elk-1*, Sap1a*) → Egr-1°, Fos°

JNK → Jun*, Elk-1*, Sap1a*

p38 MAP kinase → ATF-2*, CHOP*, MEF2C*, Sap1a*

2. AP-1 and NF-AT

ERK2 → TCFs (Elk-1*, Sap1a*) → Fos°

JNK → Jun* → Jun*/ATF-2* → Jun°

p38 MAP kinase → ATF-2* → Jun*/ATF-2*

Fos° + Jun° = AP-1 (Fos/Jun)

PLC-γ → $Ca^{2+}$ → $NF\text{-}AT_c^+$

PI 3-kinase → Akt → GSK3 → $NF\text{-}AT_c^+$

AP-1 + $NF\text{-}AT_c^+$ = NF-AT/AP-1

3. NF-κB

PLC-γ → PKC → NF-κB+ → Myc°

4. CREB

p38 MAP kinase → MAPKAP kinase 2 → CREB*

PLC-γ → PKC → CREB*

ERK → $p90^{rsk}$ → CREB*

**Fig. 5.** Transcription-factor targets of B-cell receptor (BCR) signaling. The connections between various BCR signaling pathways and the transcription factors that are direct targets of these pathways are indicated by *arrows*. Secondary targets are transcription factors whose synthesis is rapidly induced by the preceding signaling pathway/transcription factor. The *superscripted symbols* indicate the mechanisms by which the signaling pathway regulates the transcription factor. (*) phosphorylation and activation of pre-existing latent transcription factors; ($^+$) nuclear translocation of latent transcription factors; and (°) rapid synthesis of transcription factors that are immediate early gene products. The *brackets* indicate transcription factor complexes that are formed. Note that some of these signaling pathway/transcription factor connections are based on information from other cell types and have not been confirmed in B cells. *TCFs*, ternary complex factors, a family of transcription factors that dimerize with serum response factor (*SRF*) and bind to serum response elements (*SREs*) in the promoter regions of genes such as *egr-1* and c-*fos*

1991; CHILES and ROTHSTEIN 1992). As an example of higher order transcription-factor complexes that are formed during BCR signaling, AP-1 complexes interact physically with NF-$AT_c$ and cooperate to stimulate transcription from composite NF-AT/AP-1 promoter elements.

Since different transcription factors and transcription-factor complexes stimulate the transcription of different genes, the individual transcription factors can be thought of as letters in a molecular alphabet. Activating different combinations of transcription factors would result in the formation of different complexes and different sets of genes would be induced depending on which transcription factors are present and how they are combined. This combinatorial use of transcription factors would allow the BCR to induce a variety of different biological responses (anergy, apoptosis, survival, activation) provided that there were quantitative or qualitative differences in the signaling pathways that are activated under different conditions. There is now ample evidence that spectrum of signaling pathways activated by the BCR can differ depending on the development stage of the B cell and its previous antigenic encounters. This has been nicely summarized by HEALY and GOODNOW (1998) in a recent review article and, in the next section, some examples will be provided.

## 12 Combinatorial Use of Signaling Pathways to Signal Different Responses

BCR engagement can promote B-cell anergy, apoptosis, survival, or activation depending on the maturation state of the B cell, the physical nature of the antigen, the duration of the B cell's exposure to antigen, its history of previous exposure to antigen, and signals from other receptors. HEALY and GOODNOW (1998) have nicely summarized the ways in which BCR signaling could lead to so many different responses. The first possibility is that BCR signaling is always the same but that the maturation state of the B cell, its previous antigenic experiences, or signaling by other receptors causes the same signals to be interpreted differently. For example, mature and immature B cells or naive and tolerant B cells might express different transcription factors or have differences in chromatin structure so that the same set of BCR signaling pathways leads to transcription of different genes in the two cell types. Alternatively, there could be differences in BCR signaling that are specified by the cell's maturation state or antigenic experience as well as by other receptors. Effects on proximal BCR signaling components, e.g., tyrosine kinases, would lead to a global change in the amplitude of BCR signaling while effects on more downstream components would lead to selective alterations in the degree to which individual signaling pathways are activated. In either case, the net quality of the BCR signal, i.e., the set of downstream effectors that are activated, could be altered since each downstream effector may have a different threshold level of signaling required for its activation. The activation of different subsets of effectors would result in the activation of different sets of transcription factors, the induction of different sets of genes and, ultimately, different cellular responses. The large number of downstream targets of BCR signaling allows for many different combinations, each of which could promote a different cellular response. It is likely that the BCR signaling is modulated via quantitative and qualitative alterations in BCR signaling pathways as well as by nuclear changes that affect how these signals are interpreted. Below, a few examples of the different ways in which BCR signaling can be modulated are described.

One of the most obvious instances in which quantitative differences in BCR signaling can lead to different biological outcomes is the generation of tolerance to self antigens. Membrane-bound self antigens cause extensive BCR cross-linking and very strong signaling which leads to apoptosis, while soluble antigens that cause limited BCR cross-linking and lower levels of chronic BCR signaling induce anergy.

Signals from other B-cell surface proteins can also have quantitative effects on BCR signaling. CD19 and CD22 modulate the amplitude of BCR signaling and determine the amount of BCR cross-linking required for B-cell activation (reviewed by Doody et al. 1996; O'Rourke et al. 1997). The CD19/CD21 complex amplifies BCR signaling. CD21 is a complement receptor and antigens that are coated with the C3d fragment of complement can co-ligate the BCR with the CD19/CD21 complex. The cytoplasmic domain of CD19 binds Lyn and when CD19 and the BCR are co-ligated, Lyn can efficiently phosphorylate the Ig-$\alpha/\beta$ ITAMs. Tyrosine phosphorylation of the Ig-$\alpha/\beta$ ITAMs is the key event that initiates BCR signaling. In the absence of CD19 co-ligation, Ig-$\alpha/\beta$ tyrosine phosphorylation is much less efficient because only a small fraction of the BCR complexes are associated with tyrosine kinases in resting B cells. Consistent with the idea that CD19 potentiates this key initial event in BCR signaling, work by Fearon and colleagues has clearly shown that much smaller amounts of antigen are needed to activate B cells when CD19 is co-ligated with the BCR.

In contrast to CD19, CD22 limits the magnitude of BCR signaling. BCR-induced tyrosine phosphorylation of CD22 by the Lyn tyrosine kinase allows CD22 to bind the SHP1 tyrosine phosphatase which has an important role in dampening BCR signaling. Consistent with the idea that this pathway downregulates BCR signaling, B cells from mice that lack CD22, Lyn, or SHP1 are all hyperresponsive to BCR ligation (Cyster and Goodnow 1995; O'Keefe et al. 1996; Cornall et al. 1998). SHP1 presumably dephosphorylates key components involved in BCR signaling.

While CD19 and CD22 may have global effects on BCR signaling by acting on very proximal signaling components, Fc$\gamma$RIIb appears to target specific BCR signaling pathways. Co-ligation of the BCR with Fc$\gamma$RIIb by antigen-antibody complexes prevents B-cell activation and decreases BCR-induced $PIP_2$ breakdown, $Ca^{2+}$ increases, and Ras activation (reviewed by Coggeshall 1998). These effects appear to be mediated by recruitment of the 5′-inositol phosphatase SHIP to FcR$\gamma$IIb. SHIP dephosphorylates $PI(3,4,5)P_3$. While $PI(3,4)P_2$ may be still be able to activate some of the downstream effectors of PI 3-kinase, the membrane recruitment of PH domain-containing proteins by $PIP_3$ would be impaired. The failure to recruit Btk to the plasma membrane would decrease the ability of the BCR to activate PLC-$\gamma$. The inhibition of BCR-induced Ras activation by FcR$\gamma$IIb appears to be due to the ability of SHIP to bind to Shc and prevent it from binding Grb2/SOS complexes. Thus, Fc$\gamma$RIIb selectively downregulates BCR signaling pathways that are influenced by SHIP.

The maturation state of the B cell as well as prior antigen exposure can also cause qualitative and/or quantitative changes in BCR signaling. Healy et al. (1997) compared a number of BCR-induced signaling events in naive versus tolerant B cells. In naive B cells, BCR ligation increases intracellular $Ca^{2+}$ concentrations and activates NF-AT, NF-$\kappa$B, ERK, and JNK. However, in the tolerant B cells, BCR-induced $Ca^{2+}$ fluxes were greatly decreased and neither NF-$\kappa$B or JNK were activated by the BCR. The failure of the tolerant B cells to activate NF-$\kappa$B and JNK is probably a key reason why these cells are anergic and are not activated by BCR engagement. Monroe and colleagues (Chap. 8, Vol. II) have also found qualitative differences in BCR signaling between mature and immature B cells. Compared with mature B cells, BCR engagement in immature B cells causes (1) less activation of NF-$\kappa$B and subsequent expression of c-*myc*, (2) less $PIP_2$ hydrolysis, and (3) little or no $Ca^{2+}$ increases due to a defect in replenishing intracellular $Ca^{2+}$ stores. In addition, BCR ligation in the immature B cells fails to induce the expression of cyclin E which is important for G1 to S phase progression. Other BCR signaling pathways appear to be normal in immature B cells. These "imbalances" in BCR signaling, together with the fact that immature B cells express lower levels of the anti-apoptotic proteins Bcl-2, Bcl-$x_L$, and A1, may account for why immature B cells are much more sensitive to BCR-induced apoptosis than mature B cells.

In summary, there are many instances in which the net output of BCR signaling can be altered in terms of its overall amplitude or the relative amplitudes of the different signaling pathways. These differences in net signaling output are likely to be a key component in determining the cellular response to BCR signaling.

## 13 Conclusions

In this review, I have tried to give a comprehensive view of the current knowledge of BCR signaling, focusing on recent findings. In addition, I have described a number of hypothetical signaling connections that may appear in future reviews. Most importantly, I have tried to point out the gaps in our knowledge where further experimentation is needed.

If there is one take-home message, it is that the signaling pathways activated by the BCR are much more integrated than we previously thought. The notion that the PLC-$\gamma$, PI 3-kinase, and Ras pathways are separate signaling modules has been replaced by the idea that there is a high degree of connectivity between these pathways that goes beyond what we normally think of as cross-talk. For example, is now clear that the generation of $PIP_3$ by PI 3-kinase is a key initial event in many signaling pathways, helping to recruit diverse PH domain-containing signaling molecules to the plasma membrane where their substrates are located. Perhaps the activation of PLC-$\gamma$ described in Fig. 3 best illustrates the interplay between $PIP_3$-mediated membrane recruitment, SH2 domain-mediated membrane recruitment, and phosphorylation in the activation of one signaling enzyme. The complex network of events required to provide a continuous supply of $PIP_2$ for PLC-$\gamma$ (Fig. 4) also illustrates how various signaling pathways work together. Finally, I have also tried to show how a single downstream effector, for example the NF-AT transcription factor, is regulated by many different signaling pathways.

This high degree of connectivity between BCR signaling pathways means that BCR signaling is not a binary code, e.g., Ras on, PLC-$\gamma$ off, PI 3-kinase on, which specifies different cellular responses. Instead, perturbing one pathway or changing the amplitude of another will have secondary effects on the other signaling pathways, subtly changing the relative mixture of signals from the different pathways. This allows for a large spectrum of different "net" signals which are then interpreted by the B cell, which has different thresholds for the various possible responses ranging from apoptosis to activation and proliferation.

## References


Aagaard-Tillery KM, Jelinek DF (1996) Phosphatidylinositol 3-kinase activation in normal human B lymphocytes. Differential sensitivity of growth and differentiation to wortmannin. J Immunol 156:4543–4554

Abraham RT (1998) Mammalian target of rapamycin: immunosuppressive drugs uncover a novel pathway of cytokine receptor signaling. Curr Opin Immunol 10:330–336

Akimoto K, Takahashi R, Moriya S, Nishioka N, Takayanagi J, Kimura K, Fukui Y, Osada S, Mizuno K, Hirai S, Kazlauskas A, Ohno S (1996) EGF or PDGF receptors activate atypical PKClambda through phosphatidylinositol 3-kinase. EMBO J 15:788–798

Alessi DR, Andjelkovic M, Cauldwell B, Cron P, Morrice N, Cohen P, Hemmings BA (1996) Mechanism of activation of protein kinase B by insulin and IGF-1. EMBO J 15:6541–6551

Alessi DR, James SR, Downes CP, Holmes AB, Gaffney PRJ, Reese CB, Cohen P (1997) Characterization of a 3-phosphoinositide-dependent protein kinase which phosphorylates and activates protein kinase Bα. Curr Biol 7:261–269

Alessi DR, Kozlowski MT, Weng Q-P, Morrice N, Avruch J (1998) 3-Phosphoinositide-dependent kinase 1 (PDK1) phosphorylates and activates the p70 S6 kinase in vivo and in vitro. Curr Biol 8:69–81

Anderson KE, Coadwell J, Stephens LR, Hawkins PT (1998) Translocation of PDK-1 to the plasma membrane is important in allowing PDK-1 to activate protein kinase B. Curr Biol 8:684–691

Andjelkovic M, Alessi DR, Meier R, Fernandez A, Lamb NJC, Frech M, Cron P, Lucocq JM, Hemmings BA (1997) Role of translocation in the activation and function of protein kinase B. J Biol Chem 272:31515–31524

Arvidsson A-K, Rupp E, Nanberg E, Downward J, Ronnstrand L, Wennstrom S, Schlessinger J, Heldin C-H, Claesson-Welsh L (1994) Tyr-716 in the platelet-derived growth factor β-receptor kinase insert is involved in GRB2 binding and Ras activation. Mol Cell Biol 14:6715–6726

Auger KR, Cantley LC (1991) Novel polyphosphoinositides in cell growth and activation. Cancer Cells 3:263–270

Avruch J, Zhang X-F, Kyriakis JM (1994) Raf meets Ras: completing the framework of a signal transduction pathway. Trends Biochem Sci 19:279–283

Backer JM, Myers Jr. MG, Shoelson SE, Chin DJ, Sun X-J, Miralpeix M, Hu P, Margolis B, Skolnik EY, Schlessinger J, White MF (1992) Phosphatidylinositol 3′-kinase is activated by association with IRS-1 during insulin stimulation. EMBO J 11:3469–3479
Baldwin Jr. AS (1996) The NF-κB and IκB proteins: New discoveries and insights. Ann Rev Immunol 14:649–681
Barbazuk SM, Gold MR (1999) Protein kinase C-δ is a target of B cell antigen receptor signaling. J Immunol Lett in press
Baumann G, Maier D, Freuler F, Tschopp C, Baudisch K, Wienands J (1994) In vitro characterization of major ligands for Src homology 2 domains derived from protein tyrosine kinases, from the adapter protein Shc and from GTPase-activating protein in Ramos B cells. Eur J Immunol 24:1799–1807
Beals CR, Sheridan CM, Turck CW, Gardner P, Crabtree GR (1997) Nuclear export of NF-ATc enhanced by glycogen synthase kinase-3. Science 275:1930–1933
Beranger F, Goud B, Tavitian A, de Gunzburg J (1991) Association of the Ras-antagonistic Rap/Krev-1 proteins with the Golgi complex. Proc Natl Acad Sci USA 88:1606–1610
Berditchevski F, Tolias KF, Wong K, Carpenter CL, Hemler ME (1997) A novel link between integrins, transmembrane-4 superfamily proteins (CD63 and CD81), and phosphatidylinositol 4-kinase. J Biol Chem 272:2595–2598
Berridge MJ (1993) Inositol trisphosphate and calcium signaling. Nature 361:315–325
Bijsterbosch MK, Meade CJ, Turner GA, Klaus GGB (1985) B lymphocyte receptors and polyphosphoinositide degradation. Cell 41:999–1006
Bireland ML, Monroe JG (1997) Biochemistry of antigen receptor signaling in mature and developing B lymphocytes. Crit Rev Immunol 17:353–385
Boguski MS, McCormick F (1993) Proteins regulating Ras and its relatives. Nature 366:643–654
Bolland S, Pearse RN, Kurosaki T, Ravetch JV (1998) SHIP modulates immune receptor responses by regulating membrane association of Btk. Immunity 8:508–516
Bomsztyk K, Van Seuningen I, Suzuki H, Denisenko O, Ostrowski J (1997) Diverse molecular interactions of the hnRNP K protein. FEBS Lett 403:113–115
Bonnefoy-Berard N, Munshi A, Yron I, Wu S, Collins TL, Deckert M, Shalom-Barak T, Giampa L, Herbert E, Hernandez J, Meller N, Couture C, Altman A (1996) Vav: function and regulation in hematopoietic cell signaling. Stem Cells 14:250–268
Bowtell D, Fu P, Simon M, Senior D (1992) Identification of murine homologues of the *Drosophila* son of sevenless gene: Potential activators of Ras. Proc Natl Acad Sci USA 89:6511–6515
Bras A, Martinez-A. C, Baixeras E (1997) B cell receptor cross-linking prevents Fas-induced cell death by inactivating the IL-1β-converting enzyme protease and regulating Bcl-2/Bcl-x expression. J Immunol 159:3168–3177
Buday L, Downward J (1993) Epidermal growth factor regulates p21$^{ras}$ through the formation of a complex of receptor, Grb2 adapter protein, and Sos nucleotide exchange factor. Cell 73:611–620
Buhl AM, Pleiman CM, Rickert RC, Cambier JC (1997) Qualitative regulation of B cell antigen receptor signaling by CD19: Selective requirement for PI3-kinase activation, inositol-1,4,5-trisphosphate production and $Ca^{2+}$ mobilization. J Exp Med 186:1897–1910
Burgering BMT, Coffer PJ (1995) Protein kinase B (c-akt) in phosphatidylinositol 3-kinase signal transduction. Nature 376:599–602
Bustelo X, Barbacid M (1992a) Tyrosine phosphorylation of the vav proto-oncogene product in activated B cells. Science 256:1196–1199
Bustelo XR, Barbacid M (1992b) Tyrosine phosphorylation of the *vav* proto-oncogene product in activated B cells. Science 256:1196–1199
Bustelo XR (1996) The VAV family of signal transduction molecules. Crit Rev Oncogen 7:65–88
Bustelo XR, Suen K-L, Michael WM, Dreyfuss G, Barbacid M (1996) Association of the *vav* proto-oncogene product with poly(rC)-specific RNA-binding proteins. Mol Cell Biol 15:1324–1332
Campa MJ, Chang KJ, Molina V, Rep BR, Lapetina EG (1991) Inhibition of *ras*-induced germinal vesicle breakdown in Xenopus oocytes by *rap*-1B. Biochem Biophys Res Commun 174:1–5
Care A, Testa U, Bassani A, Tritarelli E, Montesoro E, Samoggia P, Cianetti L, Peschle C (1994) Coordinate expression and proliferative role of *HOXB* genes in activated adult T lymphocytes. Mol Cell Biol 14:4872–4877
Carpenter CL, Auger KR, Chanudhuri M, Yoakim M, Schaffhausen B, Shoelson S, Cantley LC (1993) Phosphoinositide 3-kinase is activated by phosphopeptides that bind to the SH2 domains of the 85-kDa subunit. J Biol Chem 268:9478–9483
Carter RH, Park DJ, Rhee SG, Fearon DT (1991) Tyrosine phosphorylation of phospholipase C induced by membrane immunoglobulin crosslinking in B lymphocytes. Proc Natl Acad Sci USA 88:2745–2749

Casillas A, Hanekom C, Williams K, Katz R, Nel AE (1991) Stimulation of B-cells via the membrane immunoglobulin receptor of with phorbol myristate 13-acetate induces tyrosine phosphorylation and activation of a 42-kDa microtubule-associated protein-2 kinase. J Biol Chem 266:19088–19094

Castresana J, Saraste M (1995) Does Vav bind to F-actin through a CH domain? FEBS Lett 374:149–151

Cerione RA, Zheng Y (1996) The Dbl family of oncogenes. Curr Opin Cell Biol 8:216–222

Chalupny NJ, Kanner SB, Schieven GL, Wee SF, Gilliland LK, Aruufo A, Ledbetter JA (1993) Tyrosine phosphorylation of CD19 in pre-B and mature B cells. EMBO J 12:2691–2696

Chen ZZ, Coggeshall KM, Cambier JC (1986) Translocation of protein kinase C during membrane immunoglobulin-mediated transmembrane signaling in B lymphocytes. J Immunol 136:2300–2304

Chiles TC, Liu J, Rothstein TL (1991) Cross-linking of surface Ig receptors on murine B lymphocytes stimulates the expression of nuclear tetradecanoyl phorbol acetate-response element-binding proteins. J Immunol 146:1730–1735

Chiles TC, Rothstein TL (1992) Surface Ig receptor-induced nuclear AP-1-dependent gene expression in B lymphocytes. J Immunol 149:825–831

Chiu R, Boyle WJ, Meek J, Smeal T, Hunter T, Karin M (1988) The c-Fos protein interacts with c-Jun/AP-1 to stimulate transcription of AP-1 response genes. Cell 54:541–552

Chong LD, Traynor-Kaplan A, Bokoch GM, Schwartz MA (1994) The small GTP-binding protein Rho regulates a phosphatidylinositol 4-phosphate-5-kinase in mammalian cells. Cell 79:507–513

Clevenger CV, Ngo W, Sokol DL, Luger SL, Gerwitz AM (1995) Vav is necessary for prolactin-stimulated proliferation and is translocated into the nucleus of a T-cell line. J Biol Chem 270:13246–13253

Coggeshall MK (1998) Inhibitory signaling by B cell FcγRIIb. Curr Opin Immunol 10:306–312

Cohen P (1997) The search for physiological substrates of MAP and SAP kinases in mammalian cells. Trends in Cell Biology 7:353–361

Collins TL, Deckert M, Altman A (1997) Views on Vav. Immunol Today 18:221–225

Cook SJ, Rubinfeld B, Albert I, McCormick (1993) RapV12 antagonizes Ras-dependent activation of ERK1 and ERK2 by LPA and EGF in Rat-1 fibroblasts. EMBO J 12:3475–3485

Cornall RJ, Cyster JG, Hibbs ML, Dunn AR, Otipoby KL, Clark EA, Goodnow CC (1998) Polygenic autoimmune traits: Lyn, CD22, and SHP-1 are limiting elements of a biochemical pathway regulating BCR signaling and selection. Immunity 8:497–508

Cory GOC, Lovering RC, Hinshelwood S, McCarthy-Morrogh L, Levinsky RJ, Kinnon C (1995) The protein product of the c-*cbl* protooncogene is phosphorylated after B cell receptor stimulation and binds the SH3 domain of Bruton's tyrosine kinase. J Exp Med 182:611–615

Coso OA, Chiariello M, Yu J-C, Teramoto H, Crespo P, Xu N, Miki T, Gutkind JS (1995) The small GTP-binding proteins Rac1 and Cdc42 regulate the activity of the JNK/SAPK signaling pathway. Cell 81:1137–1146

Crespo P, Schuebel KE, Ostrom AA, Gutkind JS, Bustelo XR (1997) Phosphorylation-dependent activation of rac-1 GDP/GTP exchange by the *vav* proto-oncogene product. Nature 385:169–172

Cross DAE, Alessi DR, Cohen P, Andjelkovic M, Hemmings BA (1995) Inhibition of glycogen synthase kinase-3 by insulin mediated protein kinase B. Nature 378:785–789

Crowley MT, Harmer SL, DeFranco AL (1996) Activation-induced association of a 145-kDa tyrosine-phosphorylated protein with Shc and Syk in B lymphocytes and macrophages. J Biol Chem 271:1145–1152

Cuenda A, Rouse J, Doza YN, Meier R, Cohen P, Gallagher TF, Young PR, Lee JC (1995) SB203580 is a specific inhibitor of a MAP kinase homologue which is stimulated by cellular stresses and interleukin-1. FEBS Lett 364:229–233

Cullen PJ, Hsuan JJ, Truong O, Letcher AJ, Jackson TR, Dawson AP, Irvine R (1996) Identification of a specific Ins(1,3,4,5)$P_4$-binding protein as a member of the GAP1 family. Nature 376:527–530

Cunningham E, Thomas GMH, Ball A, Hiles I, Cockcroft S (1995) Phosphatidylinositol transfer protein dictates the rate of inositol trisphosphate production by promoting the synthesis of $PIP_2$. Curr Biol 5:775–783

Cyster JG, Goodnow CC (1995) Protein tyrosine phosphatase 1 C negatively regulates antigen receptor signaling in B lymphocytes and determines the thresholds for negative selection. Immunity 2:13–24

D'Ambrosio D, Hippen KL, Cambier JC (1996) Distinct mechanisms mediate SHC association with the activated and resting B cell antigen receptor. Eur J Immunol 26:1960–1965

Dalby KN, Morrice N, Cauldwell FB, Avruch J, Cohen P (1998) Identification of regulatory phosphorylation sites in mitogen-activted protein kinase (MAPK)-activated protein kinase-1a/p90rsk that are inducible by MAPK. J Biol Chem 273:1496–1505

Damen JE, Liu L, Rosten P, Humphries RK, Jefferson AB, Majerus PW, Krystal G (1996) The 145-kDa protein induced to associate with Shc by mulitple cytokines is an inositol tetraphosphate and phosphatidylinositol 3,4,5 trisphosphate 5-phosphatase. Proc Natl Acad Sci USA 93:1689–1693
Datta SR, Dudek H, Tao X, Masters S, Fu H, Gotoh Y, Greenberg ME (1997) Akt phosphorylation of BAD couples survival signals to the cell-intrinsic death machinery. Cell 91:231–241
Dawson AP (1997) Calcium signalling: How do $IP_3$ receptors work? Curr Biol 7:R544–R547
De Nooij JC, Letendre MA, Hariharan IK (1996) A cyclin-dependent kinase inhibitor, Dacapo, is necessary for timely exit from the cell cycle during Drosophila embyogenesis. Cell 87:1237–1247
Deckert M, Tartare-Deckert S, Couture C, Mustelin T, Altman A (1996) Functional and physical interactions of Syk family kinases with the Vav proto-oncogene product. Immunity 5:591–604
DeFranco AL (1997) The complexity of signaling pathways activated by the BCR. Curr Opin Immunol 9:296–308
Dejgaard K, Leffers H, Rasmussen HH, Madsen P, Kruse TA, Gesser B, Nielsen H, Celis JE (1994) Identification, molecular cloning, expression and chromosome mapping of a family of transformation upregulated hnRNP-K proteins derived by alternative splicing. J Mol Biol 236:33–48
Dekker LV, Parker PJ (1994) Protein kinase C-a question of specificity. Trends Biochem Sci 19:73–77
Del Peso L, Gonzales-Garcia M, Page C, Herrera R, Nunez G (1997) Interleukin-3-induced phosphorylation of BAD through the protein kinase Akt. Science 278:687–689
Denisenko ON, O'Neill B, Ostrowski J, Van Seuningen I, Bomsztyk K (1996) Zik1, a transcriptional repressor that interacts with the heterogeneous nuclear ribonucleoprotein particle K protein. J Biol Chem 271:27701–27706
DiDonato JA, Hayakawa M, Rothwarf DM, Zandi E, Karin M (1997) A cytokine-responsive IκB kinase that activates the transcription factor NF-κB. Nature 388:548–554
Diener K, Wang XS, Chen C, Meyer CF, Keesler G, Zukowski M, Tan T-H, Yao Z (1997) Activation of the c-Jun N-terminal kinase pathway by a novel protein kinase related to human germinal center kinase. Proc Nat Acad Sci USA 94:9687–9692
Doody GM, Dempsey PW, Fearon DT (1996) Activation of B lymphocytes: integrating signals from CD19, CD22, and FcγRIIb1. Curr Opin Immunol 8:378–382
Downward J, Graves JD, Warne PH, Rayter S, Cantrell DA (1990) Stimulation of $p21^{ras}$ upon T-cell activation. Nature 346:719–723
Downward J (1998) Mechanisms and consequences of activation of protein kinase B/Akt. Curr Opin Cell Biol 10:262–267
Duchesne M, Schweighoffer F, Parker F, Clerc F, Frobert Y, Thang MN, Tocque B (1993) Identification of the SH3 domain of GAP as an essential sequence for Ras-GAP-mediated signaling. Science 259:525–528
Dudek H, Datta SR, Franke TF, Birnbaum MJ, Yao R, Cooper GM, Segal RA, Kaplan DR, Greenberg ME (1997) Regulation of neuronal survival by the serine-threonine protein kinase Akt. Science 275:661–665
Duyao MP, Buckler AJ, Sonenshein GE (1990) Interaction of an NF-κB-like factor with a site upstream of the c-myc promoter. Proc Natl Acad Sci USA 87:4727–4731
Fahey KA, DeFranco AL (1987) Cross-linking membrane IgM induces production of inositol trisphosphate and inositol tetrakisphosphate in WEHI-231 B lymphoma cells. J Immunol 138:3935–3942
Fantl WJ, Escobedo JA, Martin GA, Turck CW, del Rosario M, McCormick F, Williams LT (1992) Distinct phosphotyrosines on a growth factor receptor bind to specific molecules that mediate different signaling pathways. Cell 69:413–423
Feller SM, Knudsen B, Hanafusa H (1994) c-Abl kinase regulates the protein binding activity of c–Crk. EMBO J 13:2341–2351
Filmus J, Robles AI, Shi W, Colombo LL, Conti CJ (1994) Induction of cyclin D1 overexpression by activated ras. Oncogene 9:3627–3633
Fischer K-D, Zmuldzinas A, Gardner S, Barbacid M, Bernstein A, Guidos C (1995) Defective T-cell receptor signalling and positive selection of Vav-deficient $CD4^+CD8^+$ thymocytes. Nature 374: 474–477
Fischer K-D, Kong Y-Y, Nishina H, Tedford K, Marengere LEM, Kozieradzki I, Sasaki T, Starr M, Chan G, Gardener S, Nghiem MP, Bouchard D, Barbacid M, Bernstein A, Penninger JM (1998) Vav is a regulator of cytoskeletal reorganization mediated by the T-cell receptor. Curr Biol 8:554–562
Fluckiger A-C, Li Z, Kato RM, Wahl MI, Ochs H, Longnecker R, Kinet J-P, Witte ON, Scharenberg AM, Rawlings DJ (1998) Btk/Tec kinases regulate increases in intracellular $Ca^{2+}$ following B cell receptor activation. EMBO J 17:1973–1985

Foltz IN, Gerl RE, Weiler JS, Luckach M, Salmon RA, Schrader JW (1998) Human mitogen-activated protein kinase kinase 7 (MKK7) is a highly conserved c-Jun N-terminal kinase/stress-activated protein kinase (JNK/SAPK) activated by environmental stresses and physiological stimuli. J Biol Chem 273:9344–9351

Franke B, Akkerman J-WN, Bos J (1997) Rapid $Ca^{2+}$-mediated activation of Rap1 in human platelets. EMBO J 16:252–259

Franke TF, Yang S-I, Chan TO, Datta K, Kazlauskas A, Morrison DK, Kaplan DR, Tsichlis PN (1995) The protein kinase encoded by the *Akt* proto-oncogene is a target of the PDGF-activated phosphatidylinositol 3-kinase. Cell 81:727–736

Frech M, John J, Pizon V, Chardin P, Tavitian A, Clark R, McCormick F, Wittinghofer A (1990) Inhibition of GTPase activating protein stimulation of Ras-p21 GTPase by the Krev-1 gene product. Science 249:169–171

Frech M, Andjelkovic M, Ingley E, Reddy KK, Falck JR, Hemmings BA (1997) High affinity binding of inositol phosphates and phosphoinositides to the pleckstrin homology domain of RAC/Protein kinase B and their influence on the kinase activity. J Biol Chem 272:8474–8481

Fu C, Chan AC (1997) Identification of two tyrosine phosphoproteins, pp70 and pp68, that interact with PLCγ, Grb2, and Vav following B cell antigen receptor activation. J Biol Chem 272:27362–27368

Fu C, Turck CW, Kurosaki T, Chan AC (1998) BLNK: A central linker protein in B cell activation. Immunity, in press

Fukuda T, Kitamura D, Taniuchi I, Maekawa Y, Benhamou LE, Sarthou P, Watanabe T (1995) Restoration of surface IgM-mediated apoptosis in an anti-IgM-resistant variant of WEHI-231 lymphoma cells by HS-1, a protein-tyrosine kinase substrate. Proc Natl Acad Sci USA 92:7302–7306

Gabig TG, Crean CD, Mantel PL, Rosli R (1995) Function of wild-type or mutant Rac2 and Rap1a GTPases in differentiated HL60 cell NADPH oxidase activation. Blood 85:804–811

Gold MR, DeFranco AL (1987) Phorbol esters and dioctanoylglycerol block anti-IgM-stimulated phosphoinositide hydrolysis in the murine B cell lymphoma WEHI-231. J Immunol 138:868–876

Gold MR, Jakway JP, DeFranco AL (1987) Involvement of a guanine nucleotide-binding component in membrane IgM-stimulated phosphoinositide breakdown. J Immunol 139:3608–3613

Gold MR, Sanghera JS, Stewart J, Pelech SL (1992) Selective activation of p42 MAP kinase in murine B lymphoma cell lines by membrane immunoglobulin crosslinking. Evidence for protein kinase C-independent and -dependent mechanisms of activation. Biochem J 287:269–276

Gold MR, Aebersold RA (1994) Both phosphatidylinositol 3-kinase and phosphatidylinositol 4-kinase products are increased by antigen receptor signaling in B lymphocytes. J Immunol 152:42–50

Gold MR, Duronio V, Saxena SP, Schrader JW, Aebersold R (1994) Multiple cytokines activate phosphatidylinositol 3-kinase in hemopoietic cell lines. Association of the enzyme with various tyrosine-phosphorylated proteins. J Biol Chem 269:5403–5412

Goldschmidt-Clermont PJ, Kim JW, Machesky LM, Rhee SG, Pollard TD (1991) Regulation of phospholipase C-γ1 by profilin and tyrosine phosphorylation. Science 251:1231–1233

Gotoh T, Hattori S, Nakamura S, Kitayama H, Noda M, Takai Y, Kaibuchi K, Matsui H, Hatase O, Takahashi H, Kurata T, Matsuda M (1995) Identification of Rap1 as a Target or the Crk SH3 Domain-Binding Guanine Nucleotide-Releasing Factor C3G. Mol Cell Biol 15:6746–6753

Graves JD, Draves KE, Craxton A, Saklatvala V, Krebs EG, Clark EA (1996) Involvement of stress-activated protein kinase and p38 mitogen-activated protein kinase in mIgM-induced apoptosis of human B lymphocytes. Proc Natl Acad Sci USA 93:13814–13818

Greulich H, Hanafusa H (1996) A role for Ras in v-Crk transformation. Cell Growth and Diff 7:1443–1451

Grumont RJ, Rourke IJ, O'Reilly LA, Strasser A, Miyake K, Sha W, Gerondakis S (1998) B lymphocytes differentially use the rel and nuclear factor κB1 (NF-κB1) transcription factors to regulate cell cycle progression and apoptosis in quiescent and mitogen-activated cells. J Exp Med 187:663–674

Gulbins E, Coggeshall KM, Baier G, Katzav S, Burn P, Altman A (1993) Tyrosine kinase-stimulated guanine nucleotide exchange activity of vav in T cell activation. Nature 260:822–825

Gupta S, Barrett T, Whitmarsh AJ, Cavanagh J, Sluss HK, Derijard B, Davis RJ (1996) Selective interaction of JNK protein kinase isoforms with transcription factors. EMBO J 15:2760–2770

Han J, Jiang Y, Li Z, Kravchenko VV, Ulevitch RJ (1997) Activation of the transcription factor MEF2C by the MAP kinase p38 in inflammation. Nature 386:296–299

Han J, Luby-Phelps K, Das B, Shu X, Xia Y, Mosteller RD, Krishna UM, Falck JR, White MA, Broek D (1998) Role of substrates and products of PI 3-kinase in regulating activation of Rac-related guanosine triphosphatases by Vav. Science 279:558–560

Hariharan IK, Carthew RW, Rubin GM (1991) The Drosophila *roughened* mutation: activation of a *rap* homolog disrupts eye development and interferes with cell determination. Cell 67:717–722

Harnett MM, Klaus GGB (1988) G protein coupling of antigen receptor-stimulated polyphosphoinositide hydrolysis in B cells. J Immunol 140:3135–3139

Harwood AE, Cambier JC (1993) B cell antigen receptor cross-linking triggers rapid PKC independent activation of $p21^{ras}$. J Immunol 151:4513–4522

Hasegawa H, Kiyokawa E, Tanaka S, Nagashima K, Gotoh N, Shibuya M, Kurata T, Matsuda M (1996) DOCK180, a major Crk-binding protein, alters cell morphology upon translocation to the cell membrane. Mol Cell Biol 16:1770–1776

Hawkins PT, Eguinoa A, Qiu R-G, Stokoe D, Cooke FT, Walters R, Wennstrom S, Claesson-Welsh L, Evans T, Symons M, Stephens L (1995) PDGF stimulates an increase in GTP-Rac via activation of phosphoinositide 3-kinase. Curr Biol 5:393–403

Healy JI, Dolmetsch RE, Timmerman LA, Cyster JG, Thomas ML, Crabtree GR, Lewis RS, Goodnow CC (1997) Different nuclear signals are activated by the B cell receptor during positive versus negative signaling. Immunity 6:419–428

Healy JI, Goodnow CC (1998) Positive versus negative signaling by lymphocyte antigen receptors. Ann Rev Immunol 16:645–670

Hempel WM, Schatzman RC, DeFranco AL (1992) Tyrosine phosphorylation of phospholipase C-γ2 upon crosslinking of membrane Ig on murine B lymphocytes. J Immunol 148:3021–3027

Herrera-Velit P, Knutson KL, Reiner NE (1997) Phosphatidylinositol 3-kinase-dependent activation of protein kinase C-ζ in bacterial lipopolysaccharide-treated human monocytes. J Biol Chem 272:16445–16452

Hobert O, Jallal B, Schlessinger J, Ullrich A (1994) Novel signaling pathway suggested by SH3 domain-mediated p95vav/heterogeneous ribonucleoprotein K interaction. J Biol Chem 269:20225–20228

Hobert O, Jallal B, Ullrich A (1996) Interaction of Vav with ENX-1, a putative transcriptional regulator of homeobox gene expression. Mol Cell Biol 16:3066–3073

Holsinger LJ, Spencer DM, Austin DJ, Schreiber SL, Crabtree GR (1995) Signal transduction in T lymphocytes using a conditional allele of Sos. Proc Natl Acad Sci USA 92:9810–9814

Holsinger LJ, Graef IA, Swat W, Chi T, Bautista DM, Davidson L, Lewis RS, Alt FW, Crabtree GR (1998) Defects in actin-cap formation in Vav-deficient mice implicate an actin requirement for lymphocyte signal transduction. Curr Biol 8:563–572

Hu CD, Kariya K, Kotani G, Shirouzu M, Yokayama S, Kataoka T (1997) Coassociation of Rap1A and Ha-Ras with Raf-1 N-terminal domain region interferes with ras-dependent activation of Raf-1. J Biol Chem 272:11702–11705

Huo L, Rothstein TL (1995) Receptor-specific induction of individual AP-1 components in B lymphocytes. J Immunol 154:3300–3309

Ingham RJ, Krebs DL, Barbazuk SM, Turck CW, Hirai H, Matsuda M, Gold MR (1996) B cell antigen receptor signaling induces the formation of complexes containing the Crk adapter proteins. J Biol Chem 271:32306–32314

Ingham RJ, Holgado-Madruga M, Siu C, Wong AJ, Gold MR (1998) The Gab1 protein is a docking site for multiple proteins involved in signaling by the B cell antigen receptor. J Biol Chem 273:30630–30637

Iritani BM, Forbush KA, Farrar MA, Perlmutter RM (1997) Control of B cell development by Ras-mediated activation of Raf. EMBO J 16:7019–7031

Janknecht R, Hunter T (1997) Convergence of MAP kinase pathways on the ternary complex factor Sap-1a. EMBO J 16:1620–1627

Jones SM, Howell KE (1997) Phosphatidylinositol 3-kinase is required for the formation of constitutive transport vesicles from the TGN. J Cell Biol 139:339–349

Jugloff LS, Jongstra-Bilen J (1997) Cross-linking of the IgM receptor induces rapid translocation of IgM-associated Ig alpha, Lyn, and Syk tyrosine kinases to the membrane skeleton. J Immunol 159:1096–1106

Karin M, Liu Z-G, Zandi E (1997) AP-1 function and regulation. Curr Opin Cell Biol 9:240–246

Katzav S, Martin-Zanca D, Barbacid M (1989) vav, a novel human oncogene derived from a locus ubiquitously expressed in hematopoietic cells. EMBO J 8:2283–2290

Kauffmann-Zeh A, Thomas GMH, Ball A, Prosser S, Cunningham E, Cockcroft S, Hsuan JJ (1995) Requirement for phosphatidylinositol transfer protein in epidermal growth factor signaling. Science 268:1188–1190

Kavanaugh WM, Pot DA, Chin SM, Deuter-Reinhard M, Jefferson AB, Norris FA, Masiarz FR, Cousens LS, Majerus PW, Williams LT (1996) Multiple forms of an inositol polyphosphate 5-phosphatase form signaling complexes with Shc and Grb2 . Curr Biol 6:438–445

Kawakami Y, Miura T, Bissonnette R, Hata D, Khan WN, Kitamura T, Maeda-Yamamoto M, Hartman SE, Alt FW, Kawakami T (1997) Bruton's tyrosine kinase regulates apoptosis and JNK/SAPK kinase activity. Proc Natl Acad Sci USA 94:3938–3942

Kearns BG, Alb Jr. JG, Bankaitis VA (1998) Phosphatidylinositol transfer proteins: the long and winding road to physiological function. Trends in Cell Biology 8:276–282

Kim TJ, Kim Y-T, Pillai S (1995) Association of activated phosphatidylinositol 3-kinase with p120$^{cbl}$ in antigen receptor-ligated B cells. J Biol Chem 270:27504–27509

Kitamura D, Kaneko H, Miyagoe Y, Ariyasu T, Watanabe T (1989) Isolation and characterization of a novel human gene expressed specifically in the cells of hematopoietic lineage. Nucleic Acids Res 17:9367

Kitayama H, Sugimoto Y, Matsuzaki T, Ikawa Y, Noda M (1989) A *ras*-related gene with transformation suppressor activity. Cell 56:77–84

Klemm JD, Beals CR, Crabtree GR (1997) Rapid targeting of nuclear proteins to the cytoplasm. Curr Biol 7:638–644

Klemsz MJ, Justement LB, Palmer E, Cambier JC (1989) Induction of c-fos and c-myc expression during B cell activation by IL-4 and immunoglobulin binding ligands. J Immunol 143:1032–1039

Klippel A, Reinhard C, Kavanaugh WM, Apell G, Escobedo MA, Williams LT (1996) Membrane localization of phosphatidylinositol 3-kinase is sufficient to activate multiple signal transducing kinase pathways. Mol Cell Biol 16:4117–4127

Kohn AD, Kovacina KS, Roth RA (1995) Insulin stimulates the kinase activity of RAC-PK, a pleckstrin homology domain containing ser/thr kinase. EMBO J 4288:4295

Kon-Kozlowski M, Pani G, Pawson T, Siminovitch KA (1996) The tyrosine phosphatase PTP1 C associates with Vav Grb2, and mSOS1 in hematopoietic cells. J Biol Chem 271:3856–3862

Kroemer G, Zamzami N, Susin SA (1997) Mitochondrial control of apoptosis. Immunol Today 18:44–51

Kurachi H, Wada Y, Tsukamoto N, Maeda M, Kubota H, Hattori M, Iwai K, Minato N (1997) Human SPA-1 gene product selectively expressed in lymphoid tissues is a specfic GTPase-activating protein for Rap1 and Rap2. J Biol Chem 272:28081–28088

Kuriyama M, Harada N, Kuroda S, Yamamoto T, Nakafuku M, Iwamatsu A, Yamamoto D, Prasad R, Croce C, Canaani E, Kaibuchi K (1996) Identification of AF-6 and Canoe as putative targets for Ras. J Biol Chem 271:607–610

Kurosaki T (1997) Molecular mechanisms in B cell antigen receptor signaling. Curr Opin Immunol 9: 309–318

Lankester AC, van Schijndel GM, Rood PM, Verhoven AJ, van Lier RA (1994) B cell antigen receptor cross-linking induces tyrosine phosphorylation and membrane translocation of a multimeric Shc complex that is augmented by CD19 co-ligation. Eur J Immunol 24:2818–2825

Lavoie JN, L'Allemain G, Brunet A, Muller R, Pouyssegur J (1996) Cyclin D1 expression is regulated positively by the p42/p44$^{MAPK}$ and negatively by the p38/HOG$^{MAPK}$ pathway. J Biol Chem 271:20608–20616

Law CL, Sidorenko SP, Chandran KA, Zhao Z, Shen SH, Fischer EH, Clark EA (1996a) CD22 associates with protein tyrosine phosphatase 1C, Syk, and phospholipase C-$\gamma$1 upon B cell activation. J Exp Med 183:547–560

Law CL, Sidorenko SP, Chandran KA, Zhao ZH, Shen SH, Fischer EH, Clark EA (1996b) CD22 associates with protein tyrosine phosphatase 1C, Syk, and phospholipase C-gamma 1 upon B cell activation. J Exp Med 183:547–560

Lazarus AH, Kawauchi K, Rapoport MJ, Delovitch TJ (1993) Antigen-induced B lymphocyte activation involves the p21$^{ras}$ and ras.GAP signaling pathway. J Exp Med 178:1765–1769

Lee MH, Mori S, Raychaudhuri P (1996) Trans-activation by the hnRNP K protein involves an increase in RNA synthesis from the reporter genes. J Biol Chem 271:3420–3427

Leitges M, Schmedt C, Guinamard R, Davoust J, Schaal S, Stabel S, Tarakhovsky A (1996) Immunodeficiency in protein kinase C beta-deficient mice. Science 273:788–791

Lemmon MA, Falasca M, Ferguson KM, Schlessinger J (1997) Regulatory recruitment of signalling molecules to the cell membrane by pleckstrin-homology domains. Trends in Cell Biology 7:237–242

Lenardo M, Baltimore D (1989) NF-$\kappa$B: A pleiotropic mediator of inducible and tissue-specific gene control. Cell 58:227–229

Li HL, Forman HS, Kurosaki T, Pure E (1997a) Syk is required for BCR-mediated activation of p90Rsk, but not p70S6k, via a mitogen-activated protein kinase-independent pathway in B cells. J Biol Chem 272:18200–18208
Li X, Sandoval D, Freeberg L, Carter RH (1997b) Role of CD19 tyrosine 391 in synergistic activation of B lymphocytes by coligation of CD19 and membrane Ig. J Immunol 158:5649–5657
Liu J, Chiles TC, Sen R, Rothstein TL (1991) Inducible nuclear expression of NF-κB in primary B cells stimulated through the surface Ig receptor. J Immunol 146:1685–1691
Maltzman J, Monroe JG (1996a) A role for EGR1 in regulation of stimulus-dependent CD44 transcription in B lymphocytes. Mol Cell Biol 16:2283–2294
Maltzman J, Monroe JG (1996b) Transcriptional regulation of the ICAM-1 gene in antigen receptor and phorbol ester stimulated B lymphocytes: role for transcription factor EGR1. J Exp Med 183:1747–1759
Maly FE, Quilliam LA, Dorseuil O, Der CJ, Bokoch GM (1994) Activated or dominant inhibitory mutants of Rap1A decrease the oxidative burst of Epstein-Barr virus-transformed human B lymphocytes. J Biol Chem 269:18743–18746
Marais R, Light Y, Mason C, Paterson H, Olson MF, Marshall CJ (1998) Requirement of Ras-GTP-Raf complexes for activation of Raf1 by protein kinase C. Science 280:109–112
Marengere LEM, Mirtsos C, Kozieradzki I, Veillette A, Mak TW, Penninger JM (1997) Proto-oncoprotein Vav interacts with c-Cbl in activated thymocytes and peripheral T cells. J Immunol 159: 70–76
Marshall CJ (1996) Ras effectors. Curr Opin Cell Biol 8:197–204
Martin GA, Yatani A, Clark R, Conroy L, Polakis P, Brown AM, McCormick F (1992) GAP domains responsible for ras p21-dependent inhibition of muscarinic atrial $K^+$ channel currents. Science 255:192–194
Matsuda M, Tanaka S, Nagata S, Kojima A, Kurata T, Shibuya M (1992) Two species of human CRK cDNA encode proteins with distinct biological activities. Mol Cell Biol 12:3482–3489
Matsuda M, Hashimoto Y, Muroya K, Hasegawa H, Kurata T, Tanaka S, Nakamura S, Hattori S (1994) CRK protein binds to two guanine nucleotide-releasing proteins for the Ras family and modulates nerve growth factor-induced activation of Ras in PC12 cells. Molec Cell Biol 14:5495–5500
McLeod SJ, Ingham RJ, Bos JL, Kurosaki T, Gold MR (1998) Activation of the Rap1 GTPase by the B cell antigen receptor. J Biol Chem 273:29218–29223
McMahon SB, Monroe JG (1995) Activation of the $p21^{ras}$ pathway couples antigen receptor stimulation to induction of the primary response gene *egr*-1 in B lymphocytes. J Exp Med 181:417–422
Meier R, Alessi DR, Cron P, Andjelkovic M, Hemmings BA (1997) Mitogenic activation, phosphorylation, and nuclear translocation of protein kinase Bbeta. J Biol Chem 272:30491–30497
Melamed I, Franklin RA, Gelfand EW (1995) Microfilament assembly is required for anti-IgM-dependent MAPK and p90rsk activation in human B lymphocytes. Biochem Biophys Res Comm 209:1102–1110
Miau LH, Chang CJ, Shen BJ, Tsai WH, Lee SC (1998) Identification of heterogeneous nuclear ribonucleoprotein K (hnRNP K) as a repressor of C/EBP-beta-mediated gene activation. J Biol Chem 273:10784–10791
Michelotti EF, Michelotti GA, Aronsohn AI, Levens D (1996) Heterogeneous nuclear ribonucleoprotein K is a transcription factor. Mol Cell Biol 16:2350–2360
Minden A, Lin A, Claret F-X, Abo A, Karin M (1995) Selective activation of the JNK signaling cascade and c-Jun transcriptional activity by the small GTPases Rac and cdc42Hs. Cell 81:1147–1157
Mischak H, Kolch W, Goodnight J, Davidson WF, Rapp U, Rose-John S, Mushinski JF (1991) Expression of protein kinase C genes in hemopoietic cells is cell-type- and B cell-differentiation stage specific. J Immunol 147:3981–3987
Mittelstadt PR, DeFranco AL (1993) Induction of early response genes by cross-linking membrane Ig on B lymphocytes. J Immunol 150:4822–4832
Mizuguchi J, Yong-Yong J, Nakabayashi H, Huang K-P, Beaven MA, Chused T, Paul WE (1987) Protein kinase C activation blocks anti-IgM-mediated signaling in BAL17 B lymphoma cells. J Immunol 139:1054
Moarefi I, LaFevre-Bernt M, Sicheri F, Huse M, Lee C-H, Kuriyan J, Miller WT (1997) Activation of the Src-family tyrosine kinase Hck by SH3 domain displacement. Nature 385:650–653
Monroe JG (1988) Up-regulation of c-*fos* expression is a component of the mIg signal transduction mechanism but is not indicative of competence for proliferation. J Immunol 140:1454–1460
Monroe JG, Haldar S (1989) Involvement of a specific guanine nucleotide binding protein in receptor immunoglobulin stimulated inositol phospholipid hydrolysis. Biochim Biophys Acta 1013:273–278

Motto DS, Ross SE, Wu J, Hendricks-Taylor LR, Koretzky GA (1996) Implication of the GRB2-associated phosphoprotein SLP-76 in T cell receptor-mediated interleukin 2 production. J Exp Med 183:1937–1943

Muller G, Ayoub M, Storz P, Rennecke J, Fabbro D, Pfizenmaier (1995) PKC zeta is a molecular switch in signal transduction of TNF-$\alpha$, bifunctionally regulated by ceramide and arachidonic acid. EMBO J 14:1961

Nakanishi H, Brewer KA, Exton JH (1993) Activation of the $\zeta$ isozyme of protein kinase C by phosphatidylinositol 3,4,5-trisphosphate. J Biol Chem 268:13–16

Nel A, Wooten MW, Landreth GE, Goldschmidt-Clermont PJ, Stevenson HC, Miller PJ, Galbraith RM (1986) Translocation of phospholipid/$Ca^{2+}$-dependent protein kinase in B-lymphocytes acitvated by phorbol ester or cross-linking membrane immunoglobulin. Biochem J 233:145–149

Nimnual AS, Yatsula BA, Bar-Sagi D (1998) Coupling of Ras and Rac guanosine triphosphatases through the Ras exchange Sos. Science 279:560–563

Nishizuka Y (1995) Protein kinase C and lipid signaling for sustained cellular responses. FASEB J 9: 484–496

O'Keefe T, Williams GT, Davies SL, Neuberger MS (1996) Hyperresponsive B cells in CD22-deficient mice. Science 274:798–801

O'Rourke L, Tooze R, Fearon DT (1997) Co-receptors of B lymphocytes. Curr Opin Immunol 9:324–329

O'Rourke LM, Tooze R, Turner N, Sandoval DM, Carter RH, Tybulewicz VLJ, Fearon DT (1998) CD19 as a membrane-anchored adaptor protein of B lymphocytes: costimulation of lipid and protein kinases by recruitment of Vav. Immunity 8:635–645

Ohtsuka T, Shimizu K, Yamamori B, Kuroda S, Takai Y (1996) Activation of brain B-Raf protein kinase by Rap1B small GTP-binding protein. J Biol Chem 271:1258–1261

Panchamoorthy G, Fukazawa T, Miyake S, Soltoff S, Reedquist K, Druker B, Shoelson S, Cantley L, Band H (1996) $p120^{cbl}$ is a major substrate of tyrosine phosphorylation upon B cell antigen receptor stimulation and interacts in vivo with Fyn and Syk tyrosine kinases, Grb2 and Shc adaptors, and the p85 subunit of phosphatidylinositol 3-kinase. J Biol Chem 271:3187–3194

Philpott KL, McCarthy MJ, Klippel A, Rubin LL (1997) Activated phosphtidylinositol 3-kinase and Akt promote survival of superior cervical neurons. J Cell Biol 139:809–815

Pleiman CM, Hertz WM, Cambier JC (1994) Activation of phosphatidylinositol-3′ kinase by Src-family kinase SH3 binding to the p85 subunit. Science 263:1609–1612

Ponting CR, Benjamin DR (1996) A novel family of Ras-binding domains. Trends Biochem Sci 21:422–425

Prasad KVS, Kapeller R, Janssen O, Repke H, Duke-Cohan JS, Cantley LC, Rudd CE (1993) Phosphatidylinositol (PI) 3-kinase and PI 4-kinase binding to the CD4-$p56^{lck}$ complex: the $p56^{lck}$ SH3 domain binds to PI 3-kinase but not to PI 4-kinase. Mol Cell Biol 13:7708–7717

Pronk GJ, McGlade J, Pelicci G, Pawson T, Bos JL (1993) Insulin-induced phosphorylation of the 46- and 52-kDa Shc proteins. J Biol Chem 268:5748–5753

Rao A, Luo C, Hogan PG (1997) Transcription factors of the NFAT family: Regulation and function. Ann Rev Immunol 15:707–747

Reed JC (1997) Cytochrome c: Can't live with it – can't live without it. Cell 91:559–562

Ren XD, Bokoch GM, Traynor-Kaplan A, Jenkins GH, Anderson RA, Schwartz MA (1996) Physical association of the small GTPase Rho with a 68-kDa phosphatidylinositol-4-phosphate-5-kinase in Swiss 3T3 cells. Mol Biol Cell 7:435–442

Ridley AJ, Paterson HF, Johnston CL, Diekmann D, Hall A (1992) The small GTP-binding protein rac regulates growth factor-induced membrane ruffling. Cell 70:401–410

Rigley KP, Harnett MM, Klaus GGB (1989) Analysis of signaling via surface immunoglobulin receptors on B cells from CBA/N mice. Eur J Immunol 19:2081–2086

Robinson MJ, Cobb MH (1997) Mitogen-activated protein kinase pathways. Curr Opin Cell Biol 9:180–186

Rodriguez-Viciana P, Warne PH, Dhand R, Vanhaesebroeck B, Gout I, Fry MJ, Waterfield MD, Downward J (1994) Phosphatidylinositol-3-OH kinase as a direct target of Ras. Nature 370:527–532

Romero F, Dargemont C, Pozo F, Reeves WH, Camonis J, Gisselbrecht S, Fischer S (1996) $p95^{vav}$ associates with the nuclear protein Ku-70. Mol Cell Biol 16:37–44

Rordorf-Nikolic T, Van Horn DJ, Chen D, White MF, Backer JM (1995) Regulation of phosphatidylinositol 3′-kinase by tyrosyl phosphoproteins. Full activation requires occupancy of both SH2 domains in the 85-kDa regulatory subunit. J Biol Chem 270:3662–3666

Rozakis-Adcock M, Fernley R, Wade J, Pawson T, Bowtell D (1993) The SH2 and SH3 domains of mammalian Grb2 couple the EGF receptor to the Ras activator mSOS1. Nature 363:83–85

Rubinfeld B, Munemitsu S, Clark R, Conroy L, Crosier WJ, McCormick F, Polakis P (1991) Molecular cloning of a GTPase activating protein for the Krev-1 protein p21rap1. Cell 65:1033–1042
Salim K, Bottomley MJ, Querfurth E, Zvelebil MJ, Gout I, Scaife R, Smith CIE, Driscoll PC, Waterfield MD, Panayotou G (1996) Distinct specificity in the recognition of phosphoinositides by the pleckstrin homology domains of dynamin and Bruton's tyrosine kinase. EMBO J 15:6241–6250
Salmon RA, Foltz IN, Young PR, Schrader JW (1997) The p38 mitogen-activated protein kinase is activated by ligation of the T or B lymphocyte antigen receptors, Ras, or CD40, but suppression of kinase activity does not inhibit apoptosis induced by antigen receptors. J Immunol 159:5309–5317
Sata S, Jansen PJ, Tedder TF (1997) CD19 and CD22 expression reciprocally regulates tyrosine phosphorylation of Vav protein during B lymphocyte signaling. Proc Natl Acad Sci USA 94:13158–13162
Saxton TM, van Oostveen I, Bowtell D, Aebersold R, Gold MR (1994) B cell antigen receptor cross-linking induces phosphorylation of the Ras activators SHC and mSOS1 as well as assembly of complexes containing SHC, GRB-2, mSOS1, and a 145-kDa tyrosine-phosphorylated protein. J Immunol 153:623–636
Scheid MP, Lauener RW, Duronio V (1995) Role of phosphatidylinositol 3-OH-kinase activity in the inhibition of apoptosis in hemopoietic cells: phosphatidylinositol 3-OH-kinase inhibitors reveal a difference in signaling between interleukin-3 and granulocyte-macrophage colony stimulating factor. Biochem J 312:159–162
Schuebel KE, Bustelo XR, Nielsen DA, Song BJ, Barbacid M, Goldman D, Lee IJ (1996) Isolation and characterization of murine vav-2, a member of the vav family of proto-oncogenes. Oncogene 13:363–371
Schumacher C, Knudsen BS, Ohuchi T, Di Fiore PP, Glassman RH, Hanafusa H (1995) The SH3 domain of Crk binds specifically to a conserved proline-rich motif in Eps15 and Eps15R. J Biol Chem 270:15341–15347
Serunian LA, Haber MT, Fukui T, Kim JW, Rhee SG, Lowenstein JM, Cantley LC (1989) Polyphosphoinositides produced by phosphatidylinositol 3-kinase are poor substrates for phospholipase C from rat liver and bovine brain. J Biol Chem 264:17809–17815
Seyfert VL, Sukhatme VP, Monroe JG (1989) Differential expression of a zinc finger-encoding gene in reponse to positive versus negative signaling through receptor immunoglobulin in murine B lympohcytes. Mol Cell Biol 9:2083–2088
Seyfert VL, McMahon S, Glenn W, Cao X, Sukhatme VP, Monroe JG (1990) *Egr-1* expression in surface Ig-mediated B cell activation: kinetics and association with protein kinase C activation. J Immunol 145:3647–3653
Shepherd PR, Reaves BJ, Davidson HW (1996) Phosphoinositide 3-kinases and membrane traffic. Trends in Cell Biology 6:92–97
Sicheri F, Moarefi I, Kuriyan J (1997) Crystal structure of the Src family tyrosine kinase Hck. Nature 385:602–609
Sidorenko SP, Law C-L, Klaus SJ, Chandran KA, Takata M, Kurosaki T, Clark EA (1996) Protein kinase C μ (PKCμ) associates with the B cell antigen receptor complex and regulates lymphocyte signaling. Immunity 5:353–363
Sillman AL, Monroe JG (1995) Association of $p72^{syk}$ with the src Homology-2 (SH2) domains of PLCγ1 in B lymphocytes. J Biol Chem 270:11806–11811
Simon J (1995) Locking in stable states of gene expression: transcriptional control during Drosophila development. Curr Opin Cell Biol 7:376–385
Siranian MI, Marchetti A, Di Rocco G, Starace G, Jucker R, Nasi S (1993) Ras oncogene transformation of human B lymphoblasts is associated with lymphocyte activation and with a block of differentiation. Oncogene 8:157–163
Smit L, de Vries-Smits MM, Bos JL, Borst J (1994) B cell antigen receptor stimulation induces formation of a Shc-Grb2 complex containing multiple tyrosine-phosphorylated proteins. J Biol Chem 269:20209–20212
Smit L, van der Horst G, Borst J (1996a) Sos, Vav, and C3G participate in B cell receptor-induced signaling pathways and differentially associate with Shc-Grb2, Crk, and Crk-L adaptors. J Biol Chem 271:8654–8569
Smit L, van der Horst G, Borst J (1996b) Formation of Shc/Grb2- and Crk adaptor complexes containing tyrosine phosphorylated Cbl upon stimulation of the B-cell antigen receptor. Oncogene 13:381–389
Songyang Z, Shoelson SE, McGlade J, Olivier P, Pawson T, Bustelo XR, Barbacid M, Sabe H, Hanafusa H, Yi T, Ren R, Baltimore D, Ratnofsky S, Feldman RA, Cantley LC (1994) Specific motifs recognized by the SH2 domains of Csk, 3BP2, fps/fes, GRB-2, HCP, SHC, Syk, and Vav. Mol Cell Biol 14:2777–2785

Stephens L, Jackson TR, Hawkins PT (1993) Activation of phosphatidylinositol 4,5-bisphosphate supply by agonists and non-hydrolysable GTP analogues. Biochem J 296:481–488
Su B, Karin M (1996) Mitogen-activated protein kinase cascades and regulation of gene expression. Curr Opin Immunol 8:402–411
Sutherland CL, Heath AW, Pelech SL, Young PR, Gold MR (1996) Differential activation of the ERK, JNK, and p38 mitogen-activated protein kinases by CD40 and the B cell antigen receptor. J Immunol 157:3381–3390
Suzuki Y, Demoliere C, Kitamura D, Takeshita H, Deuschle U, Watanabe T (1997) HAX-1, a novel intracellular protein localized on mitochondria, directly associates with HS1, a substrate of Src family tyrosine kinases. J Immunol 158:2736–2744
Takata M, Kurosaki T (1996) A role for Bruton's tyrosine kinase in B cell antigen receptor-mediated activation of phospholipase C-γ2. J Exp Med 184:31–40
Tan Y, Rouse J, Zhang A, Cariati S, Cohen P, Comb MJ (1996) FGF and stress regulate CREB and ATF-1 via a pathway involving p38 MAP kinase and MAPKAP kinase-2. EMBO J 15:4629–4642
Tanaka S, Morishita T, Hashimoto Y, Hattori S, Nakamura S, Shibuya M, Matuoka K, Takenawa T, Kurata T, Nagashima K, Matsuda M (1994) C3G, a guanine nucleotide-releasing protein expressed ubiquitously, binds to the Src homology 3 domains of Crk and GRB2/ASH proteins. Proc Natl Acad Sci USA 91:3443–3447
Tanaka S, Ouchi T, Hanafusa H (1997) Downstream of Crk adaptor signaling pathway: Activation of Jun kinase by v-Crk through the guanine nucleotide exchange protein C3G. Proc Natl Acad Sci USA 94:2356–2361
Tanaka S, Hanafusa H (1998) Guanine-nucleotide exchange protein C3G activates JNK1 by a Ras-independent mechanism. J Biol Chem 273:1281–1284
Taniuchi I, Kitamura D, Maekawa Y, Fukuda T, Kishi H, Watanabe T (1995) Antigen-receptor induced clonal expansion and deletion of lymphocytes are imparied in mice lacking HS1 protein, a substrate of the antigen-receptor-coupled tyrosine kinases. EMBO J 14:3664–3678
Tarakhovsky A, Turner M, Schaal S, Mee PJ, Duddy LP, Rajewsky K, Tybulewicz VLJ (1995) Defective antigen receptor-mediated proliferation of B and T cells in the absence of Vav. Nature 374:467–470
ten Hoeve J, Morris C, Heisterkamp N, Groffen J (1993) Isolation and chromosomal localization of CRKL, a human crk-like gene. Oncogene 8:2469–2474
Tilzey JF, Chiles TC, Rothstein TL (1991) *Jun-B* gene expression mediated by the surface immunoglobulin receptor of primary B lymphocytes. Biochem Biophys Res Commun 175:77–83
Toker A, Meyer M, Reddy KK, Falck JR, Aneja R, Aneja S, Parra A, Burns DJ, Ballas LM, Cantley LC (1994) Activation of protein kinase C family members by the novel polyphosphoinositides PtdIns-3,4-$P_2$ and PtdIns-3,4,5-$P_3$. J Biol Chem 269:32358–32367
Toker A, Cantley LC (1997) Signalling through the lipid products of phosphoinositde-3-OH kinase. Nature 387:673–676
Tolias KF, Cantley LC, Carpenter CL (1995) Rho family GTPases bind to phosphoinositide kinases. J Biol Chem 270:17656–17659
Tordai A, Franklin RF, Patel H, Gardner AM, Johnson GL, Gelfand EW (1994) Cross-linking of surface IgM stimulates the Ras/Raf-1/MEK/MAPK cascade in human B lymphocytes. J Biol Chem 269:7538–7543
Treisman R (1996) Regulation of transcription by MAP kinase cascades. Curr Opin Cell Biol 8:205–215
Tridandapani S, Kelley T, Pradhan M, Cooney D, Justement LB, Coggeshall KM (1997) Recruitment and phosphorylation of SH2-containing inositol phosphatase and Shc to the B-cell Fcγ immunoreceptor tyrosine-based inhibition motif peptide motif. Mol Cell Biol 17:4305–4311
Turner M, Mee PJ, Walters AE, Quinn ME, Mellor AL, Zamoyska R, Tybulewicz VLJ (1997) A requirement for the Rho-family GTP exchange factor Vav in positive and negative selection of thymocytes. Immunity 7:451–460
Tuveson DA, Carter RH, Soltoff SP, Fearon DT (1993) CD19 of B cells as a surrogate kinase insert region to bind phosphatidylinositol 3-kinase. Science 260:986–989
Urano T, Emkey R, Feig LA (1996) Ral-GTPases mediate a distinct downstream signaling pathway from Ras that facilitates cellular transformation. EMBO J 15:810–816
van der Lugt NMT, Domen J, Linders K, van Roon M, Robanus-Maandag E, te Riele H, van der Walk M, Deschamps J, Sofroniew M, van Lohuizen M, Berns A (1994) Posterior transformation, neurological abnormalities, and severe hematopoietic defects in mice with a targeted deletion of the bmi-1 proto-oncogene. Genes Dev 8:757–769
van Lohuizen M, Verbeek S, Scheijen B, Wientjens E, van der Gulden H, Berns A (1991) Identification of cooperating oncogenes in Eμ-myc transgenic mice by provirus tagging. Cell 65:737–752

van Noesel CJM, Lankester AC, van Schijndel GMW, van Lier RAW (1993) The CR2/CD19 complex on human B cells contains the *src*-family kinase *Lyn*. Int Immunol 5:699–705

Vanhaesbroeck B, Welham MJ, Kotani K, Stein R, Warne PH, Zvelebil MJ, Higashi H, Volinia S, Downward J, Waterfield MD (1997) p110δ, a novel phosphoinositide 3-kinase in leukocytes. Proc Natl Acad Sci USA 94:4330–4335

Vavvas D, Li X, Avruch J, Zhang X-F (1998) Identification of Nore1 as a potential Ras effector. J Biol Chem 273:5439–5442

Vossler MR, Yao H, York RD, Pan MG, Rim CS, Stork PJS (1997) cAMP Activates MAP kinase and Elk-1 through a B-Raf- and Rap1-Dependent Pathway. Cell 89:73–82

Wang XZ, Ron D (1996) Stress-induced phosphorylation and activation of the transcription factor CHOP (GADD153) by p38 MAP kinase. Science 272:1347–1349

Waters SB, Holt KH, Ross SE, Syu L-J, Guan K-L, Saltiel AR, Koretzky GA, Pessin JE (1995) Desensitization of Ras activation by a feedback dissociation of the SOS-Grb2 complex. J Biol Chem 270:20883–20886

Waters SB, Pessin JE (1996) Insulin receptor substrate 1 and 2 (IRS1 and IRS2): what a tangled web we weave. Trends in Cell Biology 6:1–4

Welch H, Eguinoa A, Stephens LR, Hawkins PT (1998) PKB and Rac are activated in parallel within a PI3-kinase controlled signalling pathway. J Biol Chem 273:11248–11256

Weng WK, Jarvis L, LeBien TW (1994) Signaling through CD19 activates Vav/mitogen-activated protein kinase pathway and induces the formation of a CD19/Vav/phosphatidylinositol 3-kinase complex in human B cell precursors. J Biol Chem 269:32514–32521

Whitmarsh AJ, Yang S-H, Su MS-S, Sharrocks AD, Davis RJ (1997) Role of p38 and JNK mitogen-activated protein kinases in the activation of ternary complex factors. Mol Cell Biol 17:2360–2371

Wienecke R, Konig A, DeClue JE (1995) Identification of tuberin, the tuberous sclerosis-2 product. Tuberin possesses specific Rap1GAP activity. J Biol Chem 270:16409–16414

Wolthius RMF, Bauer B, van't Veer LJ, de Vries-Smit AMM, Cool RH, Spaargaren M, Wittinghofer A, Burgering BMT, Bos JL (1996) RalGDS-like factor (Rlf) is a novel Ras and Rap1A-associating protein. Oncogene 13:353–362

Woods D, Parry D, Cherwinski H, Bosch E, Lees E, McMahon M (1997) Raf-induced proliferation or cell cycle arrest is determined by the level of Raf activity with arrest mediated by $p21^{Cip1}$. Mol Cell Biol 17:5598–5611

Xie H, Rothstein TL (1995) Protein kinase C mediates activation of nuclear cAMP response element-binding protein (CREB) in B lymphocytes stimulated through surface Ig. J Immunol 154:1717–1723

Xing J, Ginty DD, Greenberg ME (1996) Coupling of the Ras-MAPK pathway to gene activation by rsk2, a growth factor-regulated CREB kinase. Science 273:959–963

Yamanashi Y, Okada M, Semba T, Yamori T, Umemori H, Tsunasawa S, Toyoshima K, Kitamura D, Watanabe T, Yamamoto T (1993) Identification of HS1 protein as a major substrate of protein-tyrosine kinase(s) upon B-cell antigen receptor-mediated signaling. Proc Natl Acad Sci USA 90:3631–3635

Yamanashi Y, Fukuda T, Nishizumi H, Inazu T, Higashi K-I, Kitamura D, Ishida T, Yamamura H, Watanabe T, Yamamoto T (1997) Role of tyrosine phosphorylation of HS1 in B cell antigen receptor-mediated apoptosis. J Exp Med 185:1387–1392

Yamanishi Y, Fukui Y, Wongsasant B, Kinoshita Y, Ichimori Y, Toyoshima K, Yamamoto T (1992) Activation of Src-like protien-tyrosine kinase Lyn and its association with phosphatidylinositol 3-kinase upon B-cell antigen receptor-mediated signaling. Proc Natl Acad Sci USA 89:1118–1122

Yao R, Cooper GM (1995) Requirement for phosphatidylinositol-3 kinase in the prevention of apoptosis by nerve growth factor. Science 267:2003–2006

York RD, Yao H, Dillon T, Ellig CL, Eckert SP, McCleskey EW, Stork PJ (1998) Rap1 mediates sustained MAP kinase activation induced by nerve growth factor. Nature 392:622–626

Zha J, Harada H, Yang E, Jockel J, Korsmeyer SJ (1996) Serine phosphorylation of death agonist BAD in response to survival factor results in binding to 14–3-3 not BCL-X. Cell 87:619–628

Zhang R, Alt FW, Davidson L, Orkin SH, Swat W (1995) Defective signalling through the T- and B-cell antigen receptors in lymphoid cells lacking the *vav* proto-oncogene product. Nature 374:470–473

Zhang W, Sloan-Lancaster J, Kitchen J, Trible RP, Samelson LE (1998) LAT: The ZAP-70 tyrosine kinase substrate that links T cell receptor to cellular activation. Cell 92:83–92

Zmuidzinas A, Fischer K, Lira SA, Forrester L, Bryant S, Bernstein A, Barbacid M (1995) The vav proto-oncogene is required early in embryogenesis but not for hematopoietic development in vitro. EMBO J 14:1–11

# Involvement of the Lymphocyte Cytoskeleton in Antigen-Receptor Signaling

L.A.G. da Cruz, S. Penfold, J. Zhang, A.-K. Somani, F. Shi, M.K.H. McGavin, X. Song, and K.A. Siminovitch



## 1 Introduction

Engagement of antigen receptors on lymphocytes induces a complex cascade of intracellular molecular interactions, which provides the circuitry for coupling stimulatory signals to alterations in gene expression and ultimately cell behavior. This circuitry has been intensively investigated and the results of such studies have elucidated not only the specific components of antigen-receptor-evoked signaling cascades, but also many mechanistic paradigms relevant to the signaling field as a whole. Thus, for example, the pivotal role for tyrosine phos-

Department of Medicine, University of Toronto and The Samuel Lunenfeld Research Institute, Mount Sinai Hospital, Toronto, Ontario, Canada and Department of Immunology, and Department of Molecular and Medical Genetics
e-mail: ksimin@mshri.on.ca

phorylation-based signaling interactions in initiating intracellular transmission of activation signals was highlighted by the characterization of immunoreceptor tyrosine-based activation motifs (ITAMs) within antigen-receptor components and the demonstration that Src-family-mediated phosphorylation of these motifs allows for binding and activation of ZAP-70/Syk kinases, the consequent recruitment/activation of phosphatidylinositol-3 kinase and phospholipase C (PLC)-γ, and the induction of a myriad of secondary signaling cascades (Birkeland and Monroe 1997; DeFranco 1997; Qian and Weiss 1997). Similarly, the concept of signaling thresholds for expression of given cellular responses emanated from the demonstration that signaling effectors, such as accessory receptors and cytosolic tyrosine phosphatases, can modify the outcome of antigen-receptor engagement by quantitatively modulating antigen-receptor signaling (Goodnow 1996). From these and the myriad of other insights culled from studies of antigen-receptor signaling has emerged a wealth of information on the biochemical networks that enable antigen stimulation to be translated to such diverse outcomes as proliferation, cytokine/antibody production, differentiation, death or transformation.

In contrast to the signaling events linking antigen-receptor stimulation to the nucleus and gene transcription, the biochemical and structural interactions that couple stimulation by antigen to lymphocyte cytoskeletal rearrangement are relatively undefined. The cytoskeletal apparatus has, however, been extensively studied in other cell systems and, in these latter contexts, has been shown to encompass a diverse network of cellular filaments organized as actin microfilaments, microtubules or intermediate filaments, to also include a complex array of interacting proteins, and to be integrally involved in relay of extracellular stimulatory signals to the nucleus (Bray 1995; Fowler and Vale 1996). These cumulative data indicate that the cytoskeletal apparatus imbues the cell not only with structure, but also with dynamics and the capacity to compartmentalize and redistribute proteins so as to modulate many cellular functions, including signal delivery. Thus, cytoskeletal rearrangement represents not only a response to extracellular stimuli, but also a cellular modification likely required for induction of many of the molecular interactions involved in propagating stimulatory signals through the cell.

With respect to antigen-receptor signaling, this paradigm has become increasingly appreciated consequent to the discovery that antigen-receptor-evoked signaling and cytoskeletal rearrangement are both impaired in mice genetically ablated for expression of the signaling effectors, Vav or Wiskott-Aldrich syndrome (WAS) protein. As detailed in the following pages, data garned from the analysis of these animals and from the increasing body of literature concerning both B- and T-cell cytoskeletal function provide compelling evidence of biochemical and functional liaisons between signaling effectors and cytoskeletal elements. It is now clear that the characterization of these liaisons will pave the way to understanding the mechanisms whereby lymphocyte activation and cytoarchitectural modification are co-ordinated so as to engender an appropriate immune response.

# 2 Participation of the Cytoskeletal Apparatus in Antigen-Receptor Signaling

## 2.1 Cytoskeletal Involvement in B-Cell Antigen Receptor Signaling

The potential importance of cytoskeletal rearrangement in modulating the biological sequelae of antigen-receptor engagement was initially revealed in B cells by the demonstration that cytoskeleton-disrupting agents, such as cytochalasin B, interfere with membrane immunoglobulin (mIg) internalization following antibody-mediated mIg cross-linking (LOOR et al. 1972; UNANUE et al. 1973; ROSENTHAL et al. 1973). Similarly, immunofluorescent and ultrastructural data revealing B-cell antigen receptor (BCR) engagement to be rapidly followed by receptor patching and capping (KERBEL et al. 1975) and the detection of cytoskeletal components, such as actin, myosin and tubulin in these caps (GABBIANI et al. 1977; BOURGUIGNON et al. 1978), provided initial evidence of an integral role for cytoskeletal elements in modulating antigen-receptor function. This hypothesis has since been supported by many lines of evidence. These include, for example, data revealing antigen-receptor capping and coated vesicle formation following mIg internalization to be abrogated by dihydrocytochalasin B, a drug which disrupts actin microfilament arrangement, as well as the demonstration that anti-Ig antibody treatment induces mIg association with the detergent-insoluble cell fraction (BRAUN et al. 1982).

These observations suggest that actin polymerization, i.e., the conversion of globular (G)-actin to filamentous (F)-actin, is induced by BCR ligation and is required for antigen-receptor endocytosis. While a link between mIg cross-linking and actin polymerization has not been detected in all such studies (JACKMAN and BURRIDGE 1989; WILDER and ASHMAN 1991), the cumulative data, obtained either by direct quantitation of F-actin (MELAMED et al. 1991a,b, 1995) or by the demonstration of cytochalasin-mediated inhibition of BCR capping post-receptor ligation (ROBERTS and LAVIA 1976; SCHREINER and UNANUE 1976; TETI et al. 1981; ELSON et al. 1984; SHOSAN et al. 1990), suggest that actin polymerization and F-actin microfilament rearrangement are rapidly evoked in response to BCR engagement.

In addition to induction of altered cell morphology, BCR-triggered cytoskeletal rearrangement also appears critical to the downstream transduction of activation and differentiation signals. Thus, for example, pre-treatment of B cells with cytochalasin B or, alternatively, with inhibitors of actin polymerization, such as cytochalasin D or botulism toxin, has been shown to interfere with B-cell ontogeny (HAMANO et al. 1984) and to inhibit induction of proliferation following BCR ligation (KISHIMOTO et al. 1975; MOOKERJEE et al. 1981; MOOKERJEE and JUNG 1982; MELAMED et al. 1991b). Although an inhibitory effect of cytochalasin B on BCR-evoked proliferation has not been uniformly detected (ROTHSTEIN 1985; BUCKLER et al. 1988), most of the available data pertaining to this issue support the contention of a role for cytoskeletal rearrangement in modulating the translation of

BCR engagement to cellular response. The structural basis for this link between the cellular cytoarchitecture and BCR signaling is not fully understood, but the BCR-cytoskeletal association triggered by BCR ligation is cytochalasin-D sensitive and colchicine resistant and, therefore, appears to be mediated through cortical actin microfilament and not microtubular interactions with the BCR (SALISBURY et al. 1988).

## 2.2 Cytoskeletal Involvement in T-Cell Signaling

In view of the many parallels between BCR and TCR-triggered signaling cascades, it is not surprising that T lymphocytes also undergo substantive cytoskeletal rearrangement in response to engagement of their antigen receptors. Thus, for example, in early studies of T-cell cytoarchitecture, anti-CD3 antibody-mediated cross linking of the T-cell antigen receptors (TCR) on Jurkat cells was shown to induce not only tyrosine phosphorylation, interleukin-2 (IL-2) production and growth arrest, but also the formation of a dense, actin-rich collar at the site of antibody-receptor attachment and the subsequent formation of F-actin rich pseudopods (PARSEY and LEWIS 1993). The pivotal role for cytoskeletal change in the delivery of activation signals to the nucleus was revealed in these studies by the demonstration that cytochalasin D inhibited induction of both actin polymerization and tyrosine kinase activation/growth arrest.

More recent studies of T-cell cytoarchitecture have revealed migrating T cells to manifest inherent polarity, showing both a leading edge which can ultimately interact with an antigen presenting cell (APC) and a distal uropod (NEGULESCU et al. 1996). Interaction of the T cell with an APC and, by extension, with the appropriate peptide–major histocompatability complex (MHC) complex, engenders formation of a tight interface between the two cells and polarization of the T-cell cytoskeleton toward the APC, such that actin and talin accumulate at the T-cell/APC junction (GEIGER et al. 1982; RYSER et al. 1982; KUPFER et al. 1986, 1987, 1994) and cell-surface receptors organize into an outer ring containing lymphocyte-associated antigen (LFA)-1 and an inner ring containing the TCR (MONKS et al. 1998). Polarization of the T cell toward the T-cell/APC interface is thought to serve a number of functions relevant to T-cell activation. The accumulation of F-actin and talin at this cellular interface may, for example, stabilize T-cell/APC interaction by prolonging TCR occupancy and increasing the number of intercellular contacts (VALITUTTI et al. 1995). Reorientation of the T-cell microtubule-organizing centre (MTOC) may also rearrange intracellular secretory organelles so as to favor directed secretion of cytokines and cytotoxic mediators toward the site of stimulation (KUPFER and DENNERT 1984; POO et al. 1988; KUPFER et al. 1991, 1994).

Until recently, the molecular events regulating TCR-evoked cytoskeletal reorientation have remained obscure. However, these events are beginning to be elucidated, as illustrated for example by recent data revealing actin/talin polarization and MTOC reorientation in an antigen-specific T-cell clone to be mediated by distinct mechanisms, the former, but not the latter process, requiring integrin-mediated adhesion (SEDWICK et al. 1998). Understanding of T-cell polarization at the molecular level has also been enhanced by the demonstration that TCR-triggered interaction between the CD2 adhesion protein and a novel src homology (SH)3-domain-containing CD2-associated protein plays a key role in the reorientation process, apparently providing a scaffold for CD2 clustering and T-cell cytoskeletal polarization (DUSTIN et al. 1998). These findings are consistent with previous data implicating adhesion molecules in TCR-evoked T-cell polarization (SHAW and DUSTIN 1997) and with the contention of a pivotal role for cytoskeletal rearrangement in T-cell activation.

In addition to actin and talin, several other cytoskeletal proteins have also been shown to redistribute within the T cell following antigen-receptor ligation, and the studies of these molecular rearrangements have provided further evidence of the functional connection between cytoskeletal modifications and signal transmission to the nucleus. For example, spectrin, a protein previously localized within the erythrocyte membrane cytoskeleton, aggregates in the cytosol of resting T cells, but translocates to the membrane cytoskeleton following TCR engagement, a response which is not mimicked using stimulatory maneuvers that bypass the TCR (LEE et al. 1988). Spectrin is thought to maintain "fluidity" of the phospholipid membrane (DEL BUONO et al. 1988) and, accordingly, its redistribution to the cortical cytoskeleton in response to TCR ligation suggests a role for spectrin in modulating membrane fluidity so as to expedite receptor capping and clustering.

TCR-evoked cytoskeletal changes, such as F-actin rearrangement and pseudopod formation, have also been shown to be accompanied by translocation of not only talin, but, in addition, the actin-binding proteins vinculin and α-actinin, to one pole of the cell (SELLIAH et al. 1996). As for talin, the latter two proteins are thought to subserve a linker function, providing the structural means to connect the actin cytoskeleton to integrins within the plasma membrane (HORWITZ et al. 1986; BELKIN and KOTELIANSKY 1987; WACHSSTOCK et al. 1987; JOHNSON and CRAIG 1995). The involvement of α-actinin in TCR signaling per se has also been revealed by data indicating that phospholipid-mediated activation of the calpain protease, a process which decreases the concentration of intracellular calcium required for T-cell activation, induces proteolytic cleavage of α-actinin in a TCR-dependent fashion (SAIDO et al. 1992). These as well as other data concerning calpain functions (ROCK et al. 1997) imply that calpain-mediated α-actinin cleavage modulates the threshold required for both TCR signaling and cytoskeletal rearrangement, a postulate supported by the finding that both TCR-triggered proliferation and pseudopod formation are diminished in the context of inhibiting calpain activation (SAIDO et al. 1992).

In view of the data summarized above, it is clear that the mechanisms coupling antigen-receptor engagement to both cytoskeletal change and nuclear response are complex and likely to involve a very broad amalgam of structural rearrangements and biochemical events. While characterizing these molecular interactions will be challenging, particularly in view of the difficulties inherent to studying the lymphocyte cytoskeleton, definition of the mechanisms whereby co-ordination between antigen-receptor signals and the cytoskeleton is achieved will significantly enhance understanding of the fundamental processes whereby receptor stimulation translates to changes not only in cell function but also cell dynamics.

# 3 Interactions of the Cytoskeleton with Antigen-Receptor Components

## 3.1 Cytoskeletal Links with the BCR

Translocation of selected receptor components to the cytoskeletal matrix represents one of the earliest structural rearrangements induced by antigen receptor engagement. Thus, for example, ligation of the BCR, a complex comprising mIg non-covalently associated with Igα/β heterodimers, has been shown to induce movement of a large fraction of mIg into the detergent insoluble cytoskeletal matrix (Braun et al. 1982; Woda and Woodin 1984). This phenomenon appears to occur independently of the receptor Igα/β components, as IgM mutants, which are expressed in the absence of or which fail to bind the Igα/β heterodimer, still undergo capping, internalization and cytoskeletal translocation following receptor ligation (Grupp et al. 1993; Sanchez et al. 1993; Williams et al. 1994; Hartwig et al. 1995a). By contrast, cell lines expressing these non-Igα/β-interacting mIg mutants are defective in antigen presentation, a finding which suggests that Igα/β, while not required for mIg endocytosis, is needed for directing mIg-associated antigen to appropriate endosomal compartments in the cell (Bonnerot et al. 1995; Briken et al. 1997). In any case, the capacity for mIg to associate with the cytoskeleton independently of Igα/β is also consistent with data indicating the mIg cytoskeletal translocation induced by receptor cross-linking to occur prior to and, thus, independently of the induction of protein tyrosine kinase (PTK) activation and tyrosine phosphorylation (Jugloff and Jongstra-Bilen 1997).

At present, the structural basis for mIg interactions with cytoskeletal components is not entirely defined, but appears to involve only the mIg heavy chain and most likely depends on residues comprising the mIg intracellular C-terminal segment (KVK) as well as a portion of the transmembrane domain (Hartwig et al. 1995a; Park and Jonstra-Bilen 1997). Importantly, while the Igα/β chains are not required for mIg-cytoskeletal association, these chains can also associate with the membrane cytoskeleton and the possibility that they contribute to BCR-triggered cytoskeletal modifications thus requires further investigation.

## 3.2 Cytoskeletal Links with the TCR

Activation-induced cytoskeletal translocation of antigen-receptor components has also been demonstrated in T lymphocytes. However, in contrast to B cells, in T lymphocytes it is the signaling rather than the antigen-binding elements of the antigen receptor which associate with the cytoskeleton. These signal-transducing elements include the $\gamma$, $\delta$ and $\varepsilon$ subunits of CD3 as well as TCR$\zeta$ chains, the latter of which have been shown to associate with the cytoskeleton to a minor extent during resting, but to a much more substantive extent in TCR-activated T cells (Caplan et al. 1995; Caplan and Baniyash 1996; Rozdzial et al. 1995). Similarly, other TCR subunits, most notably CD3$\varepsilon$, also appear to translocate to the cytoskeleton following TCR ligation (Rozdzial et al. 1995). Data from one study of this phenomenon suggest that cytoskeletal-sequestered TCR$\zeta$ chains do not undergo tyrosine phosphorylation in response to TCR ligation and, thus, represent a fraction of TCR$\zeta$ chains distinct from the soluble TCR$\zeta$ fraction (Caplan and Baniyash 1996). By contrast, results of two other independent studies have indicated the cytoskeletal-associated TCR$\zeta$ chains in TCR-stimulated cells to be not only tyrosine phosphorylated (Rozdzial et al. 1995), but to require Src family PTK-mediated tyrosine phosphorylation in order to translocate to the cytoskeleton (Montixi et al. 1998).

Association of TCR$\zeta$ with the cytoskeleton has also been shown to be disrupted by cytochalasin B or D and to therefore be mediated through actin microfilaments, to require at least one intact ITAM (Rozdzial et al. 1995), and to occur coincident with the cytoskeletal translocation of other tyrosine phosphorylated proteins following TCR ligation (Montixi et al. 1998). In addition, TCR$\zeta$ ITAM mutations, which disrupt the protein's association with the cytoskeleton, have been shown to be associated with a reduction in TCR-elicited IL-2 production, despite the induction of a normal pattern of protein tyrosine phosphorylation. Thus, while the full physiological significance of TCR subunit cytoskeletal recruitment following receptor ligation remains unclear, the available data suggest this translocation process to be not only induced by TCR engagement but, in fact, critical to expression of the downstream biochemical sequelae of TCR signaling, such as IL-2 production. This contention is further supported by recent data derived from analysis of non-obese diabetic mice, animals in which impairment in TCR-evoked T-cell proliferation and IL-2 production has been found to be associated with a reduction in TCR-triggered cytoskeletal translocation of TCR$\zeta$ as well as other signaling effectors (Salojin et al. 1997). Thus, while the mechanisms and significance of TCR component cytoskeletal recruitment following TCR ligation and the extent to which this phenomenon occurs in immature T cells require further investigation, the data derived from analysis of both BCR and TCR signaling suggest that cytoskeletal recruitment of selected antigen-receptor components is critical for the normal transduction of lymphocyte activation signals.

## 4 Cytoskeleton-Signaling Effector Interactions in Lymphocytes

Recognition of the integral relationship between cytoskeletal and signaling effector networks has led to an intensified effort to delineate the specific molecular interactions that connect the cytoskeleton with activation cascades. As described below and illustrated in Fig. 1, information on the cytoskeletal signaling effector inter-

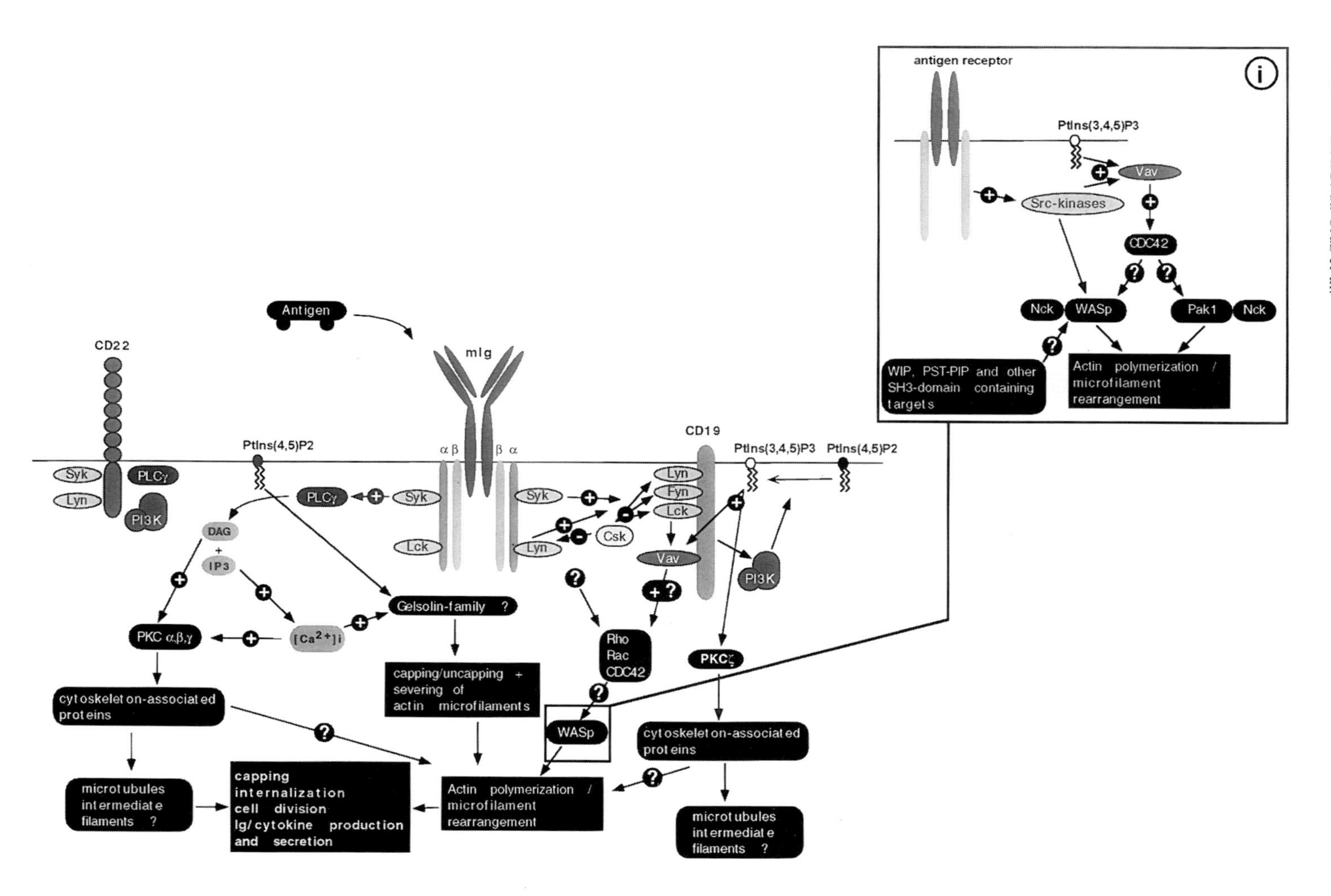

antigen receptor
PtIns(3,4,5)P3
Vav
Src-kinases
CDC42
Nck
WASp
Pak1
Nck
WIP, PST-PIP and other SH3-domain containing targets
Actin polymerization / microfilament rearrangement
i
Antigen
mIg
CD22
Syk
Lyn
PLCγ
PI3K
PtIns(4,5)P2
PLCγ
Syk
Lck
α β
β α
DAG
+
IP3
[Ca2+]i
PKC α,β,γ
Gelsolin-family ?
capping/uncapping + severing of actin microfilaments
cytoskeleton-associated proteins
microtubules intermediate filaments ?
capping internalization cell division Ig/cytokine production and secretion
Actin polymerization / microfilament rearrangement
Syk
Lyn
Csk
CD19
Lyn
Fyn
Lck
Vav
PI3K
PtIns(3,4,5)P3
PtIns(4,5)P2
Rho
Rac
CDC42
WASp
PKCζ
cytoskeleton-associated proteins
microtubules intermediate filaments ?

actions evoked by antigen-receptor engagement is rapidly increasing and facilitating definition of the mechanisms which allow antigen stimulation to co-ordinately modify both the structural and biological properties of B- and T-lymphocytes.

## 4.1 Tyrosine Phosphorylation Signaling Cascades and the Cytoskeleton

A critical role for tyrosine phosphorylation-based signaling events in antigen-receptor/cytoskeletal interactions is becoming increasingly well recognized. This relationship was identified initially in B cells by the finding that tyrosine phosphorylation of proteins associated with the detergent-insoluble membrane cytoskeleton was enhanced following mIg cross-linking (Nel et al. 1984). BCR-evoked tyrosine phosphorylation of cytoskeletal-associated proteins was subsequently demonstrated in several other studies (Hornbeck and Paul 1986; Rozenspire and Choi 1988) and shown to be required for capping (Pure and Tardelli 1992) and microfilament assembly (Melamed et al. 1991a).

More recent data, revealing BCR-triggered tyrosine phosphorylation of the related adhesion focal tyrosine kinase (RAFTK) to be decreased in cells pre-treated with cytochalasin B (Astier et al. 1997), raise the possibility that the cytoskeletal network not only contains tyrosine-phosphorylated proteins, but may, in some manner, modulate the capacity of antigen-receptor ligation to induce protein tyrosine phosphorylation. Thus, for example, antigen-receptor-evoked cytoskeletal translocation of selected PTKs might provide a mechanism for repositioning these enzymes so as to enhance their capacity to tyrosine phosphorylate receptor signaling elements and other cytosolic effectors involved in the initial stages of antigen-receptor signal delivery. This hypothesis, while currently speculative, is supported by data showing Lyn and Syk to be rapidly translocated to the cytoskeleton following BCR ligation (Jugloff and Jongstra-Bilen 1997) and indirectly by data linking the cytoskeletal translocation of Lyn and Syk to neutrophil adhesion and spreading (Yan et al. 1996, 1997) and, more recently, to fibronectin-triggered actin assembly in mast cells (Suzuki et al. 1998).

These findings as well as reports describing epidermal growth factor (EGF)-stimulated translocation of c-Src to the fibroblast membrane cytoskeleton (van der

**Fig. 1.** Mechanisms for linking antigen-receptor engagement to cytoskeletal rearrangement. B-cell antigen-receptor engagement results in the activation of protein tyrosine kinases (PTKs), such as Lyn and Syk, and the recruitment of secondary signaling effectors, such as Vav, phosphatidylinositol 3 kinase and phospholipase C-γ. Activation of these latter proteins, in turn, induces recruitment/activation of downstream effectors, including protein kinase C isoforms and the Rho-family GTPases, which induce actin polymerization and other cytoskeletal modifications. Such cytoskeletal rearrangements impact not only on cell structure, but also on transduction of signals for inducing proliferation, immunoglobulin (Ig)/cytokine production and secretion and other effector functions. The *insert* (designated as *i*) shows a model of WASp participation in antigen-receptor-induced cytoskeletal reorganization. WASp is believed to be a target for Src-family PTKs and to interact with activated cdc42, as well as WIP, Nck and other src homology (SH)3-domain containing proteins. The designations +, – and O indicate putative activating, inhibitory or uncharacterized effects, respectively, on downstream signaling effectors

Heydin 1997), TCR-triggered relocalization of Lck to the T-cell membrane cytoskeleton (Caplan and Baniyash 1996) and Lyn, Fyn, Fgr, Lck and Yes cytoskeletal translocation in activated platelets (Pestina et al. 1997), suggest that cytoskeletal localization creates a structural framework required for signaling effectors to co-ordinately modulate actin-microfilament dynamics and receptor-evoked activation cascades.

The role of PTK-cytoskeletal association in modulating cell cytoarchitecture is further evidenced by the profound cytostructural defects as well as cytoskeletal protein hyperphosphorylation, which accompany and appear to depend on enhanced Src, Lyn and Fyn activity in cells deficient for the Src family negative modulator, Csk (Nada et al. 1993; Thomas et al. 1995). Moreover, and of particular relevance to lymphocyte biology, TCR-driven cytoskeletal translocation of the TCRζ chain, an event believed to be key to the downstream relay of activation signals, has been shown to be dependent on the induction of Src family kinase activity (Montixi et al. 1998).

As for the Src-family PTKs, the signaling functions of ZAP-70, a Syk-related PTK implicated in both the early (tyrosine phosphorylation) and late [nuclear factor-activated T cells (NF-AT) activation and IL-2 synthesis] biochemical sequelae of TCR ligation (Qian et al. 1996), also appear to be mediated in a cytoskeletal context. ZAP-70's involvement in TCR signaling has been previously related to its capacity to bind phosphorylated TCRs, undergo Src-family PTK-mediated phosphorylation/activation and then phosphorylate effectors such as the LAT (linker for activation of T cells) membrane protein required for downstream propagation of the stimulatory signal (Wange et al. 1993; Iwashima et al. 1994; Chan et al. 1995; Wange et al. 1995; Zhang et al. 1998). However, recent data revealing the majority of phosphorylated ZAP-70 to be found in the membrane cytoskeleton of TCR-stimulated thymocytes (Montixi et al. 1998) suggest that the TCR signaling functions of ZAP-70 are, like those of Lck, realized at least in part in a cytoskeletal framework. Although the biochemical significance of ZAP-70 cytoskeletal translocation requires further investigation, ZAP-70 has been shown to play an integral role in the induction of specific cytoskeletal rearrangements that follow TCR ligation, antigen-receptor-evoked MTOC reorientation being inhibited by introduction of a dominant negative form of ZAP-70 (Lowin-Kropf et al. 1998). In this latter study, inhibition of ZAP-70 function was also found to block TCR-evoked activation of the NF-AT, while inhibition of mitogen-activated protein (MAP) kinase activation did not affect TCR-evoked MTOC reorientation or NF-AT activation. These latter data reinforce the complexity of the biochemical events linking antigen-receptor engagement to both cytoskeletal rearrangement and gene activation. However, taken together, the available data are consistent with the postulate of cytoskeletal involvement in antigen-receptor-evoked tyrosine phosphorylation-based signaling cascades and with the suggestion (Montixi et al. 1998) that cytoskeletal compartmentalization of selected PTKs creates a structural framework for promoting interactions between these enzymes and antigen-receptor components, thereby augmenting downstream propagation of activation signals. Parenthetically, while largely uninvestigated at the current time, a relationship between the cytoskeleton and serine/threonine kinase-mediated signaling events evoked by antigen-receptor engagement appears very likely, particularly in view of data showing protein kinase C (PKC) recruitment to the cytoskeleton following BCR ligation (CHEN et al. 1986) and demonstrating an apparent capacity of phorbol 12-myristate 13-acetate (PMA) to induce phosphorylation of some of the same cytoskeletal proteins phosphorylated consequent to mIg cross-linking (HORNBECK and PAUL 1986).

## 4.2 Associations Between PTKs and Specific Cytoskeletal Proteins

At present, the structural mechanisms whereby and sites at which signaling effectors associate with the lymphocyte cytoskeleton are largely undefined. These issues are particularly difficult to assess in lymphocytes in which the cytosolic portion of the cell is relatively small. However, as illustrated by the data cited earlier with respect to spectrin, talin and actin, there are data available, largely in relation to TCR signaling dynamics, that link antigen-receptor-evoked signal delivery to the interactions of specific cytoskeletal proteins with particular PTKs. These include, for example, recent observations concerning paxillin, a cytoskeletal protein which in fibroblasts associates with focal adhesion kinase (FAK) and localizes focal adhesions (TURNER et al. 1993). In T cells, anti-CD3 stimulation has been shown to induce paxillin tyrosine phosphorylation and its association with the SH2 domain of Lck (OSTERGAARD et al. 1998). Paxillin does not associate with FAK in this context, but has been shown to bind the closely related kinase, RAFTK/Pyk2 (GANJU et al. 1998; OSTERGAARD et al. 1998), a protein which also undergoes tyrosine phosphorylation in response to TCR stimulation and which then associates, at least in vitro, with the SH2 domains of Fyn and Grb2. Induction of RAFTK tyrosine phosphorylation is abrogated by cytochalasin D pretreatment prior to TCR stimulation, a result which again suggests that rearrangement of cytoskeletal components is required to approximate the various molecules involved in coupling TCR stimulation to PTK activation.

Another cytoskeletal protein which interacts with a PTK relevant to TCR signaling is α-tubulin, a major component of microtubules which binds to Fyn following TCR stimulation (MARIE-CARDINE et al. 1995). Again, this interaction has been shown to be SH2-domain mediated and dependent on tyrosine phosphorylation of α-tubulin. An equivalent association between α-tubulin and Lck has not been demonstrated and tyrosine phosphorylation of α-tubulin can be induced in Lck-deficient Jcam 1.6 cells. These latter findings are consistent with the distinct subcellular localizations of Lck and Fyn – Lck is found at the plasma membrane and Fyn preferentially associates with cytoskeletal microtubules (LEY et al. 1994) – and indicate the interaction of Fyn with α-tubulin to be specific and likely of functional significance. It has been suggested, for example, that TCR-mediated phosphorylation of α-tubulin engenders microtubule depolymerization and consequent redistribution of tubulin-Fyn complexes at specific sites within the plasma membrane.

In addition to the PTKs described above, several other signaling effectors that participate in antigen-receptor signaling cascades have been shown either to migrate to the cytoskeleton after cell activation and/or to interact with selected cytoskeletal proteins. These include, for example, ZAP-70 and Syk which associate with and phosphorylate α-tubulin in T- and B-cells, respectively, (HUBY et al. 1995; ISAKOV et al. 1996; PETERS et al. 1996), the SH2-domain-containing tyrosine phosphatase, SHP-1, which inhibits antigen-receptor signaling in B- and T-cells (PANI et al. 1995; PANI et al. 1996) and which translocates to the cytoskeleton in thrombin-stimulated platelets (FALET et al. 1996), and the receptor tyrosine phosphatase, CD45, which associates constitutively with the fodrin actin-binding heterodimer, but which increases its binding to fodrin and, as a consequence, its phosphatase activity following TCR stimulation (LOKESHWAR and BOURGUIGNON 1992). Thus, while definition of the specific cytoskeletal-signaling effector interactions evoked by antigen-receptor engagement is just beginning, these preliminary data provide compelling evidence that such interactions not only occur, but create the necessary framework for integrating stimulatory signals with cytoskeletal rearrangement.

## 4.3 Antigen-Receptor Co-modulator Links with Cytoskeletal Rearrangement

While activation cascades downstream of the antigen receptor on B- and T-cells are driven largely by ligand binding, the biological outcome of BCR and TCR cross-linking is substantively influenced by a spectrum of accessory molecules expressed on the lymphocyte surface. In B cells, these include for example, the FcγRIIB and CD22 proteins, which negatively modulate BCR signaling (WILSON et al. 1987; O'ROURKE et al. 1997; TEDDER et al. 1997) and CD19, a B-cell-restricted transmembrane protein which positively regulates BCR signaling functions (O'ROURKE et al. 1997; FUJIMOTO et al. 1998). The relationship of the FcγRIIB and CD22 to the cytoskeleton are currently unclear, although the capacity of CD22 to associate with the Syk, Lyn and SHP-1 proteins (CAMPBELL and KLINMAN 1995; DOODY et al. 1995; TUSCANO et al. 1996), enzymes that all undergo cytoskeletal translocation following cellular activation, raises the possibility that CD22 modulation of BCR signaling involves its interaction with cytoskeletal elements.

In contrast to CD22 and FcγRIIB, CD19 effects on BCR signaling have been more definitively linked to the cytoskeleton. CD19, which is contained within a heterotetrameric complex comprised of CD19, CD21 (or complement receptor 2), CD82 and Leu 13, has been shown by many lines of evidence to lower the threshold required for BCR signaling of a variety of cell responses (CARTER and FEARON 1992; O'ROURKE et al. 1997). CD19 is rapidly tyrosine phosphorylated following BCR cross-linking and in activated cells associates with a number of signaling effectors that appear to mediate their signaling functions, at least in part, through their cytoskeletal association (Fig. 1); these include, for example, Lyn, Lck, Fyn, Syk (CHALUPNY et al. 1993; UCKUN et al. 1993). As described earlier, these PTKs have been shown to translocate to the cytoskeleton in a variety of cell lineages and associate with cytoskeletal proteins such as paxillin (Lck) and α-tubulin (Fyn and Syk) as well as phosphatidylinositol 3 kinase (PI3-K) (Tuveson et al. 1993), the Rho GTPase guanine exchange factor Vav (Li et al. 1997), and the adaptor protein c-Cbl (Fujimoto et al. 1998). The latter three of these proteins, as described below, have all been implicated in antigen-receptor-evoked cytoskeletal rearrangement (Janmey and Stossel 1987; Meisner et al. 1995; Fischer et al. 1998). These findings, together with the capacity of anti-CD19 antibody to induce, even in the absence of BCR co-stimulation, tyrosine phosphorylation of Src-family PTKs and Vav (Weng et al. 1994), imply that CD19 co-modulatory functions are realized in part through its capacity to provide a membrane proximal docking platform for cytoskeletal-associated signaling effectors critical to BCR signal relay.

CD19 has also been more directly linked to cytoskeletal dynamics by data suggesting cross-talk between CD19 and the α4β1 (VLA-4) and α5β1 integrins on immature and mature B cells (Xiao et al. 1996). These data include, for example, the demonstration that integrin cross-linking induces tyrosine phosphorylation of CD19 and its associated Src-family kinases. Cross-linking of CD19 has also been shown to induce α4β1 integrin-mediated B-cell adhesion to tonsillar interfollicular stroma (Behr and Schriever 1995), an interaction inhibited by cytochalasin pretreatment of the cells and, thus, presumably dependent on integrity of the actin cytoskeleton. In this latter study, adhesion to the tonsillar stroma was also induced by cross-linking of CD21 but, in this context, was not cytochalasin inhibitable. These data suggest that binding of α4β1 with its ligand (fibronectin) can occur independently of the cytoskeleton, but that induction of this interaction through CD19 signaling requires CD19 physical or functional association with the cytoskeleton. Thus, while the links between BCR co-modulators such as CD19 and the cytoskeleton are not well defined, the available data identify a potential role for CD19 as a molecular conduit for linking mIg-initiated stimulatory signals to the actin cytoskeleton.

Antigen-receptor co-modulators have also been extensively studied in T cells but, as for B cells, their roles in cytoskeletal dynamics are largely undefined. However, the available data, albeit indirect, strongly suggest that the functions of a key TCR modulator, CD28, include a role in regulating cytoskeletal rearrangement. CD28 is a TCR co-stimulatory molecule, T-cell stimulation through the TCR being normally enhanced by CD28 interactions with its cognate ligands (B7-1 or B7-2 molecules) on APCs (June et al. 1994), but resulting in anergy if initiated in the absence of CD28 ligation (Mueller et al. 1989; Harding et al. 1992; Boussiotis et al. 1993; Gimmi et al. 1993). The molecules involved in CD28 signaling have not been entirely defined, but include the following proteins: the MAP kinase, stress-activated protein kinase (SAPK), which is activated by CD28 cross-linking (Su et al. 1994); the PTK ITK/EMT and the Ras-activating Grb2-Sos complex, both of which bind directly to the CD28 cytosolic tail (August et al. 1994; Raab et al. 1995; Rudd 1996); and, most notably, from the perspective of cytoskeletal interaction, PI3-K (August and Dupont 1994; Prasad et al. 1994; Truitt et al. 1994).

PI3-K has been shown to play major roles in modulating cytoskeletal organization (KAPELLER and CANTLEY 1994; KAPELLER et al. 1995; SHIMUZU and HUNT 1996). Accordingly, the capacity of PI3-K to associate with the CD28 cytoplasmic domain implies that CD28 co-stimulatory effects on the TCR may involve CD28/PI3-K-mediated recruitment of cytoskeletal elements to the membrane. This postulate is supported by recent data revealing CD28 ligation to stimulate the polymerization of actin at cell–cell contacts (KAGA et al. 1998) and showing these sites to contain not only talin, but also the Rho family GTPases, Rac and cdc42, which like CD28 stimulate SAPK signaling (BAGRODIA et al. 1995; COSO et al. 1995; MINDEN et al. 1995). As the pivotal roles for Rac and cdc42 in modulating cytoskeletal arrangement are well established (NOBES and HALL 1995), the possibility that CD28 signaling engenders activation of these enzymes suggests that CD28 effects on TCR signaling include and are potentially facilitated by cytoskeletal interaction and rearrangement.

## 4.4 Other Signaling Effector–Cytoskeletal Interactions

In addition to PTK activation and antigen-receptor tyrosine phosphorylation, which are proximal events in antigen-receptor signal transmission, a myriad of intermediary molecules also participate in the biochemical cascades mediating signal relay to the nucleus. These include, for example, the lipid kinase, PI3-K, an enzyme that is activated following antigen-receptor engagement (GOLD and AEBERSOLD 1994) and which thereby generates second messengers from inositol phopholipids. PI3-K has been linked to cytoskeletal regulation in many cell lineages (KOTANI et al. 1994; WENNSTROM et al. 1994) including T cells, in which wortmannin-mediated inhibition of PI3-K activity impairs TCR-evoked cytoskeletal polarization (STOWERS et al. 1995). The mechanisms whereby PI3-K subserves this cytoskeletal regulatory function are not well-defined, but are at least partially realized through the capacity of PI3-K to catalyze the conversion of phosphatidylinositol (PtdIns)(4,5)$P_2$ to PtdIns(3,4,5)$P_3$.

As shown in Fig. 1, PtdIns(4,5)$P_2$ has been shown to interact with the actin regulatory proteins profilin and gelsolin, the latter of which modulates microfilament rearrangement by regulating the capping/uncapping and severing of actin filaments, thereby promoting actin polymerization (STOSSEL et al. 1993). This actin-severing activity of gelsolin is calcium-dependent and generates multiple actin filament growing ends (YIN et al. 1981), which upon interaction with PtdIns(4,5)$P_2$ dissociate from gelsolin, allowing for further filament growth (JANMEY and STOSSEL 1987; HARTWIG et al. 1995b). This cytoskeletal regulatory pathway likely intersects with a second inositol signaling pathway, which is triggered by the activation of PLC-$\gamma$, and results in PtdIns(4,5)$P_2$ cleavage and formation of diacylglycerol and Ins(1,4,5)$P_3$, the latter of which induces increases in cytosolic calcium concentration that promote the actin-severing activity of gelsolin. Thus, the effects of both PI3-K and PLC-$\gamma$ activation on PtIns(4,5)$P_2$ levels may, by virtue of the regulation of gelsolin-mediated actin-filament modulation, provide one biochemical modality whereby antigen-receptor engagement translates to cytoskeletal rearrangement.

As noted above, PI3-K effects on cell cytoarchitecture are likely realized through a diversity of molecular interactions. In T cells, these interactions are believed to include the association of PI3-K with c-Cbl, a 120-kDa adaptor protein which includes an SH3-domain-interacting proline-rich region and a carboxy-terminal leucine zipper (Rivero-Lezcano et al. 1994; Meisner et al. 1995). Following TCR stimulation, c-Cbl is rapidly tyrosine phosphorylated (Donovan et al. 1994) and associates with both Lck and Fyn (Reedquist et al. 1994; Tsygankov et al. 1996); ZAP-70 (Fournel et al. 1996; Lupher et al. 1996) and the p85 subunit of PI3-K (Hartley et al. 1995; Meisner et al. 1995). In view of data implicating PI3-K activation in T-cell integrin binding to fibronectin (Shimizu et al. 1995), it has been suggested that c-Cbl-PI3-K interaction may modulate integrin activation and cell adhesion. This hypothesis is supported by data revealing a tyrosine phosphorylated pool of c-Cbl to be translocated to the cell membrane and detergent-insoluble cytoskeletal fraction after TCR ligation and the majority of PI3-K-c-Cbl complexes in activated T cells to be localized to this membrane cytoskeletal fraction (Hartley and Corvera 1996). As c-Cbl BCR ligation also induces tyrosine phosphorylation and its association with Fyn, Syk and PI3-K (Panchamoorthy et al. 1996), a role for c-Cbl in linking BCR signaling with the cytoskeleton also appears likely.

As for PI3-K, PLC-$\gamma$ also appears to invoke diverse strategies to realize cytoskeletal modulatory effects in activated lymphocytes. Thus, in addition to potentially reducing cellular levels of PtIns(4,5)$P_2$, activation of PLC-$\gamma$ also results in diacylglycerol production and the consequent activation of PKC (Fig. 1). PKC has been shown in a variety of cell lineages to associate with the cytoskeleton, localize to focal adhesions, and phosphorylate integral cytoskeletal proteins such as talin, vinculin and integrins (Keenan and Kelleher 1998). In B cells, activation of selected PKC isoforms by phorbol esters appears to induce LFA-1 capping and rearrangement of actin filaments under the cap (Haverstick et al. 1992). PKC isoforms have also been connected to the cytoskeletal structures in T cells. For example, PKC$\beta$ has been shown to associate with spectrin and ankyrin in the cytosol of resting T cells, with the complexed proteins then relocalized to the detergent-insoluble fraction following TCR or phorbol-ester stimulation (Gregorio et al. 1994). PKC$\zeta$ (Gomez et al. 1995, 1997) and PKC$\varepsilon$ (Keenan et al. 1997) have also been variably reported to associate with the cytoskeleton following TCR ligation, while recent analysis of TCR–APC interactions at the single-cell level have revealed the PKC-$\theta$ isoform to be selectively activated and translocated in conjunction with talin to the site of T-cell/APC contact during T-cell activation (Monks et al. 1997). Thus, while PKC contribution to cytoskeletal organization is not thoroughly understood, the available data suggest an important role for selected PKC isoforms in linking stimulated antigen receptors to changes in lymphocyte cytoskeletal structure and dynamics.

Among the cytosolic signaling effectors most intensively studied in relation to cytoskeletal organization are the Rho, Rac and cdc42 GTPases, members of a small subclass of the Ras superfamily. As indicated in Fig. 1, the Rho family GTP-binding proteins have been shown to play major roles in regulating the cytoskeletal rearrangements induced in various cell types by extracellular stimuli (Nobes and Hall 1995; Hall 1998). These proteins have also been implicated in the connection of antigen-receptor engagement to lymphocyte cytoskeletal and nuclear response (Moorman et al. 1996; Reif and Cantrell 1998), the cdc42 protein for example, identified as a regulator of TCR-evoked T-cell cytoskeletal polarization (Stowers et al. 1995). The mechanisms whereby Rho family GTPases subserve cytoskeletal regulatory roles are incompletely understood, but appear to include interaction with and likely activation of PI3-K (Zheng et al. 1994; Tolias et al. 1995). The related Ras protein has also been implicated in antigen-receptor-induced cytoskeletal modifications by data demonstrating BCR cross-linking to be followed by co-capping of Ras with mIg (Graziadei et al. 1990) However, as for the Rho proteins, Ras contributions to modulating the cytoarchitectural properties of lymphocytes are not well defined as of yet.

In addition to the GTPases per se, the guanine nucleotide exchange factors (GEFs) regulating activation of these enzymes, have also been identified as players in pathways connecting extracellular stimulation to activation and cytoskeletal rearrangement. In lymphocytes, this paradigm is best exemplified by Vav, a Rho GTPase expressed exclusively in hemopoietic cells and demonstrated by many lines of evidence to be key to the propagation of activation signals downstream of TCR and BCR engagement (Romero and Fischer 1997; Fischer et al. 1998a). The importance of Vav to antigen-receptor signaling has been particularly well illuminated by data derived from analysis of Vav knock-out mice, the latter of which show marked impairment in BCR- and TCR-driven lymphocyte proliferation and IL-2 production (Fischer et al. 1995; Tarakhovsky et al. 1995; Zhang et al. 1995). Despite these defects, other signaling events implicated in TCR/Vav signaling, such as ZAP-70 and SAPK/JNK activation, proceed normally in Vav-deficient T cells (Fischer et al. 1998b). By contrast, analyses of T cells from Vav-deficient mice has revealed the presence of multiple defects in TCR-evoked cytoskeletal modification. These include, for example, profound reduction not only in anti-CD3 antibody-evoked capping of TCRs, but also in TCRζ-actin association, actin polymerization and talin/vinculin recruitment to the TCR (Fischer et al. 1998b; Holsinger et al. 1998).

In addition, lack of Vav protein is associated with the failure of talin to constitutively bind the SH2-domain-containing leukocyte protein, Slp76, a Grb-2 binding adaptor which plays a pivotal role in linking TCR ligation to the induction of NF-AT and AP-1 transcriptional activity (Motto et al. 1996). Taken together, these observations suggest a pivotal role for Vav in coupling antigen-receptor signaling to the cytoskeleton and also imply that Vav is predominantly required for induction of the late and not the early activation events evoked by TCR (and potentially BCR) engagement. This latter postulate is consistent with previous observations suggesting that the capping of antigen receptors following their ligation provides a structural framework for generating antigen-receptor signals that are sufficiently sustained so as to evoke later activation events such as cytokine production (Shaw and Dustin 1997). Thus, the early and late sequelae of antigen-receptor engagement may reflect, at least in part, the outcome of biochemically and structurally distinct molecular events. However, this possibility as well as the role of Vav in divergence of the signaling cascades underlying early and late activation events require further investigation, particularly in view of the recent identification of a Vav-related protein, Vav2. Vav2 is structurally similar to Vav, but characterized by a broader range of tissue expression (Schuebel et al. 1996). As Vav2 is expressed in lymphoid cells, its antigen-receptor-signaling modulatory functions may overlap with those of Vav and, therefore, compensate for Vav deficiency in terms of some, but not all such functions. However, despite this caveat, the data described above provide compelling evidence of Vav involvement in co-ordination of antigen-receptor-evoked cytoskeletal and nuclear response.

The biochemical mechanisms enabling Vav to subserve its regulatory role are not fully understood, but are likely to relate to the capacity of Vav to interact with cytoskeletal proteins, including the focal adhesion-associated zyxin protein (Hobert et al. 1996), signaling effectors such as c-Cbl (Marengere et al. 1997) and Slp76 (Wu et al. 1996; Tuosto et al. 1996), and the Rho family GTPases, the latter of which appear to be activated by Vav (Olson et al. 1996) in response to Vav stimulation by PI3-K (Han et al. 1998). Similarly, while Vav effects on cytoskeletal dynamics have been less thoroughly examined in B- than in T cells, Vav has been shown to interact with the Slp-76 B cell homologue, Slp-65 (Wienands et al. 1998) and to associate with tyrosine phosphorylated CD19, a BCR co-modulator which also recruits PI3-K and Src-family PTKs following BCR ligation (Weng et al. 1994; Chalupny et al. 1995). These findings suggest that Vav also modulates antigen-evoked cytoskeletal rearrangements in B cells, possibly by providing an intermediary link coupling CD19 with both the BCR and the cytoskeleton.

## 5 The WAS as a Model for Delineating the Molecular Interchange Between Antigen Receptor-Signaling Cascades and the Cytoskeleton

### 5.1 The WAS Phenotype

The WAS is an X-linked recessive disease classically associated with profound immune deficiency, eczema and thrombocytopenia (Aldrich et al. 1954). The disease is clinically heterogeneous, however, and many affected boys manifest an attenuated phenotype characterized by a significantly milder clinical course than boys with the classical disease (Sullivan et al. 1994; Standen 1998). The most obvious cellular defects expressed in WAS patients involve T lymphocytes and platelets, both of which are reduced in number and show structural and functional defects. Thus, for example, platelets from affected boys are markedly reduced in size (SEMPLE et al. 1997), manifest reduced survival consequent to accelerated destruction in the spleen (GROTTUM et al. 1969; MURPHY et al. 1972; LUM et al. 1980; CORASH et al. 1985), and show impaired responses to activating stimuli (MARONE et al. 1986; SEMPLE et al. 1997). Similarly, T lymphocytes also manifest morphological defects, including reduced numbers of microvilli projections on the cell surface and altered actin organization (KENNEY et al. 1986; GALLEGO et al. 1997), as well as impaired proliferative responses to anti-CD3 antibody (MOLINA et al. 1993; GALLEGO et al. 1997) and other selected mitogenic stimuli (SIMINOVITCH et al. 1995). Several other immunological defects, such as a lack of antibody responses to polysaccharide antigens and altered levels of selected Ig isotypes, have also been variably detected among affected boys (GREER and SIMINOVITCH 1994). In addition, WAS is associated with defects in monocyte/macrophage chemotaxis (ALTMAN et al. 1974; BADOLATO et al. 1998; ZICHA et al. 1998) and has recently been shown to impact on the migratory properties of peripheral blood-derived dendritic cells, the patient cells failing to extend dendritic processes and to migrate when plated on fibronectin (BRICKELL et al. 1998). Taken together, these observations suggest a critical role for the WAS gene product in regulating and/or co-ordinating signaling cascades linking extracellular stimuli to hematopoietic cell activation and cytoskeletal rearrangement.

## 5.2 Structural Properties of the WAS Protein

The opportunity for elucidating the biochemical defects underlying the WAS phenotype arose with the positional cloning of the disease and the analysis of the structure of the 502 amino acid WAS gene product, WASp (DERRY et al. 1994). Most notably, this cytosolic protein was initially observed to be unusually proline rich, containing multiple polyproline motifs, representing potential binding sites for SH3 domains, a protein module implicated in the formation of multimeric signaling complexes (YU et al. 1994; PAWSON 1995). Structural comparisons between WASp and other proteins in published databases also reveal WASp to contain a CRIB domain, a short motif which mediates protein interactions with the Rho GTPase, cdc42, as well as two other domains present in several proline-rich proteins also implicated in cytoskeletal organization (SYMONS et al. 1996). These latter two domains, referred to as the Wiskott homology domains 1 (WH1) and 2 (WH2) are located at the N- and C-terminal regions of WASp, respectively. Either one or both of these modules are found in the vasodilator-stimulated phosphoprotein (VASP) (HAFFNER et al. 1995), verprolin (DONNELLY et al. 1993), ENA (enabled) (GERTLER et al. 1995) and its murine homologue, MENA (mammalian ENA) (GERTLER et al. 1996) and in the YSCLAS17/Bee1 *Saccharomyces cerevisiae* (LI et al. 1997), CELR144 *Caenorhabditic elegans* and N-WASP mammalian brain (MIKI et al. 1996) homologues of WASp.

The WH1 domain is of particular interest biologically as the majority of WAS gene mutations found among affected boys occur in this region (GREER et al. 1996). Based on the capacity of the N-WASP WH1 domain to interact with PtdIns(4,5) P2

(MIKI et al. 1996), it has been suggested the domain functions as a pleckstrin homology (PH) domain. By contrast, the demonstration that a WH1 domain is present in a recently identified protein, Homer, and mediates binding of the protein to the C-termini of metabotropic glutamate receptors (BRAKEMAN et al. 1997; PONTING and PHILLIPS 1997), suggests that this protein module, like PDZ domains, facilitates protein interactions with the C-termini of signal-transducing receptors. This possibility, as well as the biological functions of WH2 domain, which appear to include localization of WASp within the cytoskeleton (SYMONS et al. 1996), require further investigation. However, as outlined above, the structural properties of WASp are highly suggestive of its involvement in cytoskeletal organization and are, accordingly, consistent with the cellular defects observed in WAS patients.

Definition of the WASp sequence has led to a number of studies aimed at identifying the specific proteins that interact with WASp. As illustrated in Fig. 1 (*inset i*), data from such studies have confirmed the capacity of WASp to physically associate with activated forms of cdc42 (ASPENSTROM et al. 1996; KOLLURI et al. 1996; SYMONS et al. 1996) and also directly linked this molecular interaction to cytoskeletal organization. Thus, for example, co-expression of WASp and a dominant-negative form of cdc42 in endothelial cells has been shown to impede the normal aggregation of WASp and actin into perinuclear-localized bundles (SYMONS et al. 1996). In this latter study, WASp was also shown to inhibit the lamellipodia formation and stress fibre bundling induced by a constitutively active cdc42 protein. This result was interpreted as suggesting that WASp, by acting in a dominant-negative fashion, sequesters activated cdc42 species that would normally activate Rac and downstream cytoskeletal rearrangements.

While the mechanisms whereby WASp-cdc42 interactions regulate cytoskeletal organization remain to be defined, the postulate of this dynamic is consistent with many other lines of evidence implicating cdc42 and, by extension, cdc42 binding to WASp, in cytoskeletal rearrangement in hemopoietic cells (NOBES and HALL 1995; ALLEN et al. 1997; ROHRSCHNEIDER et al. 1997). These include, for example, data implicating cdc42 in T-cell polarization towards the site of APC/T-cell contact (STOWERS et al. 1995), and also revealing Rho and Rac activation to be required for macrophage cell motility and cdc42 for the cells to respond to chemotactic gradients (ALLEN et al. 1998). Thus, for example, macrophages microinjected with dominant-negative cdc42 protein appear to retain motility, but lose the ability to polarize and move in the direction of a chemotactic gradient. This identical behavioral phenotype has also been detected in WASp-deficient macrophages (ZICHA et al. 1998), an observation consistent with WASp and cdc42 mapping to a common functional pathway. The interdependence of WASp and cdc42 in relation to inducing selected cell behaviors provides at least one mechanistic explanation for the impairment in both T-cell responses to antigenic stimuli and macrophage responses to chemotactic stimuli detected in WAS patients.

In addition to its interaction with cdc42, WASp has also been shown to bind a number of SH3-domain-containing proteins. Most notable among these is the Nck adaptor protein, which has been shown to interact with polyproline motifs in WASp through at least two of its three SH3-domain modules (RIVERO-LEZCANO et al. 1995). This interaction has been detected in vivo in both myeloid (Rivero-Lezcano et al. 1995) and lymphoid (Siminovitch et al. unpublished data) cells. WASp has also been shown to interact with Fyn (Banin et al. 1996) and the Grb2 adapter (Miki et al. 1997) in vivo, but at present the physiological relevance of these interactions is unclear. Similarly, the biological significance of in vitro data revealing WASp binding with a spectrum of other SH3-domain-containing signaling effectors (Cory et al. 1996; Finan et al. 1996) remains uncertain.

To further elucidate the proteins that interact with WASp in vivo, our group and others have utilized yeast two-hybrid screens to isolate WASp binding partners from lymphoid cells. This work has led to our isolation of five SH3-domain-containing WASp binding proteins (McGavin et al. unpublished data), one of which includes an independently identified cytoskeletal-associated protein, PSTPIP (Spencer et al. 1997). PSTPIP has been shown by our group to not only co-localize with WASp when these proteins are overexpressed in CHO cells (Cruz et al. unpublished data), but to also redirect WASp to the cortical cytoskeleton and impede WASp actin-binding activity. Association of these proteins in vivo, however, is not detectable in hemopoietic cells. By contrast, WASp does appear to associate in vivo with another binding partner identified by a two-hybrid screen, WIP, a protein which interacts with the amino-terminal region of WASp and which appears to regulate actin distribution in lymphoid cells (Ramesh et al. 1997). Interactions between the yeast WASp homologue (Beel) and the yeast counterpart of WIP (End5p/verprolin) have very recently been shown to be critical for endocytosis (Naqvi et al. 1998). This observation suggests that this cytoskeletal-mediated function (Salisbury et al. 1980) may, in mammalian cells as well, be realized through the actin modulatory effects of WASp and WIP. This possibility is consistent with preliminary data (unpublished) from our group suggesting that four of the WASp binding proteins identified by two-hybrid screening are also involved in endocytosis.

Although WASp is also expressed in B lymphocytes, its B-cell protein binding partners have been less extensively studied in view of uncertainty as to the importance of WASp to B-cell function. WASp patients do, however, manifest defects in humoral immunity (Golding et al. 1984) and show impaired proliferative responses to BCR engagement (Simon et al. 1992) as well as defective expression of the activation antigen CD23 (Simon et al. 1993). As illustrated in Fig. 2A, our group has recently shown that WASp is constitutively tyrosine phosphorylated in resting B cells and that its tyrosine phosphorylation is enhanced after BCR crosslinking. These findings appear to reflect a physical association between WASp and one or more PTKs, as WASp immunoprecipitates from Daudi cells undergo tyrosine phosphorylation when subjected to an in vitro kinase reaction (Fig. 2B). The source of this activity is currently unknown, however, as PTK activity was not detected in anti-Lyn, anti-Fyn or anti-Lck immunoprecipitates isolated by reprecipitation from anti-WASp immunoprecipitates (Fig. 2B). By contrast, both Lyn and Btk have been shown to bind and tyrosine phosphorylate WASp in RBL-2H3 rat tumor mast cells and these interactions appear to be enhanced by cdc42 (Guinamard et al. 1998). Together, the available data suggest that WASp is a PTK

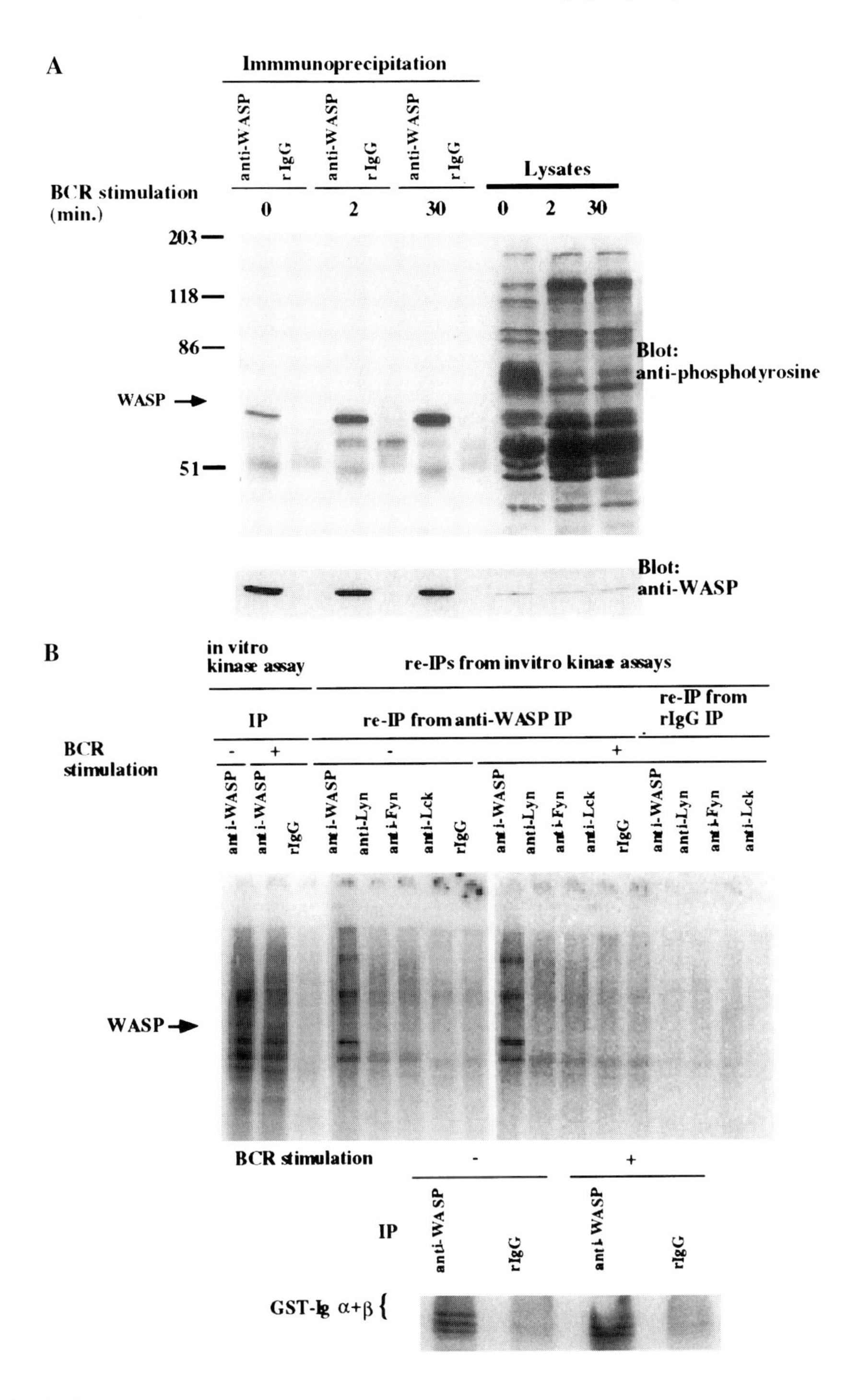
A
Immmunoprecipitation
anti-WASP
rIgG
anti-WASP
rIgG
anti-WASP
rIgG
Lysates
BCR stimulation (min.)
0
2
30
0
2
30
203
118
86
51
WASP
Blot: anti-phosphotyrosine
Blot: anti-WASP
B
in vitro kinase assay
re-IPs from invitro kinase assays
IP
re-IP from anti-WASP IP
re-IP from rIgG IP
BCR stimulation
-
+
-
+
anti-WASP
anti-WASP
rIgG
anti-WASP
anti-Lyn
anti-Fyn
anti-Lck
rIgG
anti-WASP
anti-Lyn
anti-Fyn
anti-Lck
rIgG
anti-WASP
anti-Lyn
anti-Fyn
anti-Lck
WASP
BCR stimulation
-
+
IP
anti-WASP
rIgG
anti-WASP
rIgG
GST-Ig α+β

Fig. 2A,B.

C

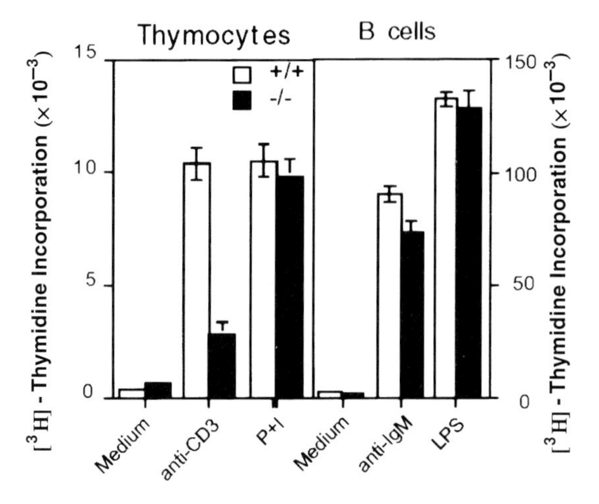


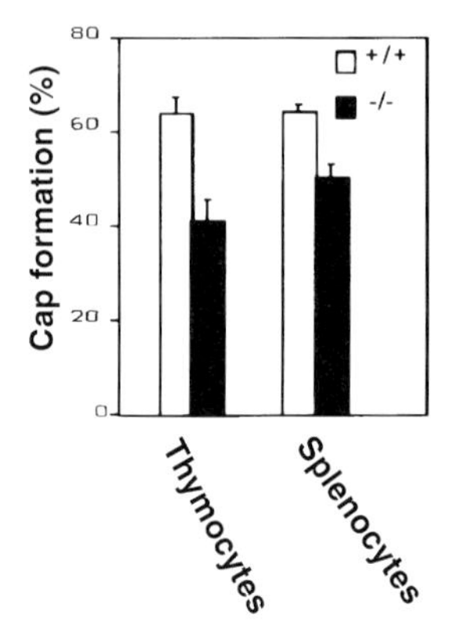


**Fig. 2A–C.** Involvement of WASp in antigen-receptor signaling, cytoskeleton change and lymphocyte effector function. **A** WASp becomes phosphorylated on tyrosine residues upon engagement of the B-cell antigen receptor. Daudi B lymphoma cells ($10^7$) were stimulated with anti-immunoglobulin (Ig) antibody [F(ab)′2 fragments], subjected to cell lysis and the lysates were then immunoprecipitated with rabbit polyclonal anti-WASp antibody. Immunoprecipitated and lysate proteins were resolved over sodium dodecyl sulfate/polyacrylamide gel electrophoresis (SDS-PAGE) and immunoblotted with anti-phosphotyrosine antibody or rabbit preimmune serum (rIgG) (*upper panel*) Blots were then re-probed with anti-WASp antibody (*lower panel*); *arrow* indicates the position of WASp. **B** WASp associates with a tyrosine kinase activity. Lysate proteins prepared from resting (–) and anti-Ig antibody-treated (+) Daudi cells were immunoprecipitated with anti-WASp antibody and the immunoprecipitates then subjected to an in vitro kinase reaction by incubation in kinase buffer containing [$\gamma$-$^{32}$P] ATP. The reaction mixture was then resolved directly over SDS-PAGE (*left panel*) or alternatively resolved on SDS-PAGE after reimmunoprecipitation with anti-WASp, anti-Lyn, anti-Fyn or anti-Lck antibodies (*upper panels* on the *right*) In vitro kinase assays of WASp immunoprecipitates from resting and activated Daudi cells were also performed in the presence of exogenous substrate (a GST-Ig$\alpha$/$\beta$ fusion protein) (*lower panel*) **C** Antigen-receptor-induced proliferative responses are reduced in WASp-deficient cells. *Upper panel*: thymocytes and splenic B cells from wild-type (+/+) and WASp-deficient (–/–) mice were stimulated for 48 h with anti-CD3 antibody (2 μg/ml) or phorbol 12-myristate 13-acetate (5 μg/ml)/ionomycin (250 ng/ml) (thymocytes) or with anti-IgM antibody (5 μg/ml) or lipopolysaccharide (5 μg/ml) (B cells) and proliferative responses then determined after a 16 h pulse with [$^3$H] thymidine. *Lower panel*: thymocyte or splenic B cells from wild-type (+/+) or WASp-deficient (–/–) mice were stimulated with biotinylated anti-TCR antibody (1 μg/ml) or anti-mouse IgD antibody (1 μg/ml), respectively, for 30 min on ice and receptor clustering then visualized by staining with streptavidin–fluoroisothiocyanate. Number of capped cells (expressed as a percentage) was determined by scoring cells from ten microsope fields per sample

target in lymphoid cells and, thus, imply that a subset of WASp interactions with other proteins involve phosphotyrosine-mediated associations.

### 5.3 Antigen-Receptor Signaling Defects in WASp-Deficient Mice

To develop a model system for in vivo analysis of WASp functions, our group has recently derived mice in which the wild-type X-chromosome-located WASp gene has been replaced by a null allele. Analyses of lymphoid cell populations from male WASp-deficient mice have revealed numbers of thymocytes and mature T- and B-cells to be markedly reduced and proliferative responses to antigen-receptor ligation to be impaired. Thus, as is shown in Fig. 2C, anti-CD3 antibody-induced thymocyte proliferation is almost abrogated in the context of WASp deficiency, while proliferative responses to PMA/ionomycin remain intact. Lack of WASp also impairs the B-cell proliferative response to BCR cross-linking, albeit only modestly; again, these cells respond normally to stimulatory maneuvers (such as lipopolysaccharide) which bypass the antigen receptor. In addition, antigen-receptor capping following cross-linking was found to be reduced in both thymocytes and splenocytes from the WASp-deficient animals (Fig. 2C). These observations indicate WASp deficiency to be associated with defects in antigen-receptor-evoked cytoskeletal change and nuclear response, a conclusion consistent with other data on WASp knock-out mice (Snapper et al. 1998) and also with the profile of lymphocyte defects present in WAS patients. Thus, as modeled in Fig. 1 (*inset i*), WASp appears to play a major role in regulating signaling cascades linking antigen stimulation to changes in cell structure and behavior. Moreover, the available data reveal WASp-deficient mice to represent a valuable model system for deciphering the complex series of molecular interactions and pathways which underpin connectivity between the cytoskeleton and transcellular signal relay.

## 6 Concluding Remarks

Cytoskeletal rearrangement is a highly dynamic facet of cell behavior recognized for many years as being critical to the modulation of cell structure and motility and a broad spectrum of cell functions including intracellular transport and relay of extracellular stimulatory signals. The complexity of the cell cytoarchitecture, however, as well as technical difficulties inherent to analyses of the cytoskeleton, have hampered definition of the mechanisms, whereby rearrangement of cytoskeletal components is co-ordinated and linked with other biochemical events induced by cell stimulation. Thus, despite the wealth of information implicating cytoskeletal reorganization in the intracellular propagation of antigen-receptor-elicited activation signals, the precise molecular interactions whereby the cytoskeleton interfaces with antigen-receptor signaling cascades so as to modulate cell behavior have not been elucidated. This situation is rapidly changing, however, in conjunction with the development of improved strategies for directly visualizing protein localization in relation to cell structure, the identification of signaling effectors involved in cytoskeletal regulation, and the generation of mice deficient for these cytoskeletal-related signaling molecules. Together, the data emerging from these studies are providing a framework for delineating the complex intracellular circuitry which allows antigen stimulation to be coincidentally translated into changes in cell morphology, dynamics and biological behavior.

*Acknowledgements.* This work was supported by grants (to K.A. Siminovitch) from the Medical Research Council of Canada, the National Cancer Institute of Canada and the Arthritis Society of Canada. K.A. Siminovitch is a Research Scientist of the Arthritis Society of Canada; Luis da Cruz is a recipient of a fellowship from the *Fundação para a Cincia e a Tecnologia* of Portugal and Ally-Khan Somani is supported by a Steve Fonyo studentship. The authors thank Drs. Andras Nagy and Josef Penninger for their assistance with some of the studies described in this manuscript.

# References

Aldrich RA, Steinberg AG, Campbell DC (1954) Pedigree demonstrating a sex-linked recessive condition characterized by draining ears, eczematoid dermatitis and bloody diarrhea. Pediatrics 13:133–139

Allen WE, Jones GE, Pollard JW, Ridley AJ (1997) Rho, Rac and Cdc42 regulate actin organization and cell adhesion in macrophages. Journal of Cell Science 110:707–720

Allen WE, Zicha D, Ridley AJ, Jones GE (1998) A Role for CDC42 in macrophage chemotaxis. Journal of Cell Biology 141:1147–1157

Altman LC, Snyderman R, Blaese RM (1974) Abnormalities of chemotactic lymphokine synthesis and mononuclear leukocyte chemotaxis in Wiskott-Aldrich syndrome. J Clin Invest 54:486–493

Aspenstrom P, Lindberg U, Hall A (1996) Two GTPases, Cdc42 and Rac, bind directly to a protein implicated in the immunodeficiency disorder Wiskott-Aldrich syndrome. Current Biology 6:70–75

Astier A, Avraham H, Manie SN, Groopman J, Canty T, Avraham S, Freedman AS (1997) The related adhesion focal tyrosine kinase is tyrosine-phosphorylated after beta1-integrin stimulation in B cells and binds to p130cas. J Biol Chem 272:228–232

August A, DuPont B (1994) CD28 of T lymphocytes associates with phosphatidylinositol 3-kinase. International Immunology 6:769–774

Badolato R, Sozzani S, Malacarne F, Bresciani S, Fiorini M, Borsatti A, Albertini A, Mantovani A, Ugazio AG, Notarangelo LD (1998) Monocytes from Wiskott-Aldrich patients display reduced chemotaxis and lack of cell polarization in response to monocyte chemoattractant protein-1 and formyl-methionyl-leucyl-phenylalanine. J Immunol 161:1026–1033

Bagrodia S, Derijard B, Davis RJ, Cerione RA (1995) Cdc42 and PAK-mediated signaling leads to Jun kinase and p38 mitogen-activated protein kinase activation. J Biol Chem 270:27995–27998

Banin S, Truong O, Katz D, Waterfield M, Brickell P, Gout I (1996) Wiskott-Aldrich syndrome protein (WASp) is a binding partner for c-Src family protein-tyrosine kinases. Current Biology 6:981–988

Behr S, Schriever F (1995) Engaging CD19 or target of an antiproliferative antibody 1 on human B lymphocytes induces binding of B cells to the interfollicular stroma of human tonsils via integrin alpha 4/beta 1 and fibronectin. J Exp Med 182:1191–1199

Belkin AM, Koteliansky VE (1987) Interaction of iodinated vinculin, metavinculin and alpha-actinin with cytoskeletal proteins. FEBS Lett 220:291–294

Birkeland ML, Monroe JG (1997) Biochemistry of antigen receptor signaling in mature and developing B lymphocytes. Crit Rev Immunol 17:353–385

Bonnerot C, Lankar D, Hanau D, Spehner D, Davoust J, Salamero J, Fridman WH (1995) Role of B cell receptor Ig alpha and Ig beta subunits in MHC class II- restricted antigen presentation. Immunity 3:335–347

Bourguignon LY, Tokuyasu KT, Singer SJ (1978) The capping of lymphocytes and other cells, studied by an improved method for immunofluorescence staining of frozen sections. J Cell Physiol 95:239–257
Boussiotis VA, Freeman GJ, Gray G, Gribben J, Nadler LM (1993) B7 but not intercellular adhesion molecule-1 costimulation prevents the induction of human alloantigen-specific tolerance. J Exp Med 178:1753–1763
Brakeman PR, Lanahan AA, O'Brien R, Roche K, Barnes CA, Huganir RL, Worley PF (1997) Homer: a protein that selectively binds metabotropic glutamate receptors [see comments]. Nature 386:284–288
Braun J, Hochman PS, Unanue ER (1982) Ligand-induced association of surface immunoglobulin with the detergent- insoluble cytoskeletal matrix of the B lymphocyte. J Immunol 128, 1198–1204
Bray N (1995) Protein molecules as computational elements in living cells. Nature 276, 307–312
Brickell PM, Katz DR, Thrasher AJ (1998) Wiskott-Aldrich syndrome: current research concepts. Br J Haematol 101:603–608
Briken V, Lankar D, Bonnerot C (1997) New evidence for two MHC class II-restricted antigen presentation pathways by overexpression of a small G protein. J Immunol 159:4653–4658
Buckler AJ, Rothstein TL, Sonenshein, GE (1988) Two-step stimulation of B lymphocytes to enter DNA synthesis: synergy between anti-immunoglobulin antibody and cytochalasin on expression of c-myc and a G1-specific gene. Mol Cell Biol 8:1371–1375
Campbell MA, Klinman NR (1995) Phosphotyrosine-dependent association between CD22 and protein tyrosine phosphatase 1 C. Eur J Immunol 25:1573–1579
Caplan S, Zeliger S, Wang L, Baniyash M (1995) Cell-surface-expressed T-cell antigen-receptor zeta chain is associated with the cytoskeleton. Proc Natl Acad Sci USA 92:4768–4772
Caplan S, Baniyash M (1996) Normal T cells express two T cell antigen receptor populations, one of which is linked to the cytoskeleton via zeta chain and displays a unique activation-dependent phosphorylation pattern. J Biol Chem 271:20705–20712
Carter RH, Fearon DT (1992) CD19: lowering the threshold for antigen receptor stimulation of B lymphocytes. Science 256:105–107
Chalupny NJ, Kanner SB, Schieven GL, Wee SF, Gilliland LK, Aruffo A, Ledbetter JA (1993) Tyrosine phosphorylation of CD19 in pre-B and mature B cells. Embo J 12:2691–2696
Chalupny NJ, Aruffo A, Esselstyn JM, Chan PY, Bajorath J, Blake J, Gilliland LK, Ledbetter JA, Tepper MA (1995) Specific binding of Fyn and phosphatidylinositol 3-kinase to the B cell surface glycoprotein CD19 through their src homology 2 domains. Eur J Immunol 25:2978–2984
Chan AC, Dalton M, Johnson R, Kong GH, Wang T, Thoma R, Kurosaki T (1995) Activation of ZAP-70 kinase activity by phosphorylation of tyrosine 493 is required for lymphocyte antigen receptor function. EMBO J 14:2499–2508
Chen ZZ, Coggeshall KM, Cambier JC (1986) Translocation of protein kinase C during membrane immunoglobulin- mediated transmembrane signaling in B lymphocytes. J Immunol 136:2300–2304
Corash L, Shafer B, Blaese RM (1985) Platelet-associated immunoglobulin, platelet size, the effect of splenectomy in the Wiskott-Aldrich syndrome. Blood 65:1439–1443
Cory G, MacCarthy-Morrogh L, Banin S, Gout I, Brickell P, Levinsky R, Kinnon C, Lovering R (1996) Evidence that the Wiskott-Aldrich syndrome protein may be involved in lymphoid signalling pathways. Journal of Immunology 157:3791–3795
Coso OA, Chiariello M, Yu JC, Teramoto H, Crespo P, Xu N, Miki T, Gutkind JS (1995) The small GTP-binding proteins Rac1 and Cdc42 regulate the activity of the JNK/SAPK signaling pathway. Cell 81:1137–1146
DeFranco AL (1997) The complexity of signaling pathways activated by the BCR. Curr Opin Immunol 9:296–308
Del Buono BJ, Williamson PL, Schlegel RA (1988) Relation between the organization of spectrin and of membrane lipids in lymphocytes. J Cell Biol 106:697–703
Derry JM, Ochs HD, Francke U (1994) Isolation of a novel gene mutated in Wiskott-Aldrich syndrome [published erratum appears in Cell 1994 Dec 2;79(5):following 922]. Cell 78:635–644
Donnelly SF, Pocklington MJ, Pallotta D, Orr E (1993) A proline-rich protein, verprolin, involved in cytoskeletal organization and cellular growth in the yeast Saccharomyces cerevisiae. Molecular Microbiology 10:585–596
Donovan JA, Wange RL, Langdon WY, Samelson LE (1994) The protein product of the c-cbl protooncogene is the 120-kDa tyrosine- phosphorylated protein in Jurkat cells activated via the T cell antigen receptor. J Biol Chem 269:22921–22924
Doody GM, Justement LB, Delibrias CC, Matthews RJ, Lin J, Thomas ML, Fearon DT (1995) A role in B cell activation for CD22 and the protein tyrosine phosphatase SHP. Science 269:242–244

Dustin ML, Otszowy MW, Holdorf AD, Li J, Bromley S, Desai N, Widder P, Rosenberger F, van der Merwe P, Allen PM, Shaw AS (1998) A novel adaptor protein orchestrates receptor patterning and cytoskeletal polarity in T-cell contacts. Cell 94:667–677

Elson EL, Pasternak C, Daily B, Young JI, McConnaughey WB (1984) Cross-linking surface immunoglobulin increases the stiffness of lymphocytes. Mol Immunol 21:1253–1257

Falet H, Ramos-Morales F, Bachelot C, Fischer S, Rendu F (1996) Association of the protein tyrosine phosphatase PTP1 C with the protein kinase c-Src in human platelets. FEBS Lett 383:165–169

Fechheimer M, Cebra JJ (1982) Phosphorylation of lymphocyte myosin catalyzed in vitro and in intact cells. J Cell Biol 93:261–268

Finan P, Soames C, Wilson L, Nelson D, Stewart D, Truong O, Hsuan J, Kellie S (1996) Identification of regions of the Wiskott-Aldrich syndrome protein responsible for association with selected Src homology 3 domains. Journal of Biological Chemistry 271:26291–26295

Fischer KD, Zmuldzinas A, Gardner S, Barbacid M, Bernstein A, Guidos C (1995) Defective T-cell receptor signalling and positive selection of Vav- deficient CD4+ CD8+ thymocytes. Nature 374:474–477

Fischer KD, Tedford K, Penninger JM (1998a) Vav links antigen-receptor signaling to the actin cytoskeleton. Semin Immunol 10:317–327

Fischer KD, Kong YY, Nishina H, Tedford K, Marengere LE, Kozieradzki I, Sasaki T, Starr M, Chan G, Gardener S, Nghiem MP, Bouchard D, Barbacid M, Bernstein A, Penninger JM (1998b) Vav is a regulator of cytoskeletal reorganization mediated by the T-cell receptor. Curr Biol 8:554–562

Fournel M, Davidson D, Weil R, Veillette A (1996) Association of tyrosine protein kinase Zap-70 with the protooncogene product p120c-cbl in T lymphocytes. J Exp Med 183:301–306

Fowler VM, Vale R (1996) Cytoskeleton. Curr Opin Cell Biol 8:1–3

Fujimoto M, Poe JC, Inaoki M, Tedder TF (1998) CD19 regulates B lymphocyte responses to transmembrane signals [In Process Citation]. Semin Immunol 10:267–277

Gabbiani G, Chaponnier C, Zumbe A, Vassalli P (1977) Actin and tubulin co-cap with surface immunoglobulins in mouse B lymphocytes. Nature 269:697–698

Gallego MD, Santamaria M, Pena J, Molina IJ (1997) Defective actin reorganization and polymerization of Wiskott-Aldrich T cells in response to CD3-mediated stimulation. Blood 90:3089–3097

Ganju RK, Hatch WC, Avraham H, Ona MA, Druker B, Avraham S, Groopman JE (1998) RAFTK, a novel member of the focal adhesion kinase family, is phosphorylated and associates with signaling molecules upon activation of mature T lymphocytes. J Exp Med 185:1055–1063

Geiger B, Rosen D, Berke G (1982) Spatial relationships of microtubule-organizing centers and the contact area of cytotoxic T lymphocytes and target cells. J Cell Biol 95:137–143

Gertler FB, Comer AR, Juang JL, Ahern SM, Clark MJ, Liebl EC, Hoffmann FM (1995) enabled, a dosage-sensitive suppressor of mutations in the Drosophila Abl tyrosine kinase, encodes an Abl substrate with SH3 domain-binding properties. Genes & Development 9:521–533

Gertler FB, Niebuhr K, Reinhard M, Wehland J, Soriano P (1996) Mena, a relative of VASP and Drosophila Enabled, is implicated in the control of microfilament dynamics. Cell 87:227–239

Gimmi CD, Freeman GJ, Gribben JG, Gray G, Nadler LM (1993) Human T-cell clonal anergy is induced by antigen presentation in the absence of B7 costimulation. Proc Natl Acad Sci USA 90:6586–6590

Gold MR, Aebersold R (1994) Both phosphatidylinositol 3-kinase and phosphatidylinositol 4-kinase products are increased by antigen receptor signaling in B cells. J Immunol 152:42–50

Golding B, Muchmore AV, Blaese RM (1984) Newborn and Wiskott-Aldrich patient B cells can be activated by TNP-Brucella Abortus X: evidence that TNP-Bruella Abortus X behaves as a T-independent type 1 antigen in humans. J. Immunol 133:2966–2971

Goodnow CC (1996) Balancing immunity and tolerance: Deleting and tuning lymphocyte repertoire. Proc Nat Acad Sci USA 93:2264–2271

Gomez J, Martinez de Aragon A, Bonay P, Pitton C, Garcia A, Silva A, Fresno M, Alvarez F, Rebollo A (1995) Physical association and functional relationship between protein kinase C zeta and the actin cytoskeleton. Eur J Immunol 25:2673–2678

Gomez J, Garcia AR-Borlado L, Bonay P, Martinez AC, Silva A, Fresno M, Carrera AC, Eicher-Streiber C, Rebollo A (1997) IL-2 signaling controls actin organization through Rho-like protein family, phosphatidylinositol 3-kinase, protein kinase C-zeta. J Immunol 158:1516–1522

Graziadei L, Riabowol K, Bar-Sagi D (1990) Co-capping of ras proteins with surface immunoglobulins in B lymphocytes. Nature 347:396–400

Greer WL, Siminovitch KA (1994) Molecular characterization of the X-linked immunodeficiency diseases. Clin Immunol News 14:17–25

Greer WL, Shehabeldin A, Schulman J, Junker A, Siminovitch KA (1996) Identification of WASP mutations, mutation hotspots and genotype-phenotype disparities in 24 patients with the Wiskott-Aldrich syndrome. Hum Genet 98:685–690

Gregorio CC, Repasky EA, Fowler VM, Black JD (1994) Dynamic properties of ankyrin in T lymphocytes: colocalization with spectrin and protein kinase C beta. J Cell Biol 125:345–358

Grottum KA, Hovig T, Holmsen H, Abrahamsen AF, Jeremic M, Seip M (1969) Wiskott-Aldrich syndrome: qualitative platelet defects and short platelet survival. Br J Haematol 17:373–388

Grupp SA, Campbell K, Mitchell RN, Cambier JC, Abbas AK (1993) Signaling-defective mutants of the B lymphocyte antigen receptor fail to associate with Ig-alpha and Ig-beta/gamma. J Biol Chem 268:25776–25779

Guinamard R, Aspenstrom P, Fougereau M, Chavrier P, Guillemot J-C (1998) Tyrosine phosphorylation of the Wiskott-Aldrich syndrome protein by Lyn and Btk is regulated by cdc42. FEBS Lett 434: 431–436

Haffner C, Jarchau T, Reinhard M, Hoppe J, Lohmann SM, Walter U (1995) Molecular cloning, structural analysis and functional expression of the proline-rich focal adhesion and microfilament-associated protein VASP. EMBO Journal 14:19–27

Hall A (1998) Rho GTPases and the actin cytoskeleton. Science 279:509–514

Hamano T, Asofsky R (1984) Functional studies on B cell hybridomas with B cell surface antigens. IV. Direct effects of cytochalasin B on differentiation. J Immunol 132:122–128

Han J, Luby-Phelps K, Das B, Shu X, Xia Y, Mosteller RD, Krishna UM, Falck JR, White MA, Broek D (1998) Role of substrates and products of PI 3-kinase in regulating activation of Rac-related guanosine triphosphatases by Vav. Science 279:558–560

Harding FA, McArthur JG, Gross JA, Raulet DH, Allison JP (1992) CD28-mediated signalling co-stimulates murine T cells and prevents induction of anergy in T-cell clones. Nature 356:607–609

Hartley D, Meisner H, Corvera S (1995) Specific association of the beta isoform of the p85 subunit of phosphatidylinositol-3 kinase with the proto-oncogene c-cbl. J Biol Chem 270:18260–18263

Hartley D, Corvera S (1996) Formation of c-Cbl. phosphatidylinositol 3-kinase complexes on lymphocyte membranes by a p56lck-independent mechanism. J Biol Chem 271:21939–21943

Hartwig JH, Jugloff LS, De Groot NJ, Grupp SA, Jongstra-Bilen J (1995a) The ligand-induced membrane IgM association with the cytoskeletal matrix of B cells is not mediated through the Ig alpha beta heterodimer. J Immunol 155:3769–3779

Hartwig JH, Bokoch GM, Carpenter CL, Janmey PA, Taylor L, Toker A, Stossel TP (1995b) Thrombin receptor ligation and activated Rac uncap actin filament barbed ends through phosphoinositide synthesis in permeabilized human platelets. Cell. 82:643–653

Haverstick DM, Sakai H, Gray LS (1992) Lymphocyte adhesion can be regulated by cytoskeleton-associated, PMA- induced capping of surface receptors. Am J Physiol 262:C916–C926

Hobert O, Schilling JW, Beckerle MC, Ullrich A, Jallal B (1996) SH3 domain-dependent interaction of the proto-oncogene product Vav with the focal contact protein zyxin. Oncogene 12:1577–1581

Holsinger LJ, Graef IA, Swat W, Chi T, Bautista DM, Davidson L, Lewis RS, Alt FW, Crabtree GR (1998) Defects in actin-cap formation in Vav-deficient mice implicate an actin requirement for lymphocyte signal transduction. Curr Biol 8:563–572

Hornbeck P, Paul WE (1986) Anti-immunoglobulin and phorbol ester induce phosphorylation of proteins associated with the plasma membrane and cytoskeleton in murine B lymphocytes. J Biol Chem 261:14817–14824

Horwitz A, Duggan K, Buck C, Beckerle MC, Burridge K (1986) Interaction of plasma membrane fibronectin receptor with talin – a transmembrane linkage. Nature 320:531–533

Huby RD, Carlile GW, Ley SC (1995) Interactions between the protein-tyrosine kinase ZAP-70, the proto-oncoprotein vav and tubulin in Jurkat T cells. J Biol Chem 270:30241–30244

Isakov N, Wange RL, Watts JD, Aebersold R, Samelson R (1996) Purification and characterization of human ZAP-70 protein tyrosine kinase from a baculovirus expression system. J Biol Chem 271:15753–15761

Iwashima M, Irving BA, van Oers NS, Chan AC, Weiss A (1994) Sequential interactions of the TCR with two distinct cytoplasmic tyrosine kinases. Science 263:1136–1139

Jackman WT, Burridge K (1989) Polymerization of additional actin is not required for capping of surface antigens in B-lymphocytes. Cell Motil Cytoskeleton 12:23–32

Janmey PA, Stossel TP (1987) Modulation of gelsolin function by phosphatidylinositol 4,5-bisphosphate. Nature 325:362–364

Johnson RP, Craig SW (1995) F-actin binding site masked by the intramolecular association of vinculin head and tail domains [see comments]. Nature 373:261–264

Jugloff LS, Jongstra-Bilen J (1997) Cross-linking of the IgM receptor induces rapid translocation of IgM- associated Ig alpha, Lyn, Syk tyrosine kinases to the membrane skeleton. J Immunol 159: 1096–1106

June CH, Bluestone JA, Nadler LM, Thompson CB (1994) The B7 and CD28 receptor families. Immunol Today 15:321–331

Kaga S, Ragg S, Rogers KA, Ochi A (1998) Stimulation of CD28 with B7–2 promotes focal adhesion-like cell contacts where Rho family small G proteins accumulate in T cells. J Immunol 160:24–27

Kapeller R, Cantley LC (1994) Phosphatidylinositol 3-kinase. Bioessays 16:565–576

Kapeller R, Toker A, Cantley LC, Carpenter CL (1995) Phosphoinositide 3-kinase binds constitutively to alpha/beta-tubulin and binds to gamma-tubulin in response to insulin. J Biol Chem 270:25985–25991

Keenan C, Volkov Y, Kelleher D, Long A (1997) Subcellular localization and translocation of protein kinase C isoforms zeta and epsilon in human peripheral blood lymphocytes. Int Immunol 9: 1431–1439

Keenan C, Kelleher D (1998) Protein kinase C and the cytoskeleton. Cell Signal 10:225–232

Kenney D, Cairns L, Remold-O'Donnell E, Peterson J, Rosen F, Parkman R (1986) Morphological abnormalities in the lymphocytes of patients with the Wiskott-Aldrich syndrome. Blood 68:1329–1332

Kerbel RS, Birbeck MS, Robertson D, Cartwright P (1975) Ultrastructural and serological studies on the resistance of activated B cells to the cytotoxic effects of anti-immunoglobulin serum. Patch and cap formation of surface immunoglobulin on mitotic B lymphocytes. Clin Exp Immunol 20:161–177

Kishimoto T, Miyake T, Nishizawa Y, Watanabe T, Yamamura Y (1975) Triggering mechanism of B lymphocytes. I. Effect of anti-immunoglobulin and enhancing soluble factor on differentiation and proliferation of B cells. J Immunol 115:1179–1184

Kolluri R, Tolias KF, Carpenter CL, Rosen FS, Kirchhausen T (1996) Direct interaction of the Wiskott-Aldrich syndrome protein with the GTPase Cdc42. Proceedings of the National Academy of Sciences of the United States of America 93:5615–5618

Kotani K, Yonezawa K, Hara K, Ueda H, Kitamura Y, Sakaye H, Ando A, Chavanieu A, Calas B, Grigorescu F, Nishiyama M, Waterfield M, Kasuga M (1994) Involvement of phosphoinositide 3-Kinase in insulin- or IGF-1 induced membrane ruffling. EMBO J 13:2313–2321

Kupfer A, Dennert G (1984) Reorientation of the microtubule-organizing center and the Golgi apparatus in cloned cytotoxic lymphocytes triggered by binding to lysable target cells. J Immunol 133:2762–2766

Kupfer A, Swain SL, Janeway CA, Jr, Singer SJ (1986) The specific direct interaction of helper T cells and antigen- presenting B cells. Proc Natl Acad Sci USA 83:6080–6083

Kupfer A, Swain SL, Singer SJ (1987) The specific direct interaction of helper T cells and antigen-presenting B cells. II. Reorientation of the microtubule organizing centre and reorganization of the membrane-associated cytoskeleton inside the bound helper T cells. J Exp Med 165:1565–1580

Kupfer A, Mosmann TR, Kupfer H (1991) Polarized expression of cytokines in cell conjugates of helper T cells and splenic B cells. Proc Natl Acad Sci USA 88:775–779

Kupfer H, Monks CR, Kupfer A (1994) Small splenic B cells that bind to antigen-specific T helper (Th) cells and face the site of cytokine production in the Th cells selectively proliferate:immunofluorescence microscopic studies of Th-B antigen- presenting cell interactions. J Exp Med 179:1507–1515

Lee JK, Black JD, Repasky EA, Kubo RT, Bankert RB (1988) Activation induces a rapid reorganization of spectrin in lymphocytes. Cell 55:807–816

Ley SC, Verbi W, Pappin DJ, Druker B, Davies AA, Crumpton MJ (1994) Tyrosine phosphorylation of alpha tubulin in human T lymphocytes. Eur J Immunol 24:99–106

Li R (1997) Beel, a yeast protein with homology to Wiskott-Aldrich syndrome protein, is critical for the assembly of cortical actin cytoskeleton. J Cell Biol 136:649–658

Li X, Sandoval D, Freeberg L, Carter RH (1997) Role of CD19 tyrosine 391 in synergistic activation of B lymphocytes by coligation of CD19 and membrane Ig. J Immunol 158:5649–5657

Lokeshwar VB, Bourguignon LYW (1992) Tyrosine phosphatase activity of lymphoma CD45 (GP180) is regulated by a direct interaction with the cytoskeleton. J Immunol 267:21551–21557

Loor F, Forni L, Pernis B (1972) The dynamic state of the lymphocyte membrane. Factors affecting the distribution and turnover of surface immunoglobulins. Eur J Immunol 2:203–212

Lowin-Kropf B, Shapiro VS, Weiss A (1998) Cytoskeletal polarization of T cells is regulated by an immunoreceptor tyrosine-based activation motif-dependent mechanism. J Cell Biol 140:861–871

Lum LG, Tubergen DG, Corash L, Blaese RM (1980) Splenectomy in the management of the thrombocytopenia of the Wiskott-Aldrich syndrome. N Engl J Med 302:892–896

Lupher ML, Jr, Reedquist KA, Miyake S, Langdon WY, Band H (1996) A novel phosphotyrosine-binding domain in the N-terminal transforming region of Cbl interacts directly and selectively with ZAP-70 in T cells. J Biol Chem 271:24063–24068

Marengere LEM, Mirtsos C, Kozieradzki I, Veillette A, Mak TW, Penninger JM (1997) Proto-oncoprotein Vav interacts with c-Cbl in activated thymocytes and peripheral T cells. J Immuno 159:70–76
Marie-Cardine A, Kirchgessner H, Eckerskorn C, Meuer SC, Schraven B (1995) Human T lymphocyte activation induces tyrosine phosphorylation of alpha-tubulin and its association with the SH2 domain of the p59fyn protein tyrosine kinase. Eur J Immunol 25:3290–3297
Marone G, Albini F, DiMartino L, Quattrin A, Poto S, Condorelli M (1986) The Wiskott-Aldrich syndrome: studies of platelets, basophils and polymorphonuclear leucocytes. Br J Haem 62:737–745
Meisner H, Conway BR, Hartley D, Czech MP (1995) Interactions of Cbl with Grb2 and phosphatidylinositol 3′-kinase in activated Jurkat cells. Mol Cell Biol 15:3571–3578
Melamed I, Downey GP, Aktories K, Roifman CM (1991) Microfilament assembly is required for antigen-receptor-mediated activation of human B lymphocytes. J Immunol 147:1139–1146
Melamed I, Downey GP, Roifman CM (1991) Tyrosine phosphorylation is essential for microfilament assembly in B lymphocytes. Biochem Biophys Res Commun 176:1424–1429
Melamed I, Franklin RA, Gelfand EW (1995) Microfilament assembly is required for anti-IgM dependent MAPK and p90rsk activation in human B lymphocytes. Biochem Biophys Res Commun 209:1102–1110
Miki H, Miura K, Takenawa T (1996) N-WASP, a novel actin-depolymerizing protein, regulates the cortical cytoskeletal rearrangement in a PIP2-dependent manner downstream of tyrosine kinases. The EMBO Journal 15:5326–5335
Miki H, Nonoyama S, Zhu Q, Aruffo A, Ochs H, Takenawa T (1997) Tyrosine Kinase Signaling Regulates Wiskott-Aldrich Syndrome Protein Function which is Essential for Megakaryocyte Differentiation. Cell Growth and Differentiation 8:195–202
Minden A, Lin A, Claret FX, Abo A, Karin M (1995) Selective activation of the JNK signaling cascade and c-Jun transcriptional activity by the small GTPases Rac and Cdc42Hs. Cell 81:1147–1157
Molina IJ, Sancho J, Terhorst C, Rosen FS, Remold-O'Donnell E (1993) T cells of patients with the Wiskott-Aldrich syndrome have a restricted defect in proliferative responses. Journal of Immunology 151:4383–4390
Monks CRF, Kupfer H, Tamir I, Barlow A, Kupfer A (1997) Selective modulation of protein kinase C-θ during T cell activation. Nature 385:83–86
Monks CRF, Freiberg BA, Kupfer H, Sciaky N, Kupfer A (1998) Three dimensional segregation of supra-molecular activation clusters in T-cells. Nature 395, in press
Montixi C, Langlet C, Bernard A-M, Thimonier J, Dubois C, Wurbel M-A, Chauvin J-P, Pierres M, He H-T (1998) Engagement of T cell receptor triggers its recruitment to low-density detergent-insoluble membrane domains. EMBO J 17:5334–5348
Mookerjee BK, Cuppoletti J, Rampal AL, Jung CY (1981) The effects of cytochalasins on lymphocytes. Identification of distinct cytochalasin-binding sites in relation to mitogenic response and hexose transport. J Biol Chem 256:1290–1300
Mookerjee BK, Jung CY (1982) The effects of cytochalasins on lymphocytes: mechanism of action of Cytochalasin A on responses to phytomitogens. J Immunol 128:2153–2159
Moorman JP, Bobak DA, Hahn CS (1996) Inactivation of the small GTP binding protein Rho induces multinucleate cell formation and apoptosis in mureine T lymphoma EL4. J Immunol 156:4146–4153
Motto DG, Ross SE, Wu J, Hendricks-Taylor LR, Koretzky GA (1996) Implication of the GRB2-associated phosphoprotein SLP-76 in T cell receptor-mediated interleukin 2 production. J Exp Med 183:1937–1943
Mueller DL, Jenkins MK, Schwartz RH (1989) An accessory cell-derived costimulatory signal acts independently of protein kinase C activation to allow T cell proliferation and prevent the induction of unresponsiveness. J Immunol 142:2617–2628
Murphy S, Oski FA, Naiman J, Lusch CJ, Goldberg S, Gardner FH (1972) Platelet size and kinetics in hereditary and acquired thrombocytopenia. New England Journal of Medicine 286:499–504
Nada S, Yagi T, Takeda H, Tokunaga T, Nakagawa H, Ikawa Y, Okada M, Aizawa S (1993) Constitutive activation of Src family kinases in mouse embryos that lack Csk. Cell 73:1125–1135
Naqvi SN, Zahn R, Mitchell DA, Stvenson BJ, Munn AL (1998) The WASp homologues Las17p functions with the WIP homologue End5p/verprolin and is essential for endocytosis in yeast. Curr Biol 8:959–962
Negulescu PA, Krasieva TB, Khan A, Kerschbaum H, Calahan M (1996) Polarity of T cell shape, motility and sensitivity to antigen. Immunity 4:421–430

Nel AE, Landreth GE, Goldschmidt-Clermont PJ, Tung HE, Galbraith RM (1984) Enhanced tyrosine phosphorylation in B lymphocytes upon complexing of membrane immunoglobulin. Biochem Biophys Res Commun 125:859–866

Nobes C, Hall A (1995) Rho, Rac, Cdc42 GTPases regulate the assembly of multimolecular focal complexes associated with actin stress fibers, lamellipodia, filopodia. Cell 81:53–62

O'Rourke L, Tooze R, Fearon DT (1997) Co-receptors of B lymphocytes. Curr Opin Immunol 9:324–329

Olson MF, Pasteris NG, Gorski JL, Hall A (1996) Faciogenital dysplasia protein (FGD1) and Vav, two related proteins required for normal embryonic development, are upstream regulators of Rho GTPases. Curr Biol 6:1628–1633

Ostergaard HL, Lou O, Arendt CW, Berg NN (1998) Paxillin phosphorylation and association with Lck and Pyk2 in anti-CD3- or anti-CD45-stimulated T cells. J Biol Chem 273:5692–5696

Panchamoorthy G, Fukazawa T, Miyake S, Soltoff S, Reedquist K, Druker B, Shoelson S, Cantley LBH (1996) p120cbl is a major substrate of tyrosine phosphorylation upon B cell antigen receptor stimulation and interacts in vivo with Fyn and Syk tyrosine kinases, Grb2 and Shc adaptors, the p85 subunit of phosphatidylinositol 3-kinase. J Biol Chem 271:3187–3194

Pani G, Kozlowski M, Cambier JC, Mills GB, Siminovitch KA (1995) Identification of the tyrosine phosphatase PTP1C as a B cell receptor-associated protein involved in the regulation of B cell signaling. J Exp Med 181:2077–2084

Pani G, Fischer K-D, Mlinaric-Rascan I, Siminovitch KA (1996) Signaling capacity of the T cell antigen receptor is negatively regulated by the PTPK tyrosine phosphatase. J Exp Med 184:839–852

Park JY, Jongstra-Bilen J (1997) Interactions between membrane IgM and the cytoskeleton involve the cytoplasmic domain of the immunoglobulin receptor. Eur J Immunol 27:3001–3009

Parsey MV, Lewis GK (1993) Actin polymerization and pseudopod reorganization accompany anti-CD3-induced growth arrest in Jurkat T cells. J Immunol 151:1881–1893

Pawson T (1995) Protein modules and signalling networks. Nature 373:573–580

Pestina TI, Stenberg PE, Druker BJ, Steward SA, Hutson NK, Barrie RJ, Jackson CW (1997) Identification of the Src family kinases, Lck and Fgr in platelets. Their tyrosine phosphorylation status and subcellular distribution compared with other Src family members. Arterioscler Thromb Vasc Biol 17:3278–3285

Peters JD, Furlong MT, Asai DJ, Harrison ML, Geahlen RL (1996) Syk, activated by cross-linking the B-cell antigen receptor, localizes to the cytosol where it interacts with and phosphorylates α-tubulin on tyrosine. J Biol Chem 271:4744–4762

Ponting CP, Phillips C (1997) Identification of homer as a homologue of the Wiskott-Aldrich syndrome protein suggests a receptor-binding function for WH1 domains. J Mol Med 75:769–771

Poo WJ, Conrad L, Janeway CA, Jr (1988) Receptor-directed focusing of lymphokine release by helper T cells. Nature 332:378–380

Prasad KV, Cai YC, Raab M, Duckworth B, Cantley L, Shoelson SE, Rudd CE (1994) T-cell antigen CD28 interacts with the lipid kinase phosphatidylinositol 3-kinase by a cytoplasmic Tyr(P)-Met-Xaa-Met motif. Proceedings of the National Academy of Sciences of the United States of America 91:2834–2838

Pure E, Tardelli L (1992) Tyrosine phosphorylation is required for ligand-induced internalization of the antigen receptor on B lymphocytes. Proc Natl Acad Sci USA 89:114–117

Qian D, Mollenauer MN, Weiss A (1996) Dominant-negative zeta-associated protein 70 inhibits T cell antigen receptor signaling. J Exp Med 183:611–620

Qian D, Weiss A (1997) T cell antigen receptor signal transduction. Curr Opin Cell Biol 9:205–212

Raab M, Cai YC, Bunnell SC, Heyeck SD, Berg LJ, Rudd CE (1995) p56Lck and p59Fyn regulate CD28 binding to phosphatidylinositol 3-kinase, growth factor receptor-bound protein GRB-2, T cell-specific protein-tyrosine kinase ITK: implications for T-cell costimulation. Proc Natl Acad Sci USA 92:8891–8895

Ramesh N, Anton IM, Hartwig JH, Geha RS (1997) WIP, a protein associated with Wiskott-Aldrich Syndrome protein, induces actin polymerization and redistribution in lymphoid cells. Proc Natl Acad Sci USA 94:14671–14676

Reedquist KA, Fukazawa T, Druker B, Panchamoorthy G, Shoelson SE, Band, H (1994) Rapid T-cell receptor-mediated tyrosine phosphorylation of p120, an Fyn/Lck Src homology 3 domain-binding protein. Proc Natl Acad Sci USA 91:4135–4139

Reif K, Cantrell DA (1998) Networking Rho family GTPases in lymphocytes. Immunity 8:395–401

Rivero-Lezcano OM, Sameshima JH, Marcilla A, Robbins KC (1994) Physical association between Src homology 3 elements and the protein product of the c-cbl proto-oncogene. J Biol Chem 269: 17363–17366

Rivero-Lezcano O, Marcilla A, Sameshima J, Robbins K (1995) Wiskott-Aldrich Syndrome Protein Physically Associates with Nck through Src Homology 3 Domains. Molecular and Cellular Biology 15:5725–5731

Roberts RL, Jr, LaVia MF (1976) Inhibition of antigen-induced B lymphocyte activation in vitro by cytochalasin B. J Immunol 117:155–159

Rock MT, Brooks WH, Roszman TL (1997) Calcium-dependent signaling pathways in T cells. Potential role of calpain, protein tyrosine phosphatase 1b, p130Cas in integrin- mediated signaling events. J Biol Chem 272:33377–33383

Rohrschneider LR, Bourette RP, Lioubin MN, Algate PA, Myles GM, Carlberg K (1997) Growth and differentiation signals regulated by the M-CSF receptor. Mol. Reprod. Dev. 46:96–103

Romero F, Fischer S (1996) Structure and function of vav. Cell Signal 8:545–553

Rosenspire AJ, Choi YS (1988) Membrane immunoglobulin: anti-immunoglobulin interactions mediate the phosphorylation of actin associated proteins in the B-lymphocyte. Life Sci 42:497–504

Rosenthal AS, Davie JM, Rosenstreich DL, Cehrs KU (1973) Antibody-mediated internalization of B lymphocyte surface membrane immunoglobulin. Exp Cell Res 81:317–329

Rothstein TL (1985) Anti-immunoglobulin in combination with cytochalasin stimulates proliferation of murine B lymphocytes. J Immunol 135:106–110

Rozdzial MM, Malissen B, Finkel TH (1995) Tyrosine-phosphorylated T cell receptor zeta chain associates with the actin cytoskeleton upon activation of mature T lymphocytes. Immunity 3:623–633

Rudd CE (1996) Upstream-downstream: CD28 cosignaling pathways and T cell function. Immunity 4:527–534

Ryser JE, Rungger-Brandle E, Chaponnier C, Gabbiani G, Vassalli P (1982) The area of attachment of cytotoxic T lymphocytes to their target cells shows high motility and polarization of actin, but not myosin. J Immunol 128:1159–1162

Saido TC, Shibata M, Takenawa T, Murofushi H, Suzuki K (1992) Positive regulation of mu-calpain action by polyphosphoinositides. J Biol Chem 267:24585–24590

Salisbury JL, Condeelis JS, Satir P (1980) Role of coated vesicles, microfilaments, calmodulin in receptor-mediated endocytosis by cultured B lymphoblastoid cells. J Cell Biol 87:132–141

Salisbury JL, Baron AT, Keller GA, Skiest D (1988) Membrane IgM: interactions with the cortical cytoskeleton in the human lymphoblastoid cell line WiL2. Cell Motil Cytoskeleton 9:140–152

Salojin K, Zhang J, Cameron M, Gill B, Arreaza G, Ochi A, Delovitch TL (1997) Impaired plasma membrane targeting of Grb2-murine son of sevenless (mSOS) complex and differential activation of the Fyn-T cell receptor (TCR)-zeta-Cbl pathway mediate T cell hyporesponsiveness in autoimmune nonobese diabetic mice. J Exp Med 186:887–897

Sanchez M, Misulovin Z, Burkhardt AL, Mahajan S, Costa T, Franke R, Bolen JB, Nussenzweig M (1993) Signal transduction by immunoglobulin is mediated through Ig alpha and Ig beta. J Exp Med 178:1049–1055

Schreiner GF, Unanue ER (1976) Calcium-sensitive modulation of Ig capping: evidence supporting a cytoplasmic control of ligand-receptor complexes. J Exp Med 143:15–31

Schuebel KE, Bustelo XR, Nielsen DA, Song BJ, Barbacid M, Goldman D, Lee IJ (1996) Isolation and characterization of murine vav2, a member of the vav family of proto-oncogenes. Oncogene 13: 363–371

Sedwick CE, Morgan MM, Jusino L, Cannon JL, Miller J, Burkhardt JK (1998) TCR, LFA-1 and CD28 play unique and complementary roles in signaling T cell cytoskeletal reorganization. J Immun (in press)

Selliah N, Brooks WH, Roszman TL (1996) Proteolytic cleavage of alpha-actinin by calpain in T cells stimulated with anti-CD3 monoclonal antibody. J Immunol 156:3215–3221

Semple JW, Siminovitch KA, Mody M, Milev Y, Lazarus AH, Wright JF, Freedman J (1997) Flow cytometric analysis of platelets from children with the Wiskott-Aldrich Syndrome reveals defects in platelet development, activation and structure. Br J Haematol 97:749–754

Shaw AS, Dustin ML (1997) Making the T cell receptor go the distance: a topologial view of T cell activation. Immunity 6:361–369

Shimizu Y, Mobley JL, Finkelstein LD, Chan AS (1995) A role for phosphatidylinositol 3-kinase in the regulation of beta 1 integrin activity by the CD2 antigen. J Cell Biol 131:1867–1880

Shimizu Y, Hunt SW (1996) Regulatory integrin-mediated adhesion: one more function for PI3-kinase. Imm Today 17:565–573

Shoshan MC, Aman P, Skog S, Florin I, Thelestam M (1990) Microfilament-disrupting Clostridium difficile toxin B causes multinucleation of transformed cells but does not block capping of membrane Ig. Eur J Cell Biol 53:357–363

Siminovitch KA, Greer WL, Novogrodsky A, Axelsson B, Somani A-K, Peacock M (1995) A diagnostic assay for the Wiskott-Aldrich syndrome and its variant forms. J Invest Med 43:159–169

Simon HU, Mills GB, Hashimoto S, Siminovitch KA (1992) Evidence for defective transmembrane signaling in B cells from patients with Wiskott-Aldrich syndrome. Journal of Clinical Investigation 90:1396–1405

Simon HU, Higgins E, Datti A, Siminovitch KA, Dennis JW (1993) Defective expression of CD23 and autocrine growth-stimulation in Epstein-Barr virus (EBV)-transformed B cells from patients with Wiskott Aldrich syndrome (WAS) Clin Exp Immunol 92:43–49

Snapper SB, Rosen FS, Mizoguchi E, Cohen P, Khan W, Liu CH, Hagemann TL, Kwan SP, Ferrini R, Davidson L, Bhan AK, Alt FW (1998) Wiskott-Aldrich syndrome protein-deficient mice reveal a role for WASP in T but not B cell activation. Immunity 9:81–91

Spencer S, Dowbenko D, Cheng J, Li W, Brush J, Utzig S, Simanis V, Lasky L (1997) PSTPIP: A Tyrosine Phosphorylates Cleavage Furrow-associates Protein that is a Substrate for a PEST Tyrosine Phosphatase. Journal of Cell Biology 138:1–16

Standen GR (1998) Wiskott-Aldrich Syndrome: new perspectives in pathogenesis and management. J Roy Coll Phys Lond 22:80–83

Stossel TP (1993) On the crawling of animal cells. Science 260:1086–1094

Stowers L, Yelon D, Berg L, Chant J (1995) Regulation of the polarization of T cells toward antigen-presenting cells by Ras-related GTPase. Proceedings of the National Academy of Sciences USA 92:5027–5031

Su B, Jacinto E, Hibi M, Kllunki T, Karin M, Neriah Y (1994) JNK is involved in signal integration during costimulation of T lymphocytes. Cell 77:727–736

Sullivan KA, Mullen CA, Blaese RM, Winkelstein J.A (1994) A multi-institutional survey of the Wiskott-Aldrich syndrome. J Pediatr 125:876–885

Suzuki T, Shoji S, Yamamato K, Nada S, Okada M, Yamato T, Honda Z (1998) Essential roles of Lyn in fibronectin-mediated filamentous actin assembly and cell motility in mast cells. J Immun 161: 3694–3701

Symons M, Derry JM, Karlak B, Jiang S, Lemahieu V, McCormick F, Francke U, Abo A (1996) Wiskott-Aldrich syndrome protein, a novel effector for the GTPase CDC42Hs, is implicated in actin polymerization. Cell 84:723–734

Tarakhovsky A, Turner M, Schaal S, Mee PJ, Duddy LP, Rajewsky K, Tybulewicz VL (1995) Defective antigen receptor-mediated proliferation of B and T cells in the absence of Vav. Nature 374:467–470

Tedder TF, Tuscano J, Sato S, Kehrl JH (1997) CD22, a B lymphocyte-specific adhesion molecule that regulates antigen receptor signaling. Annu Rev Immunol 15:481–504

Teti G, La Via MF, Misefari A, Venza-Teti D (1981) Cytochalasin A inhibits B-lymphocyte capping and activation by antigens. Immunol Lett 3:151–154

Thomas SM, Soriano P, Imamoto A (1995) Specific and redundant roles of Src and Fyn in organizing the cytoskeleton. Nature 376:267–271

Tolias KF, Cantley LC, Carpenter CL (1995) Rho family GTPases bind to phosphinositide kinase. J Biol Chem 270:17656–17659

Truitt KE, Hicks CM, Imboden JB (1994) Stimulation of CD28 triggers an association between CD28 and phosphatidylinositol 3-kinase in Jurkat T cells. J Exp Med 179:1071–1076

Tsygankov AY, Mahajan S, Fincke JE, Bolen JB (1996) Specific association of tyrosine-phosphorylated c-Cbl with Fyn tyrosine kinase in T cells. J Biol Chem 271:27130–27137

Tuosto L, Michel F, Acuto O (1996) p95vav associates with tyrosine-phosphorylated SLP-76 in antigen-stimulated T cells. J Exp Med 184:1161–1166

Turner CE, Schaller MD, Parsons JT (1993) Tyrosine phosphorylation of the focal adhesion kinase pp125FAK during development: relation to paxillin. J Cell Sci 105:637–645

Tuscano JM, Engel P, Tedder TF, Agarwal A, Kehrl JH (1996) Involvement of p72syk kinase, p53/56lyn kinase and phosphatidyl inositol-3 kinase in signal transduction via the human B lymphocyte antigen CD22. Eur J Immunol 26:1246–1252

Tuveson DA, Carter RH, Soltoff SP, Fearon DT (1993) CD19 of B cells as a surrogate kinase insert region to bind phosphatidylinositol 3-Kinase. Science 260:986–989

Uckun FM, Burkhardt AL, Jarvis L, Jun X, Stealey B, Dibirdik I, Myers DE, Tuel-Ahlgren L, Bolen JB (1993) Signal transduction through the CD19 receptor during discrete developmental stages of human B-cell ontogeny. J Biol Chem 268:21172–21184

Unanue ER, Karnovsky MJ, Engers HD (1973) Ligand-induced movement of lymphocyte membrane macromolecules. 3. Relationship between the formation and fate of anti-Ig-surface Ig complexes and cell metabolism. J Exp Med 137:675–689

Valitutti S, Dessing M, Aktories K, Gallati H, Lanzavecchia A (1995) Sustained signaling leading to T cell activation results from prolonged T cell receptor occupancy. Role of T cell actin cytoskeleton. J Exp Med 181:577–584

Van der Heyden MA, Oude Weernink PA, Van Oirschot BA, Van Bergen en Henegouwen PM, Boonstra J, Rijksen G (1997) Epidermal growth factor-induced activation and translocation of c-Src to the cytoskeleton depends on the actin binding domain of the EGF-receptor. Biochim Biophys Acta 1359:211–221

Wachsstock DH, Wilkins JA, Lin S (1987) Specific interaction of vinculin with alpha-actinin. Biochem Biophys Res Commun 146:554–560

Wange RL, Malek SN, Desiderio S, Samelson LE (1993) Tandem SH2 domains of ZAP-70 bind to T cell antigen receptor zeta and CD3 epsilon from activated Jurkat T cells. J Biol Chem 268:19797–19801

Wange RL, Guitian R, Isakov N, Watts JD, Aebersold R, Samelson LE (1995) Activating and inhibitory mutations in adjacent tyrosines in the kinase domain of ZAP-70. J Biol Chem 270:18730–18733

Weng WK, Jarvis L, LeBien TW (1994) Signaling through CD19 activates Vav/mitogen-activated protein kinase pathway and induces formation of a CD19/Vav/phosphatidylinositol 3-kinase complex in human B cell precursors. J Biol Chem 269:32514–32521

Wennstrom S, Siegbahn A, Yokote K, Arvidsson AK, Heldin CH, Mori S, Claesson WK (1994) Membrane ruffling and chemotaxis transduced by the PDGF beta-receptor require the landing site for phosphatidylinositol 3′kinase. Oncogene 9:651–660

Wienands J, Schweikert J, Wollscheid B, Jumaa H, Nielsen PJ, Reth M (1998) SLP-65: A new signaling component in B lymphocytes which requires expression of the antigen receptor for phosphorylation [In Process Citation]. J Exp Med 188:791–795

Wilder JA, Ashman RF (1991) Actin polymerization in murine B lymphocytes is stimulated by cytochalasin D but not by anti-immunoglobulin. Cell Immunol 137:514–528

Williams GT, Peaker CJ, Patel KJ, Neuberger MS (1994) The alpha/beta sheath and its cytoplasmic tyrosines are required for signaling by the B-cell antigen receptor but not for capping or for serine/threonine-kinase recruitment. Proc Natl Acad Sci USA 91:474–478

Wilson HA, Greenblatt D, Taylor CW, Putney JW, Tsien RY, Finkelman FD, Chused TM (1987) The B lymphocyte calcium response to anti-Ig is diminished by membrane immunoglobulin cross-linkage to the Fcγ receptor. J Immunol 138:1712–1718

Woda BA, Woodin MB (1984) The interaction of lymphocyte membrane proteins with the lymphocyte cytoskeletal matrix. J Immunol 133:2767–2772

Wu J, Motto DG, Koretzky GA, Weiss A (1996) Vav and SLP-76 interact and functionally cooperate in IL-2 gene activation. Immunity 4:593–602

Xiao J, Messinger Y, Jin J, Myers DE, Bolen JB, Uckun FM (1996) Signal transduction through the beta1 integrin family surface adhesion molecules VLA-4 and VLA-5 of human B-cell precursors activates CD19 receptor-associated protein-tyrosine kinases. J Biol Chem 271:7659–7664

Yan SR, Fumagalli L, Berton G (1996) Activation of Src family kinases in human neutrophils: evidence that p58C-FGR and p53/56 LYN redistributed to Triton X-100-insoluble cytoskeletal fraction, also enriched in the caveolar protein caveolin, display an enhanced kinase activity. FEBS Lett 380:198–203

Yan SR, Huang M, Berton G (1997) Signaling by adhesion in human neutrophils: activation of the $p72^{syk}$ tyrosine kinase and formation of protein complexes containing $p72^{syk}$ and Src family kinases in neutrophils spreading over fibrinogen. J Immunol 158:1902–1910

Yin HL, Hartwig JH, Maruyama K, Stossel TP (1981) $Ca^{2+}$ control of actin filament length. Effects of macrophage gelsolin on actin polymerization. J Biol Chem 256:9693–9697

Yu H, Feng J, Dalgarno S, Brauer A, Schreiber S (1994) Structural basis for the binding of proline-rich peptides to SH3 domains. Cell 76:933–945

Zhang R, Alt FW, Davidson L, Orkin SH, Swat W (1995) Defective signalling through the T- and B-cell antigen receptors in lymphoid cells lacking the vav proto-oncogene. Nature 374:470–473

Zhang W, Sloan-Lancaster J, Kitchen J, Trible RP, Samelson LE (1998) LAT: the ZAP-70 tyrosine kinase substrate that links T cell receptor to cellular activation. Cell 92:83–92

Zheng Y, Bagrodia S, Cerione RA (1994) Activation of phosphoinositide 3-kinase activity by cdc42Hs binding to p85. J. Biol. Chem. 269:18727–18730

Zicha D, Allen WE, Brickell PM, Kinnon C, Dunn GA, Jones GE, Thrasher AJ (1998) Chemotaxis of macrophages is abolished in the Wiskott-Aldrich syndrome. Br J Haematol 101:659–665

# Pax-5/BSAP: Regulator of Specific Gene Expression and Differentiation in B Lymphocytes

J. Hagman[1,2], W. Wheat[1], D. Fitzsimmons[1,2], W. Hodsdon[1], J. Negri[1], and F. Dizon[1]



## 1 Introduction

The development of B lymphocytes (Fig. 1) from committed precursor cells to antibody-producing plasma cells proceeds through multiple stages defined by the expression of distinct sets of lineage-specific genes (Rolink and Melchers 1996). The early, antigen-independent stages of B-cell differentiation were largely defined by the status of the cell's immunoglobulin (Ig) genes. In the scheme advanced by Rolink and Melchers (1996), pro-B cells are defined as cells that are committed to the B-cell lineage, but have not yet undergone any Ig gene rearrangements. At the

[1] Division of Basic Immunology, National Jewish Medical and Research Center, 1400 Jackson Street, K516, Denver, CO 80206, USA
[2] Department of Immunology, University of Colorado Health Sciences Center, Denver, CO 80262, USA
e-mail: hagmanj@njc.org

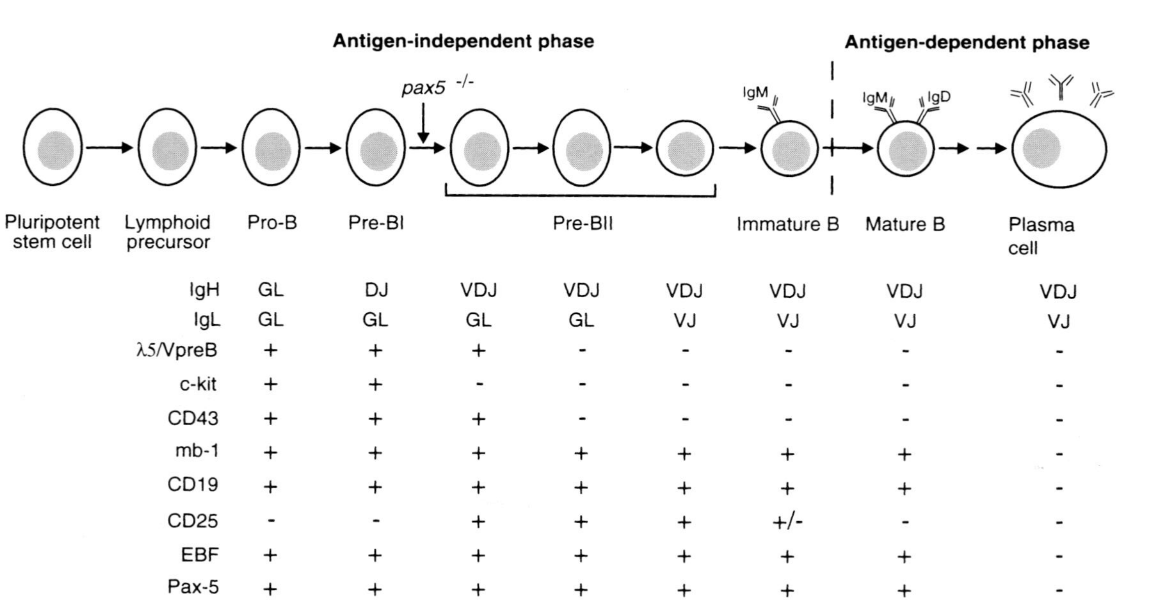


| | Pro-B | Pre-BI | Pre-BII (1) | Pre-BII (2) | Pre-BII (3) | Immature B | Mature B | Plasma cell |
|---|---|---|---|---|---|---|---|---|
| IgH | GL | DJ | VDJ | VDJ | VDJ | VDJ | VDJ | VDJ |
| IgL | GL | GL | GL | GL | VJ | VJ | VJ | VJ |
| λ5/VpreB | + | + | + | - | - | - | - | - |
| c-kit | + | + | - | - | - | - | - | - |
| CD43 | + | + | + | - | - | - | - | - |
| mb-1 | + | + | + | + | + | + | + | - |
| CD19 | + | + | + | + | + | + | + | - |
| CD25 | - | - | + | + | + | +/- | - | - |
| EBF | + | + | + | + | + | + | + | - |
| Pax-5 | + | + | + | + | + | + | + | - |

**Fig. 1.** Schematic illustration of normal murine B-cell development. Distinct stages of differentiation are characterized by the status (shown below) of immunoglobulin (Ig) genes and expression of cell surface markers (Ghia et al. 1998; Hardy et al. 1991; Rolink et al. 1994). Expression of *CD19*, *mb-1*, and the transcription factors early B-cell factor (EBF) and Pax-5 is also indicated. B-cell development is blocked at approximately the Pre-BI stage in *pax5*$^{-/-}$ knockout mice (Nutt et al. 1997). The *vertical dashed line* separates the antigen-independent (*left*) from antigen-dependent (*right*) phases of B-cell differentiation. *IgH*, status of immunoglobulin heavy-chain loci; *IgL*, status of immunoglobulin light-chain loci; *GL*, germline status (unrearranged)

early pre-B cell stage (pre-BI), cells activate their Ig gene recombination machinery including the RAG-1 and RAG-2 genes and commence rearrangements at Ig heavy-chain loci. Heavy-chain polypeptides associate with surrogate light chains encoded by the VpreB and λ5 genes and accessory polypeptides encoded by the *mb-1* (Igα/CD79a) and *B29* (Igβ/CD79b) genes to assemble the pre-B cell receptor (pre-BCR) on the plasma membrane. Following the productive rearrangement of κ or λ light chain gene loci, mature membrane-bound Ig (mIg) appears on the plasma membrane of immature B cells as the B-cell receptor (BCR) for antigen. Encounter of specific antigen by a B cell results in intracellular signaling events that can promote further differentiation, including Ig heavy-chain class switch recombination, somatic hypermutation of Ig variable-region genes, and receptor editing. Ultimately, B cells can terminally differentiate to become plasma cells, which do not express Ig on the cell surface and secrete large quantities of antibody.

B lymphopoiesis is controlled by a series of external and internal events that coordinate expression of lineage- and differentiation-stage-specific genes. In large part, these genes are activated by B-lineage-specific transcription factors (Fitzsimmons and Hagman 1996; Reya and Grosschedl, 1998). A DNA-binding protein encoded by the *pax-5* gene (Pax-5, or B-cell-specific activator protein/BSAP) plays a pivotal role in controlling the expression of genes that are required for the B cell's response to antigen. Progression from the early pre-B-cell stage to later stages of differentiation is critically dependent upon Pax-5/BSAP expression. A number of potential target genes for Pax-5/BSAP are directly associated with the BCR, or with signal transduction in B cells. In regulating these genes, Pax-5/BSAP functions either as a positive or negative regulator of transcription. Other studies show that Pax-5/BSAP regulates specific target genes, in part, through interactions with other transcription factors that enhance its DNA-binding specificity. Pax-5/BSAP may also contribute to the control of B-lymphocyte proliferation. Although our knowledge of Pax-5/BSAP functions is extensive, it is also clear that much remains to be determined. These issues and their biochemical basis are the subjects of this review.

# 2 Properties of the Pax-5/BSAP Protein and its Expression in B Cells

## 2.1 An Early B-Cell-Specific Factor Encoded by the *pax-5* Gene

The Pax-5/BSAP protein was first identified in the laboratory of M. Busslinger as a mammalian DNA-binding activity with a nucleotide-sequence specificity similar to that of tissue-specific activator protein (TSAP), a regulator of late histone gene expression in sea urchins (Barberis et al. 1990). In mammalian nuclear extracts, DNA-binding activity was only observed in tumors and cell lines representing pro-B, pre-B, and $mIg^+$ B cells from mice and humans, but not from terminally differentiated plasma cells or cells representing other tissues. Because TSAP binding sites activated transcription from a minimal promoter when transfected into early B cells, the mammalian activity was termed B-cell-specific activator protein or BSAP.

Biochemical characterization of BSAP activity showed that its DNA-binding specificity is unlike previously characterized B-lymphocyte-specific factors, suggesting that it is comprised of novel polypeptides. However, peptide sequencing of highly purified BSAP protein indicated that it is a member of the Pax, or paired domain family of DNA-binding proteins (ADAMS et al. 1992). Sequencing of cDNA clones encoding BSAP further defined it as the product of the previously cloned *pax-5* gene, which was identified in a screen for cDNA clones that cross-hybridized with sequences of the closely related *pax-1* gene (WALTHER et al. 1991). The cDNAs suggested an open reading frame encoding 391 amino acids with an estimated mass of approximately 45kDa, which is in close agreement with the experimentally determined mass of 50kDa for BSAP (the discrepancy is due to glycosylation). Like the proteins encoded by the eight other *pax* genes in mammals, the Pax-5/BSAP protein comprises a paired DNA-binding domain (Fig. 2), or paired box, and an octapeptide homology of unknown function (amino acids 179–186). The protein also comprises a separate region with homology to the homeobox (amino acids 229–251), but unlike the functional homeodomains of some other Pax proteins (e.g., Pax-6), the Pax-5/BSAP homology is too short to function as an additional DNA-binding domain.

As expected from in vitro DNA-binding data, transcripts from the *pax-5* gene were detected in early B-cell lines and tumors, but not in terminally differentiated plasma cells and most other cell types (ADAMS et al. 1992). Analysis of early stage mouse embryos by in situ hybridization also detected *pax-5* transcripts in the developing central nervous system (ASANO and GRUSS 1991; ADAMS et al. 1992). Transcripts were detected at the midbrain/hindbrain boundary and in the developing neural tube as early as day 9.5 postcoitum. Between day 11.5 and day 13.5, expression was detected in the mesencephalon and spinal cord, with no detectable expression after this period. Therefore, expression of the *pax-5* gene is temporally and spatially regulated in the CNS, in addition to expression in the B-cell lineage.

## 2.2 The Paired DNA-Binding Domain

The defining feature of the Pax family is the paired DNA-binding domain (or paired box), a 128-amino acid DNA-binding motif comprising distinct amino- and

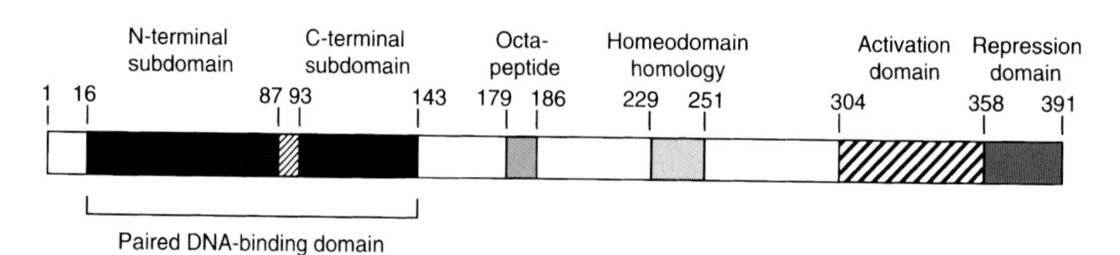


**Fig. 2.** Schematic representation of the Pax-5/BSAP protein and its functional domains. Functions of the various domains are described in the text

carboxy-terminal subdomains (Fig. 2) that bind cooperatively to specific nucleotide sequences. The bipartite structure provides the basis for the observed nucleotide sequence recognition by the Pax-5/BSAP protein (Czerny et al. 1993). In general, Pax-5/BSAP recognizes two short nucleotide sequences, or half-sites, separated by approximately one turn of the DNA helix (e.g., the sea urchin histone H2A-2.2 promoter site). These observations are consistent with DNA binding by each of the two subdomains on one face of the double helix. However, other sites are efficiently bound by Pax-5/BSAP polypeptides lacking the carboxy-terminal subdomain (e.g., the Ig switch region $\gamma$2a site; Wheat et al. 1999), suggesting that it can interact with DNA using either the amino-terminal subdomain or both subdomains.

Another important feature of paired domain/DNA interactions is their relatively relaxed nucleotide sequence specificity. Each of the Pax proteins exhibits preferences for binding different sets of nucleotide sequences that can be very degenerate. For Pax-5/BSAP, the consensus nucleotide sequence of the binding site can be described as (Morrison et al. 1998):

A   a    C  T      gACa<br>
G-GCA-TGAAGCGTGAC

Sequences that fit this consensus appear relatively frequently in a complex genome. In this regard, it has been estimated that medium- or low-affinity binding sites for Pax-5/BSAP occur every 1 kb or so throughout the mouse genome (Busslinger and Urbanek 1995). Therefore, to account for the regulation of specific genes by Pax-5/BSAP, other mechanisms have been proposed to increase its specificity for target genes in vivo. These mechanisms include interactions with partner proteins that enhance DNA-binding specificity (Sect. 5).

Structural studies of the Pax-5/BSAP paired domain are not available, but X-ray crystallographic data are available for the highly homologous DNA-binding domain of the Paired protein of Drosophila (Xu et al. 1995). Two subdomains in the Paired structure are each comprised of three $\alpha$-helices that assemble into helix-turn-helix (H-T-H) motifs typical of homeodomains and Hin recombinase. However, side chains of the recognition helix ($\alpha$3) within the amino-terminal subdomain dock into the major groove of DNA in a manner that is more reminiscent of the interaction of $\lambda$-repressor with its operator DNA. In the Paired structure, the two subdomains do not contact each other and are separated by a short linker. DNA binding is further facilitated by residues within the linker that connect the amino- and carboxy-terminal subdomains and make significant contacts with the phosphodiester backbone along the minor groove. DNA binding is also assisted by $\beta$-hairpin and type-II $\beta$-turn motifs in the amino-terminal subdomain that precede the helical regions. The $\beta$-hairpin is formed by two short antiparallel $\beta$-strands linked by a type-I $\beta$-turn. In addition to the protein/DNA contacts mentioned above, side-chain interactions were identified between various parts of the paired domain, e.g., between the $\beta$-turn motif and the linker. Together, the structure suggests that intramolecular contacts within each subdomain of the paired domain are important for its overall conformation on DNA.

### 2.3 Functional Domains for Activation/Repression of Transcription

One of the most intriguing properties of the Pax-5/BSAP protein is its ability to activate or repress transcription in different contexts. In support of its role as an activator of transcription, the carboxy-terminal 88 amino acids comprise a serine/threonine/proline-rich domain characteristic of transcriptional activation domains. Transiently expressed Pax-5/BSAP can activate transcription through specific binding sites positioned upstream of a minimal promoter (CZERNY and BUSSLINGER 1995; DÖRFLER and BUSSLINGER 1996). Using this assay and progressively truncated Pax-5/BSAP polypeptides, a potent transactivation domain was localized to amino acids 304–358. Interestingly, deletion of the carboxy-terminal 55 amino acids increased the ability of transfected Pax-5/BSAP to activate reporter gene transcription by an additional eightfold. These results suggested the presence of separable activation or repression "modules" in the carboxy-terminus of the protein, which may function differentially when the protein is bound to DNA in the context of different nucleotide sequences.

Interestingly, alternatively spliced *pax-5* mRNAs have been identified that have deleted amino acids 203–391 and, therefore, lack both the activation and repression domains (ZWOLLO et al. 1997). A new sequence encoding 42 amino acids is present in place of these segments. The functional role of putative protein(s) encoded by these transcripts is unknown.

## 3 Pax-5/BSAP Target Genes

### 3.1 Regulation of Pre-BCR/BCR Components and Associated Proteins

A large number of potential target genes for Pax-5/BSAP regulation have been identified using biochemical and genetic methods (Table 1). Most of these genes are expressed specifically in B lymphocytes and not in other cell types, although exceptions have been described (see below). The most abundant category of putative targets includes genes encoding polypeptides that comprise the pre-BCR or BCR, and associated proteins. These polypeptides include components of the Ig antigen receptor itself, signal-transduction proteins, or membrane proteins that modulate the responses of B cells to antigen stimulation. As one example, binding sites for a factor termed EBB-1 were identified in the promoters of genes encoding the surrogate light chains *VpreB1* and λ5 (OKABE et al. 1992), which comprise the pre-BCR together with Ig μ heavy chains. Expression of these genes is activated in pro-B cells and shuts off concurrently with the expression of Ig light chains in pre-BII cells (Fig. 1). The binding site for EBB-1 is important for the functional activity of the *VpreB1* promoter in transfection assays, because mutations in the EBB-1 binding site greatly reduced promoter activity in transfected pre-B cells. Later, it was determined using specific antisera that EBB-1 is Pax-5/BSAP (TIAN et al. 1997).

**Table 1.** Pax-5/BSAP target genes

| Gene | Role of Pax-5/BSAP | References |
|---|---|---|
| Pre-BCR/BCR components: | | |
| *VpreB* promoter (Ig surrogate light chain) | Activation | Okabe et al. 1992; Tian et al. 1997 |
| λ5 promoter (Ig surrogate light chain) | ? | Tian et al. 1997 |
| *Ig* heavy chain *3′Cα* enhancer | Repression | Liao et al. 1992; Singh et al. 1993; Neurath et al. 1994 |
| *Ig* switch region μ promoter | ? | Xu et al. 1992 |
| *Ig* switch region *γ2a* promoter | ? | Liao et al. 1992 |
| *Ig* α germline promoter | Repression | Qui and Stavnezer 1998 |
| *Ig* ε germline promoter | Activation | Czerny et al. 1993; Liao et al. 1994 |
| *blk* Promoter (protein tyrosine kinase) | Activation | Zwollo et al. 1994 |
| *Ig* κ*3′* enhancer | ? | Roque et al. 1996; Shaffer et al. 1997 |
| *mb-1* Promoter (Igα) | Activation | Fitzsimmons et al. 1996 |
| *J*κ promoters | ? | Tian et al. 1997 |
| Accessory proteins: | | |
| *CD19* (co-stimulatory receptor) | Activation | Kozmik et al. 1992 |
| *Ig J* chain promoter | Repression | Rinkenberger et al. 1996 |
| *CD72* promoter (co-stimulatory receptor) | Activation | Ying et al. 1998 |
| *PD-1* (activation/programmed cell death) | Repression | Nutt et al. 1998 |
| Nuclear proteins: | | |
| *p53* exon 1 | Repression | Stuart et al. 1995 |
| *hXBP-1* promoter | Repression | Reimold et al. 1996 |
| N-*myc* | Activation | Nutt et al. 1998 |
| *LEF-1* | Activation | Nutt et al. 1998 |

?, Unknown

The expression patterns of two genes, *CD19* and *mb-1*, parallel that of the *pax-5* gene itself in B cells (Fig.1). The *CD19* gene (Kozmik et al. 1992) encodes a B-cell-specific cell surface protein that reduces the threshold of BCR stimulation by antigen (Carter and Fearon 1992). Three binding sites for Pax-5/BSAP were identified in the promoter region of the *CD19* gene, including a high-affinity site in place of a recognizable TATA element at −30 bp upstream of major sites of transcription initiation (Kozmik et al. 1992). This site conferred specific activation upon a β-globin reporter gene in cells that express Pax-5/BSAP, but not in other cell types. As predicted by in vitro studies, protection of the high-affinity site from methylation was observed in vivo in a footprinting experiment using $CD19^+$ B cells, suggesting that it is occupied by Pax-5/BSAP (Riva et al. 1997). Other experiments showed that, although this site is important for the activation of transcription by Pax-5/BSAP, it is not sufficient for activation of the *CD19* promoter when introduced into non-B cells together with a vector for Pax-5/BSAP expression. Therefore, Pax-5/BSAP may interact with other factors to activate *CD19* transcription. Interactions with partner proteins are clearly required for regulation of the *mb-1* promoter (Fitzsimmons et al. 1996). The *mb-1* gene encodes the Igα polypeptide, which anchors the BCR in the cell membrane and contributes to intracellular signaling functions in response to BCR stimulation. The *mb-1* promoter encompasses functionally important sequences that bind B-cell-specific ternary complexes (BTCs) comprised of Pax-5/BSAP and at least three members of the Ets proto-oncogene family. These complexes are described in depth in Sect. 5.

Other putative targets that are expressed concurrently with Pax-5/BSAP are the Src family protein tyrosine kinase *blk* gene, which encodes a receptor-proximal kinase that becomes phosphorylated following engagement of the BCR (Tamir and Cambier 1998), and the *CD72* gene, which encodes a co-stimulatory membrane protein on B cells (Ying et al. 1998). Like *mb-1* and *CD19*, expression of the *blk* gene coincides with *pax-5* in early B cells. A site in the *blk* promoter (−68bp to −48bp) is bound efficiently by Pax-5/BSAP in vitro, and mutation of this site decreases promoter function in transfected early B-cell lines (Zwollo and Desiderio 1994). Pax-5/BSAP may regulate the *blk* promoter alternatively with NFκB polypeptides, which have been implicated in both positive and negative regulation of *blk* transcription (Zwollo et al. 1998). In parallel with the loss of Pax-5/BSAP expression, cell surface CD72 decreases when B cells differentiate into plasma cells. A site in the *CD72* promoter is important for transcriptional activity in early B cells, and mediates transactivation of the promoter by Pax-5/BSAP in transfection assays (Ying et al. 1998).

Somatic rearrangement and expression of Ig κ light chains may be regulated, in part, by the action of Pax-5/BSAP upon the κ3′ enhancer located 9 kb downstream of the murine κ constant region (Roque et al. 1996; Shaffer et al. 1997). A site within the enhancer is occupied by a Pax-5/BSAP-like factor during early stages of B-cell differentiation when the enhancer is inactive, but occupation of the site is not detected at later stages of differentiation when the enhancer is active. Therefore, it was hypothesized that Pax-5/BSAP binding may be involved in the early remodeling of κ3′ enhancer chromatin prior to its activation. Pax-5/BSAP could play an inhibitory role for expression of κ genes while it is involved in activating transcription of other genes, e.g., *CD19* and *mb-1*, in the same cells.

## 3.2 Regulation of Ig Heavy-Chain Class Switch Recombination and Terminal Differentiation

Multiple lines of evidence suggest that Pax-5/BSAP plays an important role in the control of Ig heavy-chain class switch recombination, which alters the class of Ig heavy-chain constant region expressed by B cells during the late antigen-dependent stages of differentiation. Binding sites for Pax-5/BSAP have been described in promoters and other regulatory regions associated with switching (Table 1). In support of functional roles for these sites, expression of Pax-5/BSAP affects the frequency of constant region utilization; expression of Pax-5/BSAP inhibits switching to some isotypes, while downregulation of expression increases switching to other isotypes. Inhibition of Pax-5/BSAP expression using antisense oligonucleotides decreased switching by normal B cells to produce IgG1, IgG2a, or IgG3 when cultured with bacterial lipopolysaccharide (LPS) together with interleukin-4 (IL-4), interferon-γ, or alone, respectively (Wakatsuki et al. 1994). Overexpression of Pax-5/BSAP in CH12.LX.A2 lymphoma cells reduced spontaneous switching of

IgM$^+$ cells to produce IgA, reduced the number of cells expressing high levels of Syndecan-1 (a marker of terminal differentiation), and reduced levels of expression of the late differentiation-promoting transcription factor, Blimp-1 (Usui et al. 1997). Moreover, overexpression of Pax-5/BSAP in IgM$^+$ I.29μ B-lymphoma cells reduced switching to produce IgA, but increased switching to make IgE when these cells were treated with nicotinamide, LPS, and either TGF-β1 or IL-4, respectively (Qui and Stavnezer 1998).

To account for these results, changes were observed in levels of transcription from germline switch region promoters, which are transcribed prior to recombination at a particular switch region (reviewed in Max et al. 1995). Overexpression of Pax-5/BSAP decreased transcription of the germline α promoter, which features a binding site for Pax-5/BSAP (Qiu and Stavnezer 1998). In contrast, Pax-5/BSAP expression increased transcription of the germline ε promoter (Qiu and Stavnezer 1998), which comprises a functionally important site for Pax-5/BSAP (Table 1; Rothman et al. 1991; Liao et al. 1994). Moreover, activation of the human ε promoter by IL-4 or CD40L, which promote switching to IgE, is dependent upon Pax-5/BSAP binding sites (Thienes et al. 1997). Together, these data show that Pax-5/BSAP is an important regulator of Ig class switch recombination; Pax-5/BSAP and accessory signals (e.g., cytokines) function synergistically to specify the Ig constant region target that will be expressed following switching.

As an important mechanism for the regulation of Ig class switching, downregulation of Pax-5/BSAP activates the IgH 3′α enhancer, which is located within a 40-kb cluster of control regions downstream of the heavy-chain locus in the mouse (Liao et al. 1992; Singh and Birshtein 1993; Neurath et al. 1994). The enhancer comprises part of a "locus control region" that can influence the rearrangement and expression of Ig heavy-chain genes spanning some 200kb (reviewed in Pettersson et al. 1995). Targeted mutation of the enhancer resulted in reduced expression of germline transcripts, reduced isotype class switching, and deficient or reduced secretion of IgG2a, IgG2b, IgG3, IgE, and IgA (Cogné et al. 1994). The enhancer includes two Pax-5/BSAP binding sites that repress enhancer function in B cells (Singh and Birshtein 1993; Neurath et al. 1994; Singh and Birshtein, 1996). Interestingly, a site near the Pax-5/BSAP binding sites (αP) is only occupied when Pax-5/BSAP is downregulated, implying that an (unknown) activator of enhancer function may bind alternatively with Pax-5/BSAP (Neurath et al. 1995). Moreover, in response to extracellular signaling through the OX-40 ligand on CD40L-stimulated B cells, Pax-5/BSAP expression is reduced with a commensurate increase in levels of Ig secretion (Stüber et al. 1995). The downregulation of Pax-5/BSAP expression is correlated with a decrease in protection of a Pax-5/BSAP binding site in the enhancer, suggesting that reduced occupancy in vivo is linked to the activation of enhancer function. Together, these data suggest that levels of Pax-5/BSAP expression are important for the control of late differentiation, with high expression promoting some types of isotype switching (e.g., IgE) and proliferation (possibly in the context of germinal centers), and low expression promoting secretion of Ig.

Another example of a Pax-5/BSAP target gene expressed during late differentiation is the *Ig J chain* gene, which encodes a polypeptide required for secretion of pentamer IgM by antibody-secreting cells. A binding site for a B-cell-specific factor (NF-JC) that represses transcription was identified in the *J chain* promoter (RINKENBERGER et al. 1996). The factor, which was identified as Pax-5/BSAP, blocks activation of the promoter by other B-cell-specific activators including the Ets protein PU.1. Downregulation of Pax-5/BSAP occurs in B cells in response to treatment with IL-2, which activates *J chain* gene transcription in B-cell lines.

### 3.3 Other Pax-5/BSAP-Regulated Genes

Two widely expressed genes encoding nuclear proteins were identified as Pax-5/BSAP targets. The *p53* tumor suppressor gene encodes a protein that can arrest cells in the G1 phase of the cell cycle, activate apoptosis, or act as a differentiation signal in response to a host of stimuli (reviewed in GIACCIA and KASTAN 1998). The *p53* gene was identified as a target for suppression by Pax-5/BSAP in human astrocytomas (STUART et al. 1995). A high affinity site within the first exon mediates the downregulation of *p53* expression in response to overexpression of Pax-5/BSAP. Other Pax proteins (Pax-2, Pax-6, and Pax-8) can bind this site in vitro. The authors of this study hypothesized that Pax proteins downregulate *p53* expression during stages of cellular development that feature rapid cell growth, including the early stages of B-cell differentiation (e.g., pro-B cells). Moreover, the abrogation of *p53* expression may contribute to oncogenic activity observed with overexpression of Pax proteins (MAULBECKER and GRUSS 1993).

The other putative Pax-5-regulated nuclear protein is human X-box-binding protein-1 (hXBP-1), a basic region leucine zipper (b-zip) protein that may function as a regulator of major histocompatibility complex (MHC) class-II gene promoters (REIMOLD et al. 1996). The *hXBP-1* promoter is downregulated when co-transfected with a plasmid for Pax-5/BSAP expression. Therefore, expression of hXBP-1 and Pax-5/BSAP are inversely correlated in early B cells. Together, these data suggest that Pax-5/BSAP may play at least an indirect role in regulating expression of MHC class-II genes.

## 4 Lessons from *pax-5*$^{-/-}$ Knockout Mice

### 4.1 *pax-5*$^{-/-}$ Mice Lack Functional B Cells

As expected from its pattern of expression and putative target genes in early B lymphocytes, targeted deletion of genes encoding Pax-5/BSAP in mice resulted in a complete lack of functional B cells due to a block of differentiation (Fig. 1) at an early pre-B cell-like stage (URBÁNEK et al. 1994). Normal B lymphopoiesis occurs in heterozygous *pax-5*$^{+/-}$ mice. In contrast, homozygous *pax-5*$^{-/-}$ knockout mice exhibit only very early stage-B cells in their bone marrow. Within 2 weeks of birth, the majority of cells that express B220 and CD43 in bone marrow (characteristic of pro-B and early pre-B cells) also express c-kit, heat-stable antigen (HSA), IL-7 receptor (IL-7R), and λ5 surrogate light chains, and lack markers characteristic of pre-BII cells (e.g. CD25 and BP-1) and mIg$^{+}$ cells (Nutt et al. 1997). Analysis of Ig gene rearrangements in these pro/pre-BI-like cells detected only partially rearranged Ig heavy-chain loci (DJ rearrangements). B cells with functionally rearranged heavy chain (VDJ) genes are only detected in greatly reduced numbers (1/50th of normal levels). Rearrangements of Ig light-chain loci were not detected, and the mutant mice lack detectable serum antibody. Interestingly, early B cells detected in mutant bone marrow were not similarly detected in fetal liver, suggesting that one or more Pax-5/BSAP-regulated genes is required differentially for B lymphopoiesis in the latter tissue. Together, these observations show that Pax-5/BSAP is crucial for promoting B lymphopoiesis. In addition, the lack of Pax-5/BSAP protein results in alterations of the posterior midbrain with a resulting abnormal motor reflex; clasping of the hind legs is observed when *pax-5*$^{-/-}$ mutant mice are raised by the tail.

Effects of the lack of Pax-5/BSAP on the expression of specific genes are clearly discernable in mutant B cells (Nutt et al. 1997). Expression of the *CD19* gene was not detected in knockout B-lineage cells, confirming that its promoter is a target for regulation by Pax-5. Strikingly, it was noted that expression of many putative target genes, including the surrogate light chains, *blk* tyrosine kinase, and the *p53* tumor-suppressor gene, was unaffected in *pax-5*$^{-/-}$ mice. Expression of the recombination-activating genes RAG-1 and RAG-2, and germline transcripts (Iμ) that are observed prior to Ig heavy-chain gene rearrangements was also unaffected. In agreement with the hypothesis that Pax-5/BSAP acts downstream of other regulators of early gene expression in B cells (reviewed by Singh 1996), expression of E2A and EBF was also unaffected. The block to B-cell differentiation induced in the knockout mouse precedes stages at which many putative Pax-5/BSAP target genes are expressed (e.g. Ig κ light chains) and, therefore, it is not possible to assess the contributions of Pax-5/BSAP in their regulation.

Further studies of the knockout phenotype confirmed the status of some genes as Pax-5/BSAP targets and, in addition, revealed effects on the expression of previously unsuspected target genes. Using an elegant experimental system, Nutt and colleagues (1998) showed that expression of a small number of genes is either increased or decreased when Pax-5/BSAP expression is reconstituted in knockout pre-B cells. Pre B cells from *pax-5*$^{-/-}$ knockout mice were expanded from bone marrow using IL-7 and transduced with a retroviral vector that allows for expression of a fusion protein comprised of either full-length Pax-5/BSAP or its paired domain tethered to the ligand-binding domain of the estrogen receptor (ER). In the fusion protein, subcellular localization and, therefore, DNA-binding activity of Pax-5:ER is blocked in the absence of estrogen. With the addition of estrogen (17β-estradiol), the fusion protein can translocate to the nucleus, and effects on specific genes can be measured using specific ribonuclease protection assays.

Three types of effects were noted following activation of functional Pax-5/BSAP. First, the majority of genes tested were not affected by the presence or absence of Pax-5/BSAP, including the surrogate light chains *VpreB* and λ5, *blk*, *XBP-1*, and *p53* genes. A second group includes genes that are downregulated in *pax-5*$^{-/-}$ pre-B cells and are strongly activated by expression of exogenous Pax-5/BSAP. *CD19* expression, which is undetectable in the mutant pre-B cells, is partially reactivated when Pax-5/BSAP expression is restored. Levels of *CD19* expression were dependent on the levels of Pax-5:ER fusion protein. Expression of the *mb-1* gene, which is reduced to one-tenth that of wild-type pre-B cells in knockout pre-B cells, is reactivated by Pax-5:ER expression to nearly wild-type levels. Expression of the transcription factor genes N-*myc* and *lymphoid enhancer factor-1* (*LEF-1*) was also strongly reactivated by Pax-5:ER fusion protein. Activation of each of these genes was observed in the presence of an inhibitor of protein synthesis (cycloheximide), suggesting that they are direct targets of Pax-5:ER. Reduced expression of $bcl\text{-}x_L$ gene expression was also observed in knockout pre-B cells and upregulated by Pax-5:ER, but data indicate that this gene is not a direct target for regulation by Pax-5/BSAP. The third class of targets are upregulated in the absence of Pax-5/BSAP expression and are strongly downregulated by expression of the Pax-5:ER fusion protein. The product of the *PD-1* gene is an Ig superfamily protein that comprises a variant immunoreceptor tyrosine-based activation motif (ITAM) in its putative cytoplasmic domain (Ishida et al. 1992). Normally, expression of *PD-1* is induced in response to cross-linking of mIgM on splenic B cells (Agata et al. 1996). *PD-1* may be an inducer of programmed cell death and/or other functions in B cells. *PD-1* expression is strongly upregulated in Pax-5/BSAP knockout pre-B cells (Nutt et al. 1998), and reconstitution of Pax-5/BSAP expression downregulates *PD-1* transcripts by three- to fourfold. The other gene is that of the cell cycle regulator c-*myc*, which is strongly upregulated in the absence of Pax-5/BSAP expression. Increased expression of the c-*myc* gene in these cells is likely due to the very low levels of N-*myc* expression; it was shown previously that c-*myc* and N-*myc* levels are coordinately controlled in a reciprocal fashion (Dildrop et al. 1989; Rosenbaum et al. 1989).

Analysis of polypeptide sequences required for the activation of target genes by Pax-5:ER revealed an important difference between gene targets identified in this study (Nutt et al. 1998). Reactivation of *CD19* and *N-myc* required expression of intact Pax-5/BSAP in mutant pre-B cells. In contrast, reactivation of the *mb-1* or *LEF-1* genes required a fusion protein comprising only the paired domain of Pax-5/BSAP fused with ER, which lacks sequences required for transcriptional activation in other contexts. These data support other studies showing that Pax-5/BSAP activates its promoter through cooperative interactions between its paired domain and partner proteins (Fitzsimmons et al. 1996). *LEF-1* may be regulated in a similar fashion.

How can discrepancies between target genes identified by biochemical means and *pax-5*$^{-/-}$ knockout mice be resolved? The lack of changes in transcriptional efficiency in *pax-5*$^{-/-}$ knockout B cells may be due to functionally redundant proteins in B cells. However, it is possible that the physiological relevance of

Pax-5/BSAP binding may have been assigned incorrectly due to fortuitous binding of the protein in vitro. It is notable that transcriptional activation or repression by Pax-5/BSAP is highly dependent upon its concentration in vivo (DÖRFLER and BUSSLINGER 1996; WALLIN et al. 1998), a characteristic that could explain conflicting determinations of the protein's function at a particular site.

## 4.2 Why is B Lymphopoiesis Blocked in *pax-5*$^{-/-}$ Mice?

A number of genes are expressed at altered levels in the *pax-5*$^{-/-}$ knockout mouse, but the complete picture of how these changes in gene expression block B lymphopoiesis is unclear. The lack of *CD19* expression is unlikely to be a major contributing factor to the mutant phenotype, because mice with mutated *CD19* genes possess mIg$^{+}$ B cells in normal numbers (ENGEL et al. 1995; RICKERT et al. 1995). An obvious source for the block to differentiation is defective BCR assembly, because Ig heavy and light chains are lacking in these mice. Therefore, it is expected that reconstitution of normal heavy-chain expression should restore at least part of normal B lymphopoiesis. In mutant mice with RAG gene knockouts, or in severe combined immune-deficient (SCID) mice, a functionally rearranged Ig μ transgene can partially reconstitute the progression of B lineage cells by providing Ig heavy chains (SPANOPOULOU et al. 1994; YOUNG et al. 1994; CHANG et al. 1995). In contrast, expression of μ heavy chains did not initiate further differentiation in *pax-5*$^{-/-}$ mice (THEVENIN et al. 1998). Inhibition of apoptosis (which occurs at high rates in *pax-5*$^{-/-}$ pre-B cells in the absence of a growth signal) with a transgene directing expression of the survival protein bcl-2 also did not significantly enhance differentiation in the presence of the Ig μ transgene. Together, these results indicate that the block of differentiation is not due solely to either a rearrangement defect or to increased cell death from a lack of survival signals.

The lack of rescued differentiation in Ig heavy-chain-supplemented *pax-5*$^{-/-}$ knockout mice is not likely due to a lack of Igα function resulting from reduced *mb-1* gene expression. Mice were reconstituted with a transgene encoding an Ig heavy chain:Igβ cytoplasmic domain fusion protein (THEVENIN et al. 1998), which can advance B lymphopoiesis to late (pre-BII) pre-B cells in RAG-deficient mice (PAPAVASILIOU et al. 1995a,b; TEH and NEUBERGER 1997). The chimeric protein did not affect B lymphopoiesis in *pax-5*$^{-/-}$ knockout mice, even though pre-BCR including the chimeric polypeptide were detected on the cell surface. The authors concluded that additional defects in normal B-cell signaling responses are revealed in these mice.

Together, these data suggest that the early block of differentiation in these mice is due to pleiotropic effects of the lack of Pax-5/BSAP, including the absence of functional Ig heavy- and light chains. In addition, altered expression of a previously unidentified polypeptide that is crucial for normal pre-BCR signaling may be a factor, because expression of other signaling components of the BCR (or pre-BCR) is unaffected in *pax-5*$^{-/-}$ mutant mice (e.g., surrogate light chains, terminal deoxynucleotidyl transferase, and the protein-tyrosine kinases Lyn, Fyn, Blk, Btk, and Syk).

More recently, IL-7Rα$^{-/-}$ knockout mice confirmed the link between Pax-5/BSAP and VDJ recombination (Corcoran et al. 1998). In these mice, the lack of IL-7Rα chains abrogates the response of B cells to IL-7, with a resulting impairment of VDJ recombination. Further analysis demonstrated reduced levels of *pax-5* and Pax-5/BSAP target gene transcripts (*CD19*), which suggest that Pax-5/BSAP levels are maintained by IL-7 signaling in early B cells. The authors also proposed that reduced expression of Pax-5/BSAP mediates the decreased rearrangement of Ig genes in these mice.

## 5 Regulation of *mb-1* (Igα) Gene Transcription

### 5.1 The mb-1 Promoter Controls Igα Protein Expression in Early B Cells

Regulation of the *mb-1* (Igα; CD79a) gene is a particularly significant model system because protein–protein interactions play an important role for the control of its promoter by Pax-5/BSAP. The *mb-1* gene is expressed during the early stages of B-cell differentiation that precede and include expression of mIg, but is shut off in terminally differentiated plasma cells, which lack mIg and function largely as secretory cells (Fig. 1) (Sakaguchi et al. 1988). The pattern of *mb-1* expression parallels the functions of the Igα protein, a transmembrane protein that acts as scaffolding for the BCR in the plasma membrane, and as a mediator of transmembrane signaling from the BCR to the inside of the cell (Campbell and Cambier 1990; Hombach et al. 1990). The Igα protein mediates signaling functions through its cytoplasmic tail, which includes an ITAM that acts as a docking site for protein tyrosine kinases and other proteins of the BCR-activated signaling cascade (reviewed in Tamir and Cambier 1998).

The *mb-1* promoter (Fig. 3A) has served as a useful model system for identifying mechanisms that control early B-cell-specific gene expression. Preliminary analysis of the *mb-1* gene did not identify an upstream TATA element, and detected utilization of multiple sites for transcriptional initiation (Travis et al. 1991; Feldhaus et al. 1992). In transient transfection assays, 323 bp of genomic DNA encompassing the *mb-1* promoter region correctly initiate mRNA transcripts in early B cells that express endogenous *mb-1* mRNA transcripts, but not in *mb-1*-negative plasma cells or non-lymphoid cells. Therefore, the genomic segment comprises necessary information to confer an early B lymphocyte-specific transcriptional activity in appropriate cells.

Analysis of promoter DNA using progressive deletions and mutations identified proximal and distal regulatory sequences within the promoter fragment (Fig. 3A). Activity of the promoter is dependent on two adjacent nucleotide sequences within the proximal region (–60bp to +70bp) that, when mutated, greatly

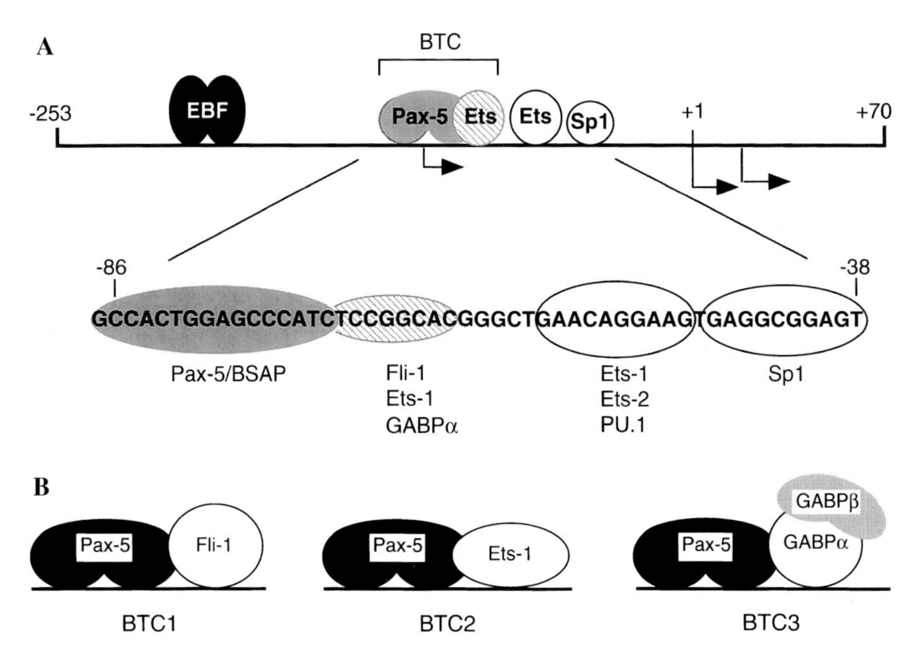


**Fig. 3A,B.** Transcription factor interactions on the *mb-1* promoter. **A** Factor binding and transcription initiation sites. *Above*: Functionally important factor binding sites and proteins. *BTC*, B-cell-specific ternary complex. *+1*, the 5′-most major site of transcription initiation (Travis et al. 1991). *Arrows*, nucleotide sequence interval utilized for initiation. An upstream site of initiation coincides with the Pax-5/BSAP binding site (Travis et al. 1991; Feldhaus et al. 1992). *Below*: Nucleotide sequence from −38 bp to −86 bp and approximate binding sites for regulatory factors (Hagman et al. 1991; Travis et al. 1991; Hagman and Grosschedl 1992; Fitzsimmons et al. 1996). **B** Composition of B-cell-specific ternary complexes (Fitzsimmons et al. 1996)

decrease promoter function. These sites specifically bind proteins in nuclear extracts from B lymphocytes and other cell types in vitro in DNAseI footprinting and electrophoretic mobility shift assays (EMSA). One of the proximal sites is bound by the ubiquitous activator Sp1 (Travis et al. 1991). The other is recognized by proteins of the Ets proto-oncogene family, including Ets-1, Ets-2, and PU.1 (Hagman and Grosschedl 1992).

Although the binding of multiple proteins to the proximal region was detected in vitro, none of these proteins exhibits the early B-cell-specific pattern of activity of the promoter itself. However, complexes that specifically assembled with a site (−160bp to −180bp) within the distal part of the promoter (−253bp to −61bp) exhibited a distribution in nuclear extracts similar to that of transcripts of the *mb-1* gene itself. These complexes were shown to be comprised of early B-cell factor (EBF), which is expressed in early B cells, but not in terminally differentiated plasma cells (Hagman et al. 1991; Feldhaus et al. 1992). Subsequent purification and cloning of cDNAs identified EBF as a member of a novel family of DNA-binding proteins (Hagman et al. 1993). Targeted deletion of genes encoding EBF resulted in a lack of functional B lymphocytes due to a block in early differentiation at a pro-B-like transitional stage (Lin and Grosschedl 1995).

## 5.2 Regulation by Pax-5/BSAP and Ets Proteins

Although EBF was strongly implicated as a regulator of *mb-1* transcription, other data suggested that the promoter is a target for other early B-cell-specific factors. Promoter constructs with mutated EBF binding sites retained their early B-cell specificity, although with reduced activity (Hagman et al. 1991). To address this problem, a screen for factor binding was carried out using DNaseI footprinting and EMSA assays. DNA binding by an early B-cell-specific activity was detected that, based on its mobility in EMSA, was clearly not EBF, but was likely to be Pax-5/BSAP (Fitzsimmons et al. 1996). Studies using recombinant Pax-5/BSAP showed that it specifically binds the *mb-1* promoter in vitro (Fig. 3A). The binding site showed only very limited identity with previously identified binding sites, and competition assays showed that the *mb-1* promoter site is bound only one-eighth as well as is the high affinity site of the *CD19* promoter.

Because Pax-5/BSAP exhibits a degenerate DNA-binding specificity, protein–protein interactions with other transcription factors are necessary for functional DNA binding in vivo. In this regard, it is significant that complexes assembled on the *mb-1* promoter with Pax-5/BSAP were found to include interacting partner proteins (Fitzsimmons et al. 1996). Analysis using EMSA detected complexes comprised of Pax-5(BSAP), and three B-cell-specific ternary complexes or BTCs (termed BTC1, BTC2, or BTC3), which have different mobilities in nondenaturing polyacrylamide gels. Alignment of the *mb-1* promoter with the recognition sequence of the Ets proto-oncogene family protein Elk-1 (Treisman 1992) identified a highly homologous region adjacent to the Pax-5/BSAP binding site (antisense sequence is 5′GCCGGA<u>G</u>AT; only the underlined base is a mismatch with the Elk-1 consensus). The presence of Ets proteins in the slower migrating complexes was strongly suggested by competition for their assembly using oligonucleotides comprising an optimized Ets-1 binding site. As a more stringent test, specific antisera detected Pax-5/BSAP in each of the four complexes, while antisera specific for individual Ets proteins affected the three slower complexes selectively. The composition of BTCs was shown to include Pax-5(BSAP) and different Ets proteins – Fli-1 in BTC1, Ets-1 in BTC2, and GABPα in BTC3 (Fig. 3B). In addition to GABPα, BTC3 also includes the ankyrin repeat protein GABPβ.

BTC sites are important for activating *mb-1* transcription (Fitzsimmons et al. 1996). Mutation of either the Pax-5/BSAP site or the Ets site similarly reduced (by four- to fivefold) transcription from the promoter in transfected early B cells, while mutation of both sites had only a slightly greater effect, indicating that the two sites function together in vivo. Interestingly, mutation of either site reduced promoter function in a manner similar to mutations in the upstream EBF site or to deletions of distal region sequences including all of these sites (J. Hagman, unpublished data). Together, these data show that Pax-5:Ets complexes enhance transcription and suggest that early B-cell-specific factors (EBF and Pax-5) and Ets proteins activate transcription synergistically in early B cells. In accord with in vitro studies, expression of the *mb-1* gene is reduced by 90% in the *pax-5*$^{-/-}$ knockout mouse (Nutt et al. 1998), and *mb-1* transcription is restored to nearly normal levels by expression of Pax-5/BSAP in *pax-5*$^{-/-}$ knockout pre-B cells. As one caveat to these studies, direct evidence has not been obtained to implicate one (or more) of the previously identified BTCs as the complex that participates in *mb-1* regulation.

In vitro studies using recombinant proteins showed that Pax-5/BSAP recruits its Ets partners to bind a suboptimal DNA sequence through cooperative interactions on specific DNA sequences (FITZSIMMONS et al. 1996). In addition to the three Ets components detected in BTCs (Fli-1, Ets-1, and GABPα), other recombinant Ets proteins can be recruited by Pax-5/BSAP to bind the promoter in vitro, including Ets-2, v-Ets, Net, and Elk-1 (D. Fitzsimmons and J. Hagman, unpublished data). Only SAP1a binds detectably by itself under normal conditions, but SAP1a (and PU.1) does not bind the promoter cooperatively with Pax-5. Analysis of DNA binding by progressively truncated Pax-5/BSAP and Ets polypeptides showed that minimal sequences for cooperative DNA binding are included in their respective DNA-binding motifs, and that BTC complexes are comprised of monomers of each of the two components.

As an important point, the last G in the Elk-1 binding-site homology (5′GCCGGA$\underline{\text{G}}$AT) is contacted by Pax-5/BSAP in the ternary complex. Mutation of the G to an A greatly decreases Pax-5/BSAP binding (to 5% of wild-type binding), while the affinity of Ets DNA binding to the mutated site is increased by more than 100-fold. Moreover, Ets proteins (e.g., Ets-1) efficiently recruit Pax-5/BSAP to bind the mutated site. The recruitment of partner proteins in a reciprocal fashion on wild-type versus mutated sites strongly suggests that the two polypeptides contact each other in the ternary complex.

Candidate amino acids that may participate in Pax-5/BSAP:Ets protein–protein interactions have been identified. The ability of Ets proteins to be recruited by Pax-5/BSAP coincides with the presence of a specific aspartic acid (Asp398 in Ets-1) within the ETS DNA-binding domain (FITZSIMMONS et al. 1996). The aspartic acid is just carboxy-terminal to an α-helix that recognizes the major groove of DNA (reviewed in GRAVES and PETERSEN 1998). The aspartic acid is conserved among the Ets proteins that are recruited by Pax-5, but not among those that are not recruited (valine or lysine are substituted for aspartic acid in SAP1a or PU.1, respectively). Substitution of aspartic acid for valine in SAP1a allows for its efficient recruitment by Pax-5/BSAP to bind the promoter. This relationship strongly suggests that the aspartic acid participates directly in Pax-5:Ets interactions. More recently, it was observed that aspartic or glutamic acid, but not other amino acids, allows for recruitment of Ets proteins by Pax-5/BSAP (D. Fitzsimmons, unpublished data).

To make predictions concerning sequences within Pax-5/BSAP that interact with Ets proteins on specific DNA, a structural model (Fig. 4) was prepared using previously determined X-ray crystallographic and nuclear magnetic resonance (NMR) structures of related proteins bound to their cognate sites (WHEAT et al. 1999). Because structural determinations were not available for Pax-5/BSAP itself, a structure determined for the highly homologous paired domain (77% identity) of Drosophila Paired bound to DNA was utilized. In this crystal structure, the amino-terminal sub-domain and linker of the paired domain contact its optimized site, while the carboxy-terminal sub-domain does not interact with DNA. Because preliminary

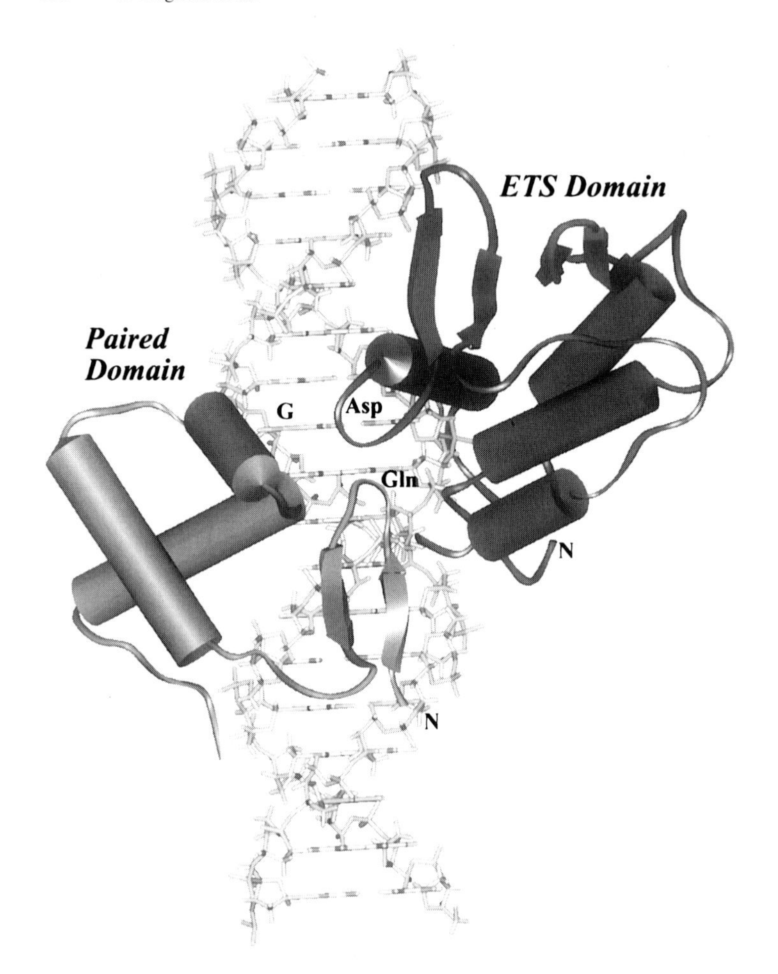


**Fig. 4.** Model for Pax-5/BSAP:Ets interactions. The model was generated using the graphic program SETOR (Evans 1993) to represent (a) the nuclear-magnetic-resonance (NMR)-derived structure of the murine Ets-1 ETS domain and flanking carboxy-terminal α-helix (Donaldson et al. 1996; Werner et al. 1997); *Brookhaven Protein Data Base (PDB) code*: 1ETC docked on idealized B-DNA in an orientation similar to that observed in the crystal structure of the PU.1 ETS domain/DNA complex (Kodandapani et al. 1996); *dark grey*, PDB code: 1PUE, and (b) the amino-terminal sub-domain of drosophila paired (Xu et al. 1995); *light grey*, PDB code: 1PDN docked by alignment of the *mb-1* promoter and optimized Paired binding-site sequences. Relative positioning and orientation of the ETS domain was suggested by the obvious relationship between the *mb-1* promoter and consensus Ets-1 binding sites (5′CCGGAG and 5′CCGGAA/T, respectively). Positioning of the amino-terminal sub-domain of Paired was based on the assumption that the G common to the Ets-1 and Pax-5/BSAP sites (5′CCGGAG) is, by analogy with Paired DNA binding, contacted by the first residue of its DNA recognition α-helix (α3). Small distortions of the DNA observed in the structures of the Paired-DNA and ETS domain-DNA complexes were neglected for this modeling

data suggested that the amino-terminal sub-domain of Pax-5/BSAP is oriented toward the Ets binding site, the amino-terminal sub-domain of Paired was docked onto an ideal B-form DNA corresponding to the *mb-1* sequence. For the ETS domain, the recently determined structures of the Ets-1 (Fig. 4) and PU.1 DNA-binding domains bound to their recognition sites were readily appropriated for the ternary complex model (DONALDSON et al. 1996; KODANDAPANI et al. 1996; WERNER et al. 1997).

The model predicts that Pax/BSAP and Ets proteins can bind the *mb-1* promoter in very close proximity to one another. Furthermore, the model is consistent with previous studies indicating that an aspartic acid (Asp398) of Ets-1 is important for interactions with Pax-5/BSAP, and suggests that this amino acid may contact the paired domain (FITZSIMMONS et al. 1996). If correct, recruitment of Ets proteins is a function of the amino-terminal sub-domain alone. The model cannot address whether the carboxy-terminal sub-domain is required for high affinity binding of the *mb-1* promoter by Pax-5; however, the contribution of this sub-domain in the ternary complex may be the stabilization of Pax-5/BSAP binding through additional contacts with DNA, and not through contacts with the ETS domain.

As predicted by the model, recruitment of Ets-1 was localized to sequences within the amino-terminal sub-domain of Pax-5/BSAP (WHEAT et al. 1999). Amino acids 12–84 of Pax-5/BSAP can recruit the Ets-1 ETS domain, but only when tethered to a heterologous DNA-binding domain bound to DNA. A key amino acid for Ets-1 recruitment is an invariant glutamine residue (Gln22 in Pax-5/BSAP) present in the β-hairpin motif near the amino-termini of all paired domains. Conversion of this amino acid to alanine resulted in a 75–80% decrease in the ability of Pax-5/BSAP to recruit Ets-1 to bind DNA, but had no effect on DNA binding by Pax-5/BSAP alone. In the ternary complex model, this residue is in close proximity to the ETS domain DNA recognition α-helix and the key aspartic acid (Asp398) and could play a direct role in interactions between these two proteins; however, an indirect structural role cannot be excluded.

Although much is clear now concerning how Pax-5/BSAP and Ets proteins interact with each other on the *mb-1* promoter, a number of questions remain to be addressed. As reported previously, autoinhibitory sequences flanking either side of the ETS domain of Ets-1 decrease its affinity for DNA (HAGMAN and GROSSCHEDL 1992; LIM et al. 1992; NYE et al. 1992; WASYLYK et al. 1992). Formation of the Pax-5:Ets-1 ternary complex may lead to de-repression of Ets-1 DNA binding by protein–protein interactions with residues within its amino- or carboxy-terminal inhibitory domains. Apparently, interaction with Pax-5/BSAP overcomes obstacles to Ets-1 DNA binding by stabilizing both proteins in the ternary complex (J. Hagman and W. Wheat, unpublished data), but it is not clear how this is accomplished. It is also not understood how other polypeptides that associate with BTCs, e.g., GABPβ, participate in ternary complex assembly. Further structural and biochemical studies will be required to define precise mechanisms for assembly of Pax-5:Ets ternary complexes.

## 6 Control of B-Cell Proliferation

Proliferation of B lymphocytes occurs at important developmentally controlled points during early B lymphopoiesis (MELCHERS 1997). Studies have implicated

Pax-5/BSAP as a regulator of proliferation in B cells. Pax-5/BSAP DNA-binding activity increases in normal ex vivo B cells treated with mitogens (e.g., LPS), BCR cross-linking with anti-IgM, or cross-linking of CD40 through co-culture with CD40L-expressing cells (Wakatsuki et al. 1994). In the same study, addition of mitogens to resting B-cell cultures increased Pax-5/BSAP DNA-binding activity by three- to fourfold. Proliferation of these cultures correlated positively with Pax-5/BSAP levels. Conversely, incubation of LPS-treated splenic B cells with *pax-5* mRNA-specific antisense oligonuleotides inhibited proliferation by more than 100-fold. Specific inhibition was observed following treatment of Pax-5/BSAP-expressing lymphoma cells, but not of cells that do not express Pax-5. It was concluded that Pax-5/BSAP regulates proliferation in a manner similar to regulation by other cellular proto-oncogenes, e.g., c-*fos*, but does so in a tissue-specific manner. Other data that link Pax-5/BSAP to oncogenesis in B cells (Busslinger et al. 1996; Iida et al. 1996) and the brain (Kozmik et al. 1995; Stuart et al. 1995) support the hypothesis that Pax-5/BSAP levels are crucial for regulating the proliferation of normal cells in a tissue-specific fashion (reviewed in Morrison et al. 1998).

We performed an experiment to assess the effects of LPS stimulation on the expression of B-cell-specific ternary complexes, which we identified as important for regulating *mb-1* gene transcription (Fitzsimmons et al. 1996). Nuclear extracts were prepared from B lymphoid tumor cells (as controls), from normal resting splenic B cells, or from splenic B cells incubated for 24 h with LPS. Nuclear extract proteins were tested for the presence of BTCs in an EMSA, using $^{32}$P-labeled probes (Fig. 5). Unlike pre-B (70Z/3 and 38B9) or mIg$^{+}$ (WEHI-231) B-lymphoid tumor cells, normal resting B cells express almost no detectable Pax-5/BSAP by itself or complexed with Ets proteins as BTCs. After overnight incubation with LPS, levels of BTCs increased by nine- to tenfold. The greater induction of Pax-5/BSAP binding relative to previous studies (Wakatsuki et al. 1994) may be due to use of the *mb-1* gene probe, which is bound cooperatively by Pax-5/BSAP and Ets proteins. In contrast, detection of Ets proteins using an Ets consensus probe detected relatively constant levels of these proteins in the various extracts (Fig. 5). We conclude that BTC expression in B cells is low in resting cells and increased in proliferating cells treated with LPS. The data suggest that BTC levels are primarily controlled by modulating levels of Pax-5/BSAP, but further studies are needed to clarify the status of Ets proteins in these cells.

## 7 Future Directions

A great deal has been learned concerning the roles of the Pax-5/BSAP protein as a regulator of tissue-specific gene expression and differentiation. In addition to the problems addressed above, a number of other questions remain to be addressed. One of the major mysteries concerning Pax-5/BSAP concerns its activities as either

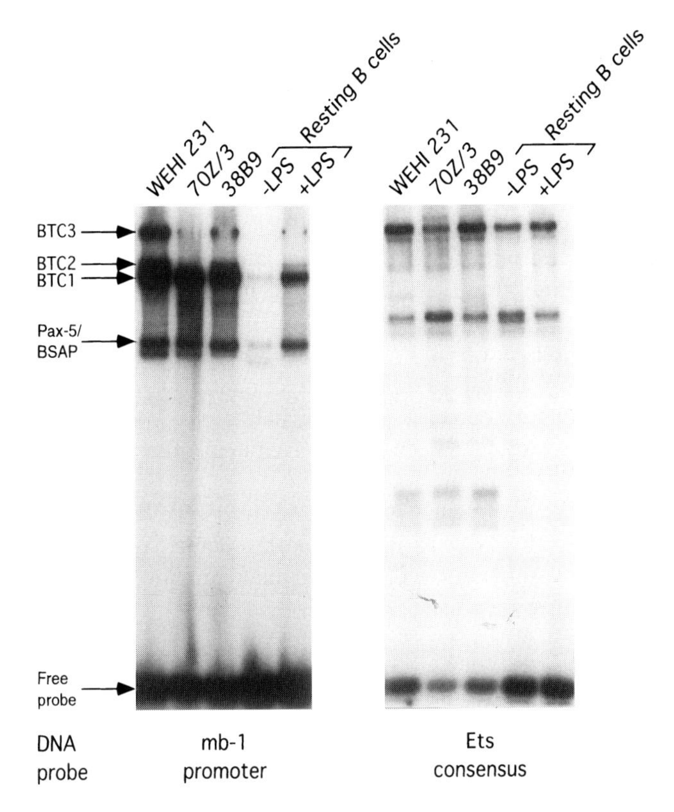


**Fig. 5.** Detection of B-cell-specific ternary complexes in nuclear extracts of B lymphoid tumor cells, and normal resting or lipopolysaccharide-activated B cells. Electrophoretic mobility shift assays (EMSA), $^{32}$P-labeled double-stranded oligonucleotide probes (indicated below), and nuclear extract preparation were all described previously (Fitzsimmons et al. 1996). For normal B-cell extracts, B cells were isolated from normal AKR spleens following red cell lysis and depletion of T cells using monoclonal antibodies T24 (anti-Thy1.1/1.2; gift from U. Staerz), HO2.2 (anti-CD8), GK1.5 (anti-CD4), and baby rabbit complement. Resting B cells were isolated from remaining cells by density gradient centrifugation on Percoll gradients and collection from the 1.085/1.079 interface (Julius et al. 1988). Purity of B cells was assessed as more than 90% B220/Ia$^{+}$ by flow cytometry (FACS). Cells were used to make nuclear extracts immediately, or cultured overnight in the absence or presence of 50μg/ml lipopolysaccharide (*E. coli* 0111:B4)

an activator or repressor in different contexts, e.g., of the *CD19* vs *PD-1* genes. In part, the differential activities of the protein may be related to consequences of binding different sites, e.g., DNA-induced conformational changes that selectively expose activation or repression domains, but it is not clear how this would be accomplished. One study has suggested that activator sites are higher-affinity binding sites, while lower-affinity binding is a context for repression (Wallin et al. 1998), but this may be an oversimplification. "Context" may actually reflect binding to sequences that promote interactions with different partner proteins.

Interactions with partners are clearly important for increasing the DNA-binding specificity of Pax-5/BSAP, and may also affect the functional outcome(s) of

DNA binding on transcription. Partnerships with Ets proteins, which are nuclear effectors of the Ras-MAP-kinase signaling pathway (Wasylyk et al. 1998), suggest that BTCs mediate changes in gene expression as an endpoint of signaling pathways, but more work is needed to resolve this intriguing possibility. The identification of other proteins that interact with Pax-5/BSAP will continue to be a major area of interest.

Another problem concerns the roles of Pax-5/BSAP at later stages of B-cell differentiation. The protein is clearly implicated as a regulator of Ig heavy-chain class switching, proliferation, and terminal differentiation, but further studies in these areas have been hampered by the lack of a useful mouse model system. Moreover, more information is needed to understand how the *pax-5* gene itself is controlled in cells that possess the ability to differentiate in multiple ways, e.g., in germinal centers. The use of mice (or cells) that can delete *pax-5* genes in a lineage- and differentiation-stage-specific fashion will facilitate answering these questions.

In summary, the Pax-5/BSAP protein has proved to be a versatile regulator of specific gene expression and development of B lymphocytes. These studies have provided important insights into the control of cellular differentiation and functional properties of Pax transcription factors.

*Acknowledgements.* The authors wish to thank Drs. Lawrence P. McIntosh and Lisa N. Gentile for generating the Pax:Ets model, Drs. M. Karen Newell and John Freed for assistance with the preparation of normal resting B cells, and Boyd Anderson for assistance with illustrations. This work was supported by generous awards to J. Hagman from the American Cancer Society (DB-8309), the National Institutes of Health (R01 AI37574 and P01 AI22295), and by a Local Chapter Grant from the Rocky Mountain Chapter of the Arthritis Foundation. W. Wheat was supported by a fellowship from the Cancer League of Colorado and by NJMRC. D. Fitzsimmons is supported by a fellowship from NJMRC.

## References

Adams B, Dorfler P, Aguzzi A, Kozmik Z, Urbánek P, Maurer-Fogy I, Busslinger M (1992) Pax-5 encodes the transcription factor BSAP and is expressed in B lymphocytes, the developing CNS, adult testis. Genes Dev 6:1589–1607

Agata Y, Kawasaki A, Nishimura H, Ishida Y, Tsubata T, Yagita H, Honjo T (1996) Expression of the PD-1 antigen on the surface of stimulated mouse T and B lymphocytes. Int Immunol 8:765–772

Asano M, Gruss P (1991) Pax-5 is expressed at the midbrain-hindbrain boundary during mouse development. Mech Dev 39:29–39

Barberis A, Widenhorn K, Vitelli L, Busslinger M (1990) A novel B-cell lineage-specific transcription factor present at early but not late states of differentiation. Genes Dev 22:37–43

Busslinger M, Klix N, Pfeffer P, Graninger P, Kozmik Z (1996) Deregulation of Pax-5 by translocation of the Eμ enhancer of the IgH locus adjacent to two alternative Pax-5 promoters in diffuse large-cell lymphoma. Proc Natl Acad Sci USA 93:6129–6134

Busslinger M, Urbanek P (1995) The role of BSAP (Pax-5) in B-cell development. Curr Opin Genet Dev 5:595–601

Campbell KS, Cambier JC (1990) B lymphocyte antigen receptors (mIg) are non-covalently associated with a disulfide linked, inducibly phosphorylated glycoprotein complex. EMBO J 9:441–448

Carter RH, Fearon DT (1992) CD19: Lowering the threshold for antigen receptor stimulation of B lymphocytes. Science 256:105–107

Chang Y, Bosma GC, Bosma MJ (1995) Development of B cells in scid mice with immunoglobulin transgenes, implications for the control of VDJ recombination. Immunity 2:607–616
Cogné M, Lansford R, Bottaro A, Zhang J, Gorman J, Young F, Cheng H-L, Alt FW (1994) A class switch control region at the 3′ end of the immunoglobulin heavy chain locus. Cell 77:737–747
Corcoran AE, Riddell A, Krooshoop D, Venkitaraman AR (1998) Impaired immunoglobulin gene rearrangement in mice lacking the IL-7 receptor. Nature 391:904–907
Czerny T, Busslinger M (1995) DNA-binding and transactivation properties of Pax-6, three amino acids in the paired domain are responsible for the different sequence recognition of Pax-6 and BSAP (Pax-5). Mol Cell Biol 15:2858–2871
Czerny T, Schaffner G, Busslinger M (1993) DNA sequence recognition by Pax proteins, bipartite structure of the paired domain and its binding site. Genes Dev 7:2048–2061
Dildrop R, Ma A, Zimmermann K, Hsu E, Tesfaye A, DePinho R, Alt FW (1989) IgH enhancer-mediated deregulation of N-*myc* gene expression in transgenic mice, generation of lymphoid neoplasias that lack c-*myc* expression. EMBO J 8:1121–1128
Donaldson LW, Petersen JM, Graves BJ, McIntosh LP (1996) Solution structure of the ETS domain from murine Ets-1, a winged helix-turn-helix DNA binding motif. EMBO J 15:125–134
Dörfler P, Busslinger M (1996) C-terminal activating and inhibitory domains determine the transactivation potential of BSAP (Pax-5), Pax-2, Pax-8. EMBO J 15:1971–1982
Engel P, Zhou L-J, Ord DC, Sato S, Koller B, Tedder TF (1995) Abnormal B lymphocyte development, activation, and differentiation in mice that lack or overexpress the CD19 signal transduction molecule. Immunity 3:39–50
Evans S (1993) SETOR, Hardware lighted three-dimensional solid representation of macromolecules. J Mol Graphics 11:134–138
Feldhaus A, Mbangkollo D, Arvin K, Klug C, Singh H (1992) BlyF, a novel cell-type- and stage-specific regulator of the B-lymphocyte gene *mb-1*. Mol Cell Biol 12:1126–1133
Fitzsimmons D, Hagman J (1996) Regulation of gene expression at early stages of B-cell and T-cell differentiation. Curr Opin Immunol 8:166–174
Fitzsimmons D, Hodsdon W, Wheat W, Maira S-M, Wasylyk B, Hagman J (1996) Pax-5 (BSAP) recruits Ets proto-oncogene family proteins to form functional ternary complexes on a B-cell-specific promoter. Genes Dev 10:2198–2211
Giaccia AJ, Kastan MB (1998) The complexity of p53 modulation, emerging patterns from divergent signals. Genes Dev 12:2973–2983
Graves BJ, Petersen J M (1998) Specificity within the ets family of transcription factors. Adv Cancer Res 75:1–55
Hagman J, Belanger C, Travis A, Turck CW, Grosschedl R (1993) Cloning and functional characterization of early B-cell factor, a regulator of lymphocyte-specific gene expression. Genes Dev 7:760–73
Hagman J, Grosschedl R (1992) An inhibitory carboxyl-terminal domain in Ets-1 and Ets-2 mediates differential binding of ETS family factors to promoter sequences of the *mb-1* gene. Proc Natl Acad Sci USA 89:8889–8893
Hagman J, Travis A, Grosschedl R (1991) A novel lineage-specific nuclear factor regulates *mb-1* gene transcription at the early stages of B cell differentiation. EMBO J 10:3409–17
Hardy RR, Carmack CE, Shinton SA, Kemp JD, Hayakawa K (1991) Resolution and characterization of pro-B and pre-pro-B cell stages in normal mouse bone marrow. J Exp Med 173:1213–1225
Hombach J, Tsubata T, Leclerq L, Stappert H, Reth M (1990) Molecular components of the B-cell antigen receptor complex of the IgM class. Nature 343:760–762
Iida S, Rao PH, Nallasivam P, Hibshoosh H, Butler M, Louie DC, Dyomin V, Ohno H, Chaganti RSK, Dalla-Favera R (1996) The t(9;14)(p13;q32) chromosomal translocation associated with lymphoplasmacytoid lymphoma involves the PAX-5 gene. Blood 88:4110–4117
Ishida Y, Agata Y, Shibahara K, Honjo T (1992) Induced expression of PD-1, a novel member of the immunoglobulin gene superfamily, upon programmed cell death. EMBO J 11:3887–3895
Julius MH, Janusz M, Lisowski J (1988) A colostral protein that induces the growth and differentiation of resting B lymphocytes. J Immunol 140:1366–1371
Kodandapani R, Pio F, Ni C-Z, Piccialli G, Klemsz M, McKercher S, Maki RA, Ely KR (1996) A new pattern for helix-turn-helix recognition revealed by the PU1 ETS-domain-DNA complex. Nature 380:456–460
Kozmik Z, Sure U, Ruedi D, Busslinger M, Aguzzi A (1995) Deregulated expression of PAX5 in medulloblastoma. Proc Natl Acad Sci USA 92:5709–5713
Kozmik Z, Wang S, Dörfler P, Adams B, Busslinger M (1992) The promoter of the *CD19* gene is a target for the B-cell-specific transcription factor BSAP. Mol Cell Biol 12:2662–2672

Liao F, Birshtein BK, Busslinger M, Rothman P (1994) The transcription factor BSAP (NF-HB) is essential for immunoglobulin germ-line ε transcription J Immunol 152:2904–2911
Liao F, Giannini S, Brishtein B (1992) A nuclear DNA-binding protein expressed during early stages of B cell differentiation interacts with diverse segments within and 3′ of the Ig H chain gene cluster J Immunol 148:2909–2917
Lim F, Kraut N, Frampton J, Graf T (1992) DNA binding by c-Ets-1, but not v-Ets, is repressed by an intramolecular mechanism. EMBO J 11:643–652
Lin H, Grosschedl R (1995) Failure of B-cell differentiation in mice lacking the transcription factor EBF. Nature 376:263–267
Maulbecker CC, Gruss, P (1993) The oncogenic potential of Pax genes. EMBO J 12:2361–2367
Max EE, Wakatsuki Y, Neurath MF, Strober W (1995) The role of BSAP in immunoglobulin isotype switching and B-cell proliferation. Curr Top Micro Immunol 194:449–458
Melchers F (1997) Control of the sizes and contents of precursor B cell repertoires in bone marrow Ciba Found Symp 204:172–182
Morrison AM, Nutt SL, Thevenin C, Rolink A, Busslinger M (1998) Loss- and gain-of-function mutations reveal an important role of BSAP (Pax-5) at the start and end of B cell differentiation. Semin Immunol 10:133–142
Neurath MF, Max EE, Strober W (1995) Pax5 (BSAP) regulates the murine immunoglobulin 3′ alpha enhancer by suppressing binding of NF-alpha P, a protein that controls heavy chain transcription. Proc Natl Acad Sci USA 92:5336–5340
Neurath MF, Strober W, Wakatsuki Y (1994) The murine immunoglobulin 3′α enhancer is a target site with repressor function for the B-cell lineage-specific transcription factor BSAP (NF-HB, Sα-BP) J Immunol 153:730–742
Nutt SL, Morrison AM, Dörfler P, Rolink A, Busslinger M (1998) Identification of BSAP (Pax-5) target genes in early B-cell development by loss- and gain-of-function experiments. EMBO J 17:2319–2333
Nutt SL, Urbánek P, Rolink A, Busslinger M (1997) Essential functions of Pax5 (BSAP) in pro-B cell development, difference between fetal and adult B lymphopoiesis and reduced V-to-DJ recombination at the IgH locus. Genes Dev 11:476–491
Nye JA, Petersen JM, Gunther CV, Jonsen MD, Graves BJ (1992) Interaction of murine Ets-1 with GGA-binding sites establishes the ETS domain as a new DNA-binding motif. Genes Dev 6:975–990
Okabe T, Watanabe T, Kudo A (1992) A pre-B and B cell-specific DNA-binding protein, EBB-1 which binds to the promoter of the VpreB1 gene. EMBO J 12:2753–2772
Papavasiliou F, Jankovic M, Suh H, Nussenzweig MC (1995a) The cytoplasmic domains of immunoglobulin (Ig) α and Ig β can independently induce the precursor B cell transition and allelic exclusion. J Exp Med 182:1389–1394
Papavasiliou F, Misulovin Z, Suh H, Nussenzweig MC (1995b) The role of Ig beta in precursor B cell transition and allelic exclusion. Science 268:408–411
Pettersson S, Grant P, Faxen M, Samuelsson A, Skogberg M, Arulampalam V (1995) The 3′ end of the IgH locus, an important regulator of B-lymphoid development. The Immunologist 3:146–151
Qiu G, Stavnezer J (1998) Overexpression of BSAP/Pax-5 inhibits switching to IgA and enhances switching to IgE in the I29μ B cell line. J Immunol 161:2906–2918
Reimold AM, Ponath PD, Li Y-S, Hardy RR, David CS, Strominger JL, Glimcher LH (1996) Transcription factor B cell lineage-specific activator protein regulates the gene for human X-box binding protein 1. J Exp Med 183:393–401
Reya T, Grosschedl R (1998) Transcriptional regulation of B-cell differentiation. Curr Opin Immunol 10:158–165
Rickert RC, Rajewsky K, Roes J (1995) Impairment of T-cell-dependent B-cell responses and B-1 cell development in CD19-deficient mice. Nature 376:352–355
Rinkenberger JL, Wallin JJ, Johnson KW, Koshland ME (1996) An interleukin-2 signal relieves BSAP (Pax5)-mediated repression of the immunoglobulin J chain gene. Immunity 5:377–386
Riva A, Wilson GL, Kehrl JH (1997) In vivo footprinting and mutational analysis of the proximal CD19 promoter reveal important roles for an SP1/Egr-1 binding site and a novel site termed the PyG box. J Immunol 159:1284–1292
Rolink A, Grawunder U, Winkler TH, Karasuyama H, Melchers F (1994) IL-2 receptor α chain (CD25, TAC) expression defines a crucial stage in pre-B cell development. Int Immunol 6:1257–1264
Rolink A, Melchers F (1996) B-cell development in the mouse. Immunol Lett 54:157–161
Roqué, MC, Smith PA, Blasquez VC (1996) A developmentally modulated chromatin structure at the mouse immunoglobulin κ3′ enhancer. Mol Cell Biol 16:3138–3155

Rosenbaum H, Webb E, Adams JM, Cory S, Harris AW (1989) N-*myc* transgene promotes B lymphoid proliferation, elicits lymphomas, and reveals cross-regulation with c-*myc*. EMBO J 8:749–755

Rothman P, Li SC, Gorham B, Glimcher L, Alt F, Boothby M (1991) Identification of a conserved lipopolysaccharide-plus-interleukin-4-responsive element located at the promoter of germ line ε transcripts. Mol Cell Biol 11:5551–5561

Sakaguchi N, Kashiwamura S-I, Kimoto M, Thalmann P, Melchers F (1988) B lymphocyte lineage-restricted expression of *mb-1*, a gene with CD-3 like structural properties. EMBO J 7:3457–3464

Shaffer AL, Peng A, Schlissel MS (1997) In vivo occupancy of the κ light chain enhancers in primary pro- and pre-B cells, a model for κ locus activation. Immunity 6:131–143

Singh H (1996) Gene targeting reveals a hierarchy of transcription factors regulating specification of lymphoid cell fates. Curr Opin Immunol 8:160–165

Singh M, Birshtein BK (1996) Concerted repression of an immunoglobulin heavy-chain enhancer, 3′ alpha E(hs1,2). Proc Natl Acad Sci USA 93:4392–4397

Singh M, Birshtein BK (1993) NF-HB (BSAP) is a repressor of the murine immunoglobulin heavy-chain 3′α enhancer at early stages of B-cell differentiation. Mol Cell Biol 13:3611–3622

Spanopoulou E, Roman CA, Corcoran LM, Schlissel MS, Silver DP, Nemazee D, Nussenzweig MC, Shinton SA, Hardy RR, Baltimore D (1994) Functional immunoglobulin transgenes guide ordered B-cell differentiation in Rag-1-deficient mice. Genes Dev 8:1030–1042

Stuart ET, Haffner R, Oren M, Gruss P (1995) Loss of p53 function through PAX-mediated transcriptional repression. EMBO J 14:5638–5645

Stuart ET, Kioussi C, Aguzzi A, Gruss P (1995) PAX5 expression correlates with increasing malignancy in human astrocytomas. Clin Cancer Res 1:207–214

Stüber E, Neurath M, Calderhead D, Fell HP, Strober W (1995) Cross-linking of OX-40 ligand, a member of the TNF/NGF cytokine family, induces proliferation and differentiation in murine splenic B cells. Immunity 2:507–521

Tamir I, Cambier JC (1998) Antigen receptor signaling, integration of protein tyrosine kinase functions. Oncogene 17:1353–1364

Teh YM, Neuberger MS (1997) The immunoglobulin (Ig)α and Igβ cytoplasmic domains are independently sufficient to signal B cell maturation and activation in transgenic mice. J Exp Med 185:1753–1758

Theines CP, De Monte L, Monticelli S, Busslinger M, Gould HJ, Vercelli D (1997) The transcription factor B cell-specific activator protein (BSAP) enhances both IL-4- and CD40-mediated activation of the human ε germline promoter. J Immunol 158:5874–5882

Thevenin C, Nutt SL, Busslinger M (1998) Early function of Pax5 (BSAP) before the pre-B cell receptor stage of B lymphopoiesis. J Exp Med 188:735–744

Tian J, Okabe T, Miyazaki T, Takeshita S, Kudo A (1997) Pax-5 is identical to EBB-1/KLP and binds to the VpreB and lambda5 promoters as well as the KI and KII sites upstream of the Jkappa genes. Eur J Immunol 27:750–755

Travis A, Hagman J, Grosschedl R (1991) Heterogeneously initiated transcription from the pre-B- and B-cell-specific *mb-1* promoter: analysis of the requirement for upstream factor-binding sites and initiation site sequences. Mol Cell Biol 11:5756–5755

Treisman R (1992) The serum response element. Trends Biochem Sci 17:423–426

Urbánek P, Wang Z-Q, Fetka I, Wagner EF, Busslinger M (1994) Complete block of early B cell differentiation and altered patterning of the posterior midbrain in mice lacking Pax5/BSAP. Cell 79:901–912

Usui T, Wakatsuki Y, Matsunaga Y, Kaneko S, Kosek H, Kita T (1997) Overexpression of B cell-specific activator protein (BSAP/Pax-5) in a late B cell is sufficient to suppress differentiation to a Ig high producer cell with plasma cell phentoype. J Immunol 158:3197–3204

Wakatsuki Y, Neurath M, Max E, Strober W (1994) The B cell-specific transcription factor BSAP regulates B cell proliferation. J Exp Med 179:1099–1108

Wallin JJ, Gackstetter ER, Koshland ME (1998) Dependence of BSAP repressor and activator functions on BSAP concentration. Science 279:1961–1964

Walther C, Guenet JL, Simon D, Deutsch U, Jostes B, Goulding MD, Plachov D, Balling R, Gruss P (1991) Pax, a murine multigene family of paired box-containing genes. Genomics 11:424–434

Wasylyk B, Hagman J, Gutierrez-Hartmann A (1998) Ets transcription factors: nuclear effectors of the Ras-MAP-kinase signaling pathway. Trends Bioc Sci 23:213–216

Wasylyk C, Kerckaert J-P, Wasylyk B (1992) A novel modulator domain of Ets transcription factors. Genes Dev 6:965–974

Werner MH, Clore GM, Fisher CL, Fisher RJ, Trinh L, Shiloach J, Gronenborn AM (1997) Correction of the NMR structure of the Ets1/DNA complex. J Biomol NMR 10:317–328

Wheat W, Fitzsimmons D, Lennox H, Krautkramer SR, Gentile LN, McIntosh LP, Hagman J (1999) The highly conserved β-hairpin of the paired DNA-binding domain is required for the assembly of Pax:Ets ternary complexes. Mol Cell Biol (in press)

Xu W, Rould MA, Jun S, Desplan C, Pabo CO (1995) Crystal structure of a paired domain-DNA complex at 25 Å resolution reveals structural basis for Pax developmental mutations. Cell 80:639–650

Ying H, Healy JI, Goodnow CC, Parnes JR (1998) Regulation of mouse *CD72* gene expression during B lymphocyte development. J Immunol 161:4760–4767

Young F, Ardman B, Shinkai Y, Lansford R, Blackwell TK, Mendelsohn M, Rolink A, Melchers F, Alt FW (1994) Influence of immunoglobulin heavy- and light-chain expression on B-cell differentiation. Genes Dev 8:1043–1057

Zwollo P, Arrieta H, Ede K, Molinder K, Desiderio S, Pollock R (1997) The Pax-5 gene is alternatively spliced during B-cell development. J Biol Chem 272:10160–10168

Zwollo P, Desiderio S (1994) Specific recognition of the *blk* promoter by the B-lymphoid transcription factor B-cell-specific activator protein. J Biol Chem 269:15310–15317

Zwollo P, Rao S, Wallin JJ, Gackstetter ER, Koshland ME (1998) The transcription factor NF-kappaB/p50 interacts with the *blk* gene during B cell activation. J Biol Chem 273:18647–18655

# Receptor Modulators of B-Cell Receptor Signalling – CD19/CD22

K.G.C. Smith and D.T. Fearon



## 1 Introduction

Activation of the B cell by stimulation through its antigen receptor, along with co-operation from a cognate T cell, forms the basis of the T-cell-dependent humoral immune response. The amount of information that can be transmitted to the B cell by this simple interaction with membrane immunoglobulin (mIg) is, however, limited. Optimally, B-cell responses should be maximised if the B cell is in an appropriate microenvironment, if the antigen is likely to be dangerous to the host, and if an adequate immune response to the antigen has not already been made. There is increasing evidence that activation through the B-cell receptor (BCR) can

The Wellcome Trust Immunology Unit, Department of Medicine, University of Cambridge School of Clinical Medicine, Addenbrooke's Hospital, Cambridge CB2 2SP, England
e-mail: kgcs2@cam.ac.uk

be modulated by a number of co-receptors, providing the B cell with information of this kind. Two of these co-receptors, CD22 and CD19, will be discussed in this review.

# 2 CD22

## 2.1 CD22 Expression, Structure and Function

CD22 is a cell-surface glycoprotein expressed only on B cells (Dörken et al. 1986; Stamenkovic and Seed 1990). It was first identified on mouse B cells by a monoclonal antibody recognising the Lyb 8.2 alloantigen (Symington et al. 1982); the alternative allotype of CD22, Lyb 8.1, is expressed by only a few mouse strains. The expression of CD22 was originally thought to be restricted to mature B cells (Campana et al. 1985; Dörken et al. 1986; Erickson et al. 1996), but more recent studies have demonstrated expression, albeit at lower levels, on pre B and immature B cells (Nitschke et al. 1997; Stoddart et al. 1997). All subsets of mature murine B cells appear to express CD22 with the exception of plasma cells; CD22 has a similar distribution in the human (reviewed by Tedder et al. 1997a).

The CD22 glycoprotein is a member of the subclass of the Ig superfamily known as the sialoadhesins. The gene for human CD22 has 15 exons and is located at q13.1 of chromosome 19. A splice variant of CD22 makes up a small proportion of the total, failing to express exons 6 and 7 which encode Ig domains 3 and 4 (Wilson et al. 1993). Mouse CD22 is 62% homologous to human CD22 at the amino acid level (Torres et al. 1992), and its gene is located on chromosome 7 (Law et al. 1993). The two allotypes of mCD22, Lyb 8.1 and 8.2, differ by only 3% at the amino acid level and most of this difference is in the two amino-terminal Ig domains, to which the allotypic antibody binds. Both human and mouse CD22 molecules are made up of an extracellular region comprised of seven Ig domains, a transmembrane region and a cytoplasmic domain which contains six conserved tyrosine residues.

Early studies of COS cells expressing CD22 demonstrated its ability to adhere to a number of cell types, in particular both B and T lymphocytes (reviewed by Tedder et al. 1997b). This binding was found to be mediated by the two amino-terminal Ig domains (Engel et al. 1995a; Law et al. 1995) which bind to Siaα2-6Galβ1-4GlcNAc. This sialic acid derivative is widespread on cell-surface glycoproteins, but is expressed particularly on cells expressing α2,6-sialyltransferase, which sialylates a number of cell surface proteins and promotes CD22 binding but, paradoxically, can sialylate CD22 itself and inhibit ligand binding (Braesch-Anderson and Stamenkovic 1994; Razi and Varki 1998). Appropriate sialic acid residues for binding CD22 exist on a number of cell surface molecules, but in particular on CD45 (Sgroi et al. 1995), IgM and CD22 itself. Use of a CD22-Ig fusion protein has demonstrated preferential binding to B rather than T cells

(Law et al. 1995) with less binding to nonhematopoietic cells, though endothelial cells express CD22 ligands when exposed to inflammatory cytokines (Hanasaki et al. 1994).

CD22, being a member of the sialoadhesin family, was initially thought to act predominantly as an adhesion molecule. It soon became clear, however, that CD22 was involved in B-cell activation, and it was thought that CD22 may act as a co-receptor enhancing mIg-induced B-cell activation. This was suggested by the increase in B-cell proliferation noted to occur when CD22 was separately ligated by anti-CD22 antibodies (Pezzutto et al. 1987, 1988; Tuscano et al. 1996a), and by the findings that CD22 could be co-precipitated with the antigen receptor (Leprince 1993; Peaker and Neuberger 1993) and was also associated with Lyn, syk and PI-3 kinase (Law et al. 1996; Tuscano et al. 1996b). Finally the observation that CD22 became phosphorylated after stimulation of the B cell through the antigen receptor (Schulte et al. 1992; Peaker and Neuberger 1993; Williams et al. 1994) seemed to confirm the role of CD22 as a co-activator.

More recently it has become clear that the predominant effect of CD22 on B-cell activation involves negative regulation. The first suggestions of this were raised by the fact that after antigen-receptor stimulation CD22 became associated with SHP-1 (Campbell and Klinman 1995; Doody et al. 1995), a protein tyrosine phosphatase which negatively regulates signalling through the surface receptor (Cyster and Goodnow 1995; Pani et al. 1995) and a deficiency of which is responsible for the phenotype of motheaten mice (Kozlowski et al. 1993; Schulz et al. 1993; Tsui et al. 1993). SHP-1 also mediates negative regulation of cell signalling through killer-inhibitory receptors on NK cells (Burshtyn et al. 1996; Ono et al. 1997) and PIR-B receptors on B cells (Kubagawa et al. 1997; Bléry et al. 1998).

This circumstantial evidence suggesting a negative regulatory role for CD22 in B-cell activation was supported by Doody et al. (1995), who ligated CD22 to beads coated with anti-CD22 antibody. This sequestration of CD22 away from the antigen receptor resulted in increased B-cell activation in response to mIg stimulation. Conversely, ligation of CD22 to mIg reduced mitogen-activated protein (MAP) kinase activation (Tooze et al. 1997) and intracellular calcium flux (Smith et al. 1998). The conclusion that CD22 is an inhibitory co-receptor was unambiguously confirmed by the production of mice deficient in CD22 (O'Keefe et al. 1996; Otipoby et al. 1996; Sato et al. 1996a; Nitschke et al. 1997). These mice demonstrated increased flux of intracellular calcium in response to cross linking of the antigen receptor and down-modulation of surface IgM on B cells (known to follow antigen engagement). They had increased plasma cells and serum IgM and IgA, but reduced circulating mature B cells, perhaps due to their predisposition to activation and, therefore, to subsequent apoptosis.

This evidence indicates that the primary role for CD22 is in the negative regulation of B-cell activation (Fig. 1A). This does not, however, exclude an additional positive stimulatory role. SHP-1 ligation to CD22 and subsequent activation is likely to be mediated by one or more of the three immunoreceptor tyrosine-based inhibitory motifs (ITIMs) of the cytoplasmic domain. Three addi-

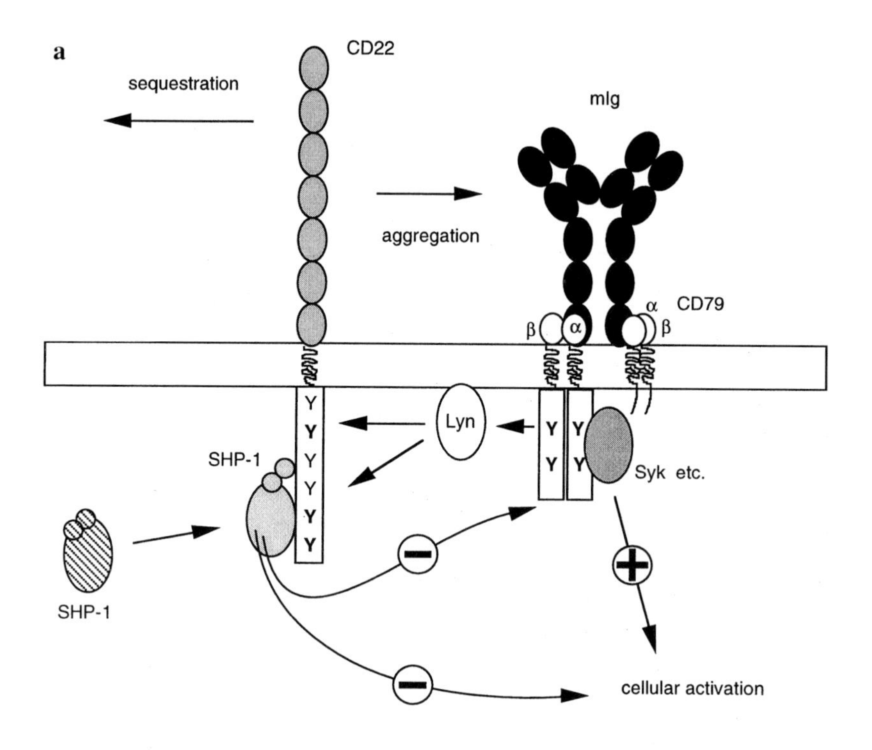
a
CD22
sequestration
mIg
aggregation
α CD79
β
α
β
Y
Y
Y
Y
Y
Y
Lyn
Y
Y
Y
Y
SHP-1
Syk etc.
SHP-1
cellular activation

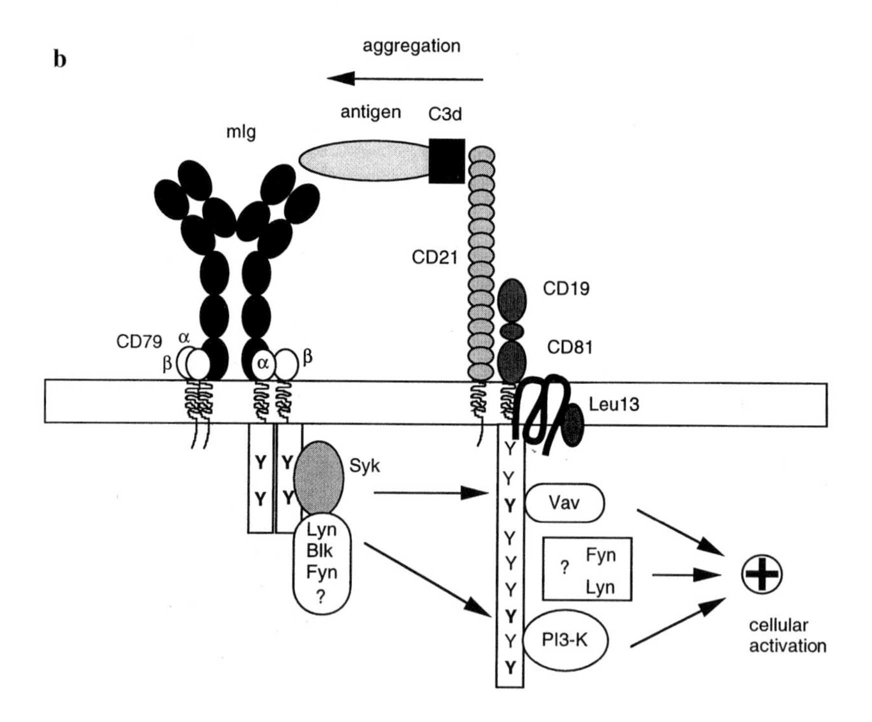
b
aggregation
antigen
C3d
mIg
CD21
CD19
CD79
α
β
α
β
CD81
Leu13
Syk
Lyn
Blk
Fyn
?
Vav
? Fyn
Lyn
PI3-K
cellular
activation

tional tyrosines also exist in the cytoplasmic domain, phosphorylation of which may well enhance activation, though evidence supporting this possibility is yet to be produced.

## 2.2 Intracellular Signalling Pathways Involved in Negative Regulation by CD22

Surface IgM is constitutively associated with a proportion of CD22 molecules (LEPRINCE et al. 1993; PEAKER and NEUBERGER 1993). Phosphorylation of CD22 increases markedly after stimulation of the B cell through the antigen receptor (SCHULTE et al. 1992; LEPRINCE et al. 1993; PEAKER and NEUBERGER 1993; WILLIAMS et al. 1994) as does the subsequent association with SHP-1 (CAMPBELL and KLINMAN 1995; DOODY et al. 1995). Doody et al. (1995) characterised the SHP-1 binding sites in CD22, demonstrating that phosphorylation of tyrosines associated with three ITIM motifs having the consensus sequence, allowed the binding and activation of SHP-1.

The tyrosine kinase Lyn has been demonstrated to associate both with the BCR-associated signal transduction molecule CD79 (YAMANASHI et al. 1991) and with CD22 (TUSCANO et al. 1996b). This together with some similarities in B-cell phenotype displayed by mice deficient in Lyn (HIBBS et al. 1995; NISHIZUMI et al. 1995; CHAN et al. 1997), SHP-1 (SCHULZ et al. 1993; TSUI et al. 1993), and CD22 (O'KEEFE et al. 1996; OTIPOBY et al. 1996; SATO et al. 1996a; NITSCHKE et al 1997), suggested that Lyn, CD22 and SHP-1 may be all part of the same signalling process, subserving CD22's negative regulatory function. In support of this, Lyn has recently been demonstrated to be required for constitutive and mIg-induced phosphorylation of the ITIM motifs of CD22 and the subsequent association of CD22 with SHP-1 (CHAN et al. 1998; CORNALL et al. 1998; Smith et al. 1998).

SHP-1 has a negative regulatory effect in a number of biological systems (SCHARENBERG and KINET 1996), and two specific downstream signalling pathways influenced by CD22 have been defined. MAP kinases form part of the signalling cascade which translate signals from the cell surface into alterations in gene transcription. Stimulation through the antigen receptor is known to modestly increase activation of three such MAP kinases, ERK2, JNK and p38, and this effect is markedly increased by the co-ligation of CD19 (TOOZE et al. 1997). Ligation of CD22 to the antigen receptor suppresses activation of all three MAP kinases, while

**Fig. 1a,b.** B-cell co-receptors and signalling through mIg. **a** CD22. The position of CD22 in the membrane relative to membrane immunoglobulin (mIg) is likely to be controlled by the relative proximity of its ligand to antigen. Signalling through mIg results in the phosphorylation of tyrosines 783, 843 and 863, denoted by *Y*, which mediates binding with SHP-1, which in turn negatively regulates cellular activation. **b** CD19/CD21/CD81 complex. CD19 and the three molecules which complex with it are shown aggregated to mIg by antigen and C3d. This co-ligation results in the phosphorylation of a number of tyrosines (*Y*) in the intracytoplasmic domain of CD19. Tyrosine 391 associates with Vav while tyrosines 482 and 513 associate with PI3-K (shown in *bold*). The phosphotyrosines mediating interactions with tyrosine kinases, such as Fyn and Lyn, are less well defined. Interaction with and activation of these molecules leads to downstream signalling events culminating in an enhancement of cellular activation sequestrating CD22 away from the antigen receptor increases their activation, presumably by releasing the tonic negative regulatory effects of CD22 and SHP-1 (Tooze et al. 1997). CD22 also acts to suppress the release of intracellular free calcium by stimulation of the antigen receptor (Smith et al. 1998), perhaps by suppressing the tyrosine phosphorylation of phospholipase C (PLC)-γ (Doody and Fearon unpublished observations).

## 2.3 CD22 and Regulation of the Immune Response In Vivo

CD22 inhibits B-cell activation when associated with mIg, and regulation of this spatial association in the membrane is likely to be important in controlling CD22 function in vivo. This is probably largely controlled by interactions between CD22 and its ligand(s). This regulation is likely to be complex in the in vivo immune response, as the sialic acid derivative which binds CD22 can be expressed on a number of glycoproteins and on a number of cell types, and this expression itself can be modified by the action of α2,6-sialyltransferase. Mutant mice deficient in CD22 or its ligand are informative, but more approaches are needed to dissect the complex regulatory interactions of the B-cell immune response. Details of the role of CD22 in this response, therefore, remain somewhat speculative.

### 2.3.1 Sequestration by Ligand on Adjacent Cells

CD22 may provide the B cell with information on its microenvironment, helping ensure that activation occurs where this environment is most appropriate. Evidence for this is provided by experiments demonstrating CD22 sequestration on the surface of beads results in B-cell hyperactivity (Doody et al. 1995). In addition, the CD22 ligand is most highly expressed on B and T lymphocytes (Law et al. 1995). These observations suggest that when a B cell finds itself in a lymphoid environment, the neighbouring lymphocytes could ligate CD22, sequestering it from the antigen receptor and thus making the B cell more prone to activation. Conversely, when outside a lymphoid microenvironment, where the B cell is less likely to find either appropriate T-cell help or B-cell follicles to allow induction of an effective immune response, the absence of CD22 ligand would allow increased association of CD22 with mIg, thus restraining B-cell activation until a more appropriate anatomical site is found. Convincing evidence supporting this hypothesis has been provided by the study of mice deficient in α2,6-sialyltransferase and thus unable to make Siaα2-6Galβ1-4GlcNAc, the major CD22 ligand. These mice have normal B-cell number and development, but markedly impaired B-cell responses to both T-dependent and T-independent antigens in vivo, and CD40 ligand, lipopolysaccharide (LPS) and anti-IgM in vitro (Hennet et al. 1998). In addition, decreased serum IgM and decreased intracellular calcium flux and tyrosine phosphorylation in response to mIg cross-linking was observed. Increased association of CD22 with the B-cell receptor following the removal of its ligand from adjacent cells would explain much of the phenotype of these mice.

In addition to modulating immune responsiveness, contact between CD22 and its ligand may assist in promoting appropriate intercellular interaction in lymphoid organs. While the apparent normality of lymphoid microarchitecture in CD22 deficient and α2,6-sialyltransferase-deficient mice suggests that this effect is not critical for the overall structure of lymphoid organs, it may nonetheless be important at a cellular level. For example, it is possible that the interaction between CD22 and its ligand plays a role in slowing the passage of B cells through lymphoid organs or specific areas within lymphoid organs, perhaps enhancing their chances of finding appropriate T-cell help. CD22 may also play a role in B-cell homing to areas of inflammation, as suggested by the upregulation of CD22 ligand on endothelial cells treated with inflammatory cytokines (Hanasaki et al. 1994).

### 2.3.2 CD22 and IgM

Another major glycoprotein which can function as a CD22 ligand is IgM. Interaction between CD22 and surface IgM on the same cell may contribute to the constitutive association of CD22 with the antigen receptor, though whether an appropriate sialic acid exists on monomeric, as distinct from pentameric, IgM has not been confirmed. Nonetheless, if such an interaction did exist, competition between the B-cell receptor and other CD22 ligands on the surface of the same cell, for example CD45, may play an important role in the regulation of B-cell reactivity.

The binding of CD22 to circulating pentameric IgM may also be of physiological significance. Circulating IgM may cross link CD22, sequestrating it from the antigen receptor. This may be important in the early augmentation of the immune response, when very rapid IgM production by the antibody-forming cell foci (Jacob et al. 1991; Smith et al. 1996) might result in high IgM levels in the immediate vicinity, thus increasing sequestration of CD22 locally and amplifying such a response. Conversely if IgM was bound to the antigen receptor, it could act to coligate CD22 and decrease B-cell activation. This might provide a form of negative feedback once antigen-IgM complexes had been formed, with the presence of antigen allowing targeting of such a complex, along with CD22 to the B-cell receptor. Germinal center B cells express low levels of SHP-1, perhaps rendering them insensitive to such negative regulation through CD22 (Delibrias et al. 1997). Coligation of CD22 by IgM might also be a mechanism by which the production of anti-idiotypic antibodies (i.e. anti-Ig autoantibodies or rheumatoid factor) is prevented, as previously suggested by Cyster and Goodnow (1997).

### 2.3.3 Other Possible Mechanisms of CD22 Regulation

The activity of α2,6-sialyltransferase, and therefore the density of appropriate CD22 ligands, on the B cell itself may also be altered to control CD22 regulation. It has been suggested that much CD22 on resting B cells exists bound to ligands on the same cell, or has its binding-site blocked by sialylation, and is therefore unavailable to bind CD22 on adjacent cells. Activation of B cells appears to "unmask" CD22 making it accessible to ligation by other cells (Razi and Varki 1998).

Control of CD22 expression may also be used to regulate its function. CD22 is rapidly trafficked to and from the cell surface and has reported to be expressed at low levels on human germinal centre (GC) B cells (Dörken et al. 1986; Ling et al. 1987), perhaps reducing their level of negative suppression and promoting GC formation and function. Probably more importantly, SHP-1, the key protein tyrosine phosphatase required for CD22 negative regulation, is also expressed at particularly low levels in the GC in both mice and humans (Delibrias et al. 1998). This too is likely to contribute to the appropriate hyperactivity of this cell population and may be necessary for GC development.

At least two other negative regulatory molecules in addition to CD22 are known to exist on the surface of B cells. The function of the recently described PIR-B receptors, which also recruit SHP-1, has not been fully determined (Bléry et al. 1998). The role of FcγRIIB1, which interacts with SHIP, is more apparent. It seems likely that, as discussed above, a major role of CD22 is to indicate whether the B cell is within a lymphoid microenvironment, predisposing it to activation by antigen in a context where T-cell help can readily be found. Later in the immune response FcγRIIB1 provides an antigen-specific negative signal as it is ligated to mIg by IgG specific for antigen, thus providing a means of shutting down the successful, isotype-switched, antigen-specific immune response.

## 2.4 CD22 and Autoimmunity

Negative regulation of B-cell activation through CD22 may be important in maintaining tolerance and thus avoiding autoimmune disease, and there is some suggestion that defects in CD22-mediated negative regulation of B cells may contribute to the development of such disease.

Homozygous defects in Lyn or SHP-1 (both expressed in B and non-B cells) result in autoimmune disease with some similarities to systemic lupus erythematosus (SLE) (Schulz et al. 1993; Hibbs et al. 1995; Nishizumi et al. 1995; Pani et al. 1995; Wang et al. 1996; Chan et al. 1997). The phenotype of CD22-deficient mice is less severe, but these mice nonetheless develop the isotype-switched anti-dsDNA antibodies typical of SLE (O'Keefe et al. 1996; Otipoby et al. 1996; Sato et al. 1996a; Nitschke et al. 1997). Heterozygous defects in all three of these components of the CD22 pathway also combine to cause a defect in B-cell regulation, implying that even partial defects in CD22 function could contribute to the development of polygenic autoimmune disease (Cornall et al. 1998).

A number of mouse models of autoimmune disease have hyperreactive B cells, with a tendency to polyclonal activation in vitro and autoantibody production in vivo. Prominent among them is the NZB/W model of SLE which has three strong recessive alleles associated with the development of glomerulonephritis, as determined by study of a derived inbred strain particularly susceptible to SLE, NZM/Aeg2410 (Rudofsky et al. 1993). One of these (*Sle3*) maps to an area of chromosome 7 containing the CD22 gene (Law et al. 1993; Morel et al. 1994) and is particularly associated with autoantibody production (Mohan et al. 1997).

There are also features of human SLE which point towards a failure of B-cell negative regulation. Spontaneously active polyclonal antibody-forming cells are found in the circulation of those with SLE (Budman et al. 1977; Fauci et al. 1980), and their peripheral blood B cells demonstrate an exaggerated intracellular calcium response to BCR stimulation, similar to that seen in CD22-deficient mice (Liossis et al. 1996). Early in the course of SLE, many patients have hyper-IgM (Budman et al. 1977; Fauci et al. 1980), again similar to mice in which the CD22 pathway has been disrupted. Thus, defects in the CD22 pathway may well contribute to autoimmunity in both mice and humans. Such defects might be due to genetic polymorphisms, or loss of function mutations, in *Lyn*, *Cd22* or *Shp-1*.

# 3 CD19

## 3.1 CD19 Expression and Structure

CD19 is a cell-surface glycoprotein of the Ig superfamily expressed only on B cells. Its gene is located at chromosome 16 p11.2, and it is expressed throughout the B-cell lineage in both mouse and man from the pro-B cell to the mature B cell, though it is not expressed at significant levels on plasma cells (reviewed by Tedder et al. 1994). It exists in a complex with CD21 (complement receptor 2: CR2), CD81 (TAPA-1) and Leu 13.

The extracellular region of CD19 contains three domains, two of which are of the Ig type. The contribution of this extracellular region to the function of CD19 is not known, other than its mediation of the association with CD21 and CD81 (Bradbury et al. 1993; Matsumoto et al. 1993). The cytoplasmic domain of CD19 is more conserved between human, mice and guinea pigs and contains nine highly conserved tyrosine residues. Phosphorylation of these residues allows interaction of the cytoplasmic domain of CD19 with intracellular signalling molecules, including phosphatidylinositol-3 kinase (PI-3 kinase) and Vav.

## 3.2 The CD19/CD21/CD81 Complex

CD19 exists as part of a complex also containing CD21, CD81 and Leu 13 (Matsumoto et al. 1991; Bradbury et al. 1992). CD21 is a cell-surface molecule with an extracellular domain comprised of 15–16 short consensus repeats and a 34 amino acid cytoplasmic domain. The extracellular domain mediates binding with three products of C3 cleavage, iC3b, C3dg and C3d. The association of CD19 with CD21 is necessary for the function of the former, as mice deficient in either of these molecules have similar phenotypes (see below). CD81 is a member of the trans-membrane 4 superfamily expressed on many cell types (reviewed by Levy et al.

1998). It associates with a number of cell-surface molecules in addition to CD19, though the phenotype of CD81 knockout mice suggests CD81 may be necessary for membrane expression of CD19. Leu 13 is a 16-kDa protein which often associates with CD81 (Evans et al. 1990) though its functional role in the CD19/CD21/CD81 complex has not been defined.

CD19 modulates the proliferative and activation signals delivered to the B cell by mIg. Co-ligation of the CD19/CD21/CD81 complex to the antigen receptor results in a 100- to 1000-fold decrease in both the proliferation and calcium thresholds of B cells stimulated through surface IgM (Carter et al. 1991; Carter and Fearon 1992; Dempsey et al. 1996). The results of these initial in vitro studies were extended by the generation of mice deficient in CD19 (Engel et al. 1995b; Rickert et al. 1995). Although these mice had only modest decreases in proliferative responses to mIg stimulation and CD40 ligation in vitro, more striking abnormalities were observed in vivo. The CD19 deficiency was associated with a 75% reduction in serum IgM, IgG1 and IgG2a levels, markedly diminished T-dependent immune responses, decreased GC formation, a reduction in affinity maturation, and a reduction in the number of B1 cells to less than 25% of normal (Engel et al. 1995b; Rickert et al. 1995). These abnormalities contrasted with relatively unimpaired responses to type-2 T-independent antigens.

Mice in which human CD19 was overexpressed in B cells have a reciprocal phenotype. They show an increase in in vitro proliferation in response to antigen-receptor cross-linking and increases in serum Ig titres and B1 cell numbers (Engel et al. 1995b; Sato 1996b). In addition, they displayed decreased production of mature B cells, perhaps due to increased deletion of immature B cells in the bone marrow as a result of enhanced signalling through the antigen receptor. It therefore appears that the effect of CD19 co-stimulation in B cells is to enhance both the sensitivity of the B cell for activation and the magnitude of the subsequent response. The biological implications of this will be discussed below.

CD21 acts as a ligand-binding subunit of the CD19/CD21/CD81 complex. Mice deficient in CD21 have a similar, though less severe, phenotype than those deficient in CD19. They have decreased T-dependent B-cell responses, but normal levels of serum IgM and IgG. GCs are detectable but may be reduced in size (Ahearn et al. 1996; Molina et al. 1996). Reductions in T-dependent B-cell responses in these mice are due to the absence of CD21 from B cells and not FDCs (Ahearn et al. 1996; Croix et al. 1996). In addition, CD21 may provide a necessary survival signal for GC B cells (Fischer et al. 1998).

The wide distribution of CD81 explains why mice deficient in CD81 have a more complex phenotype than those deficient in either CD19 or CD21 alone. Nonetheless, these mice exhibit reduced membrane expression of CD19, decreased proliferation in response to stimulation through the antigen receptor, lower levels of most subclasses of Ig, and a reduced number of B1 cells. This phenotype is consistent with deficient activity of the CD19/CD21/CD81 complex, most likely due to its decreased expression in CD81-deficient mice (Maecker and Levy 1997; Miyazaki et al. 1997; Tsitsikov et al. 1997) and suggesting a role for CD81 in the trafficking of CD19 to the plasma membrane.

## 3.3 Intracellular Signalling Pathways Downstream of CD19

CD19 complex acts by synergistically increasing signalling through mIg (Fig. 1B). CD19 co-activation reduces the number of antigen receptors which need to be stimulated to result in increases in intracellular calcium concentration and activation of MAP kinases (CARTER et al. 1991; DEMPSEY et al. 1996; TOOZE et al. 1997). CD19 mediates a similar reduction in the threshold required for B-cell proliferation (CARTER and FEARON 1992).

Intracellular signalling arising from the CD19/CD21/CD81 complex is independent of both CD21 and CD81 (MATSUMOTO et al. 1993) and requires the cytoplasmic domain of CD19 (SATO et al. 1997a). This cytoplasmic domain contains nine conserved tyrosine residues which, when phosphorylated, allow association with PI3-kinase and Vav (TUVESON et al. 1993; VAN NOESEL et al. 1993; WENG et al. 1994; CHALUPNY et al. 1995). The roles of three phosphotyrosines have been defined: pY482 and pY513 bind PI3-K through its p85 subunit (TUVESON et al. 1993), and pY391 recruits Vav (O'ROURKE et al. 1998). The mechanism of association between CD19 and the tyrosine kinases lyn and fyn is not known (VAN NOESEL et al. 1993; CHALUPNY et al. 1995).

Co-activation through CD19 involves at least two major types of signalling pathway. The first of these is the co-stimulation of intracellular calcium flux. Ligation of CD19 alone results in the production of inositol 1,4,5-trisphosphate, but this production is greatly enhanced by co-ligation of both CD19 and mIg. These effects of CD19 do not occur through an effect on the tyrosine phosphorylation of PLC-$\gamma$ (CARTER et al. 1991), which is induced by mIg (1991), but by increasing the availability of the substrate phosphatidylinositol 4,5-bisphosphate ($PI(4,5)P_2$), a rate limiting component in inositol lipid hydrolysis (STEPHENS et al. 1993). The mechanism by which CD19 induces the synthesis of $PI(4,5)\ P_2$ requires the recruitment of Vav, which is a guanine nucleotide-exchange protein for the Rho family of small GTPases. The activation of Rac, or possibly Cdc42, by Vav leads to activation of phosphatidylinositol 4-phosphate 5-kinase and synthesis of $PI(4,5)\ P_2$ (O'ROURKE et al. 1998). This interaction is required in the maintenance of intracellular calcium elevation after the initial stimulation of the B cell through mIg. These effects of CD19 would be anticipated to regulate the transcription factor nuclear factor-activated T cells (NF-AT).

The other major pathway of signalling co-stimulated by CD19 is that of the MAP kinases ERK, JNK and p38, which transactivate several transcription factors. Membrane Ig stimulation modestly activates ERK2 (CASILLAS et al. 1991; GOLD et al. 1992) and there is even less of an effect on JNK and p38 (LI et al. 1996; SUTHERLAND et al. 1996). CD19 co-ligation with mIg, however, markedly augments the activation of all three MAP kinases (LI et al. 1997; TOOZE et al. 1997), with co-stimulation of JNK but not ERK2 being mediated by Vav (O'ROURKE et al. 1998). Thus, CD19, when co-ligated with mIg, should cause the expression of genes that are regulated by NF-AT, ERK, JNK and p38. These genes remain to be identified.

## 3.4 CD19 and Regulation of the Immune Response In Vivo

### 3.4.1 CD19 and Complement

The complement system is comprised of a number of plasma proteins which are sequentially activated in response to microbial and other inflammatory insults. When the complement cascade is activated a number of complement proteins are cleaved and one, C3, attaches covalently to the target of complement activation while also serving as a ligand for cellular receptors, including CD21.

The depletion of C3 by cobra venom factor has long been known to impair the humoral immune response (Pepys 1974). The importance of complement in T-cell-dependent responses has been subsequently extended by the generation of mice deficient in complement components (Fischer et al. 1996). It seems likely that a major mechanism by which complement contributes to humoral immunity is via the CD19/CD21/CD81 complex. It has been shown that antigen coupled to C3d causes co-localisation of the antigen receptor and of the CD19/CD21/CD81 complex via interaction of C3d with CD21. This lowers the threshold for B-cell activation by approximately one order of magnitude for each C3d molecule attached to a model antigen in vitro, with even greater reductions of up to 10,000-fold seen in vivo (Dempsey et al 1996). The CD19/CD21/CD81 complex therefore provides a link between innate and acquired immunity. C3d "tags" antigens associated with microbial infection, allowing recruitment of the CD19/CD21/CD81 complex and thus reducing the threshold for activation of the B cell. Thus, the complement system and the CD19/CD21/CD81 complex are providing the B cell with qualitative information about the potential microbial origin of a given antigen, promoting maximal immune responses where most appropriate.

### 3.4.2 The Possible Role of a CD19 Ligand

The position of the CD19/CD21/CD81 complex relative to the antigen receptor can be controlled by the binding of antigen-associated C3d to CD21. That the phenotype of mice deficient in CD19 is more severe than that of those deficient in CD21 (see above) suggests, however, that factors in addition to CD21 might regulate CD19 signalling, such as a ligand for CD19. Although CD77 has been suggested to function as a ligand for CD19, the biological significance of this observation is unclear (Maloney and Lingwood 1994).

### 3.4.3 CD19 and the Maintenance of B1 and Memory B Cells

The presence of the CD19/CD21/CD81 complex is important for the maintenance of the B1 cell population. B1 cells, many of which express CD5, are the predominant B cell found in the peritoneal cavity and represent a lineage of B cells distinct from conventional (B2) B cells, that, in some circumstances, produce autoantibodies (Murakami and Honjo 1995). The B1 cell population is reduced in mice deficient in either CD19 or CD21 (see above) or chronically treated with anti-CD19 antibody (KROP et al. 1996), and is expanded in mice overexpressing CD19 (ENGEL et al. 1995b; SATO 1996b). These findings imply that the CD19/CD21/CD81 complex may be required for lowering B1 cell thresholds and facilitating a basal level of turnover.

Persistence of memory B2 cells requires the presence of antigen, which may cause their slow turnover (GRAY and SKARVALL 1988; SCHITTEK and RAJEWSKY 1990). The possible analogous role of the CD19/CD21/CD81 complex for the maintenance of B2 cell memory is suggested by the recent demonstration that, while GCs can be formed in mice deficient in CD19, B-cell memory does not develop (FEHR et al. 1998).

# 4 Conclusion

Activation of the B cell is influenced by co-receptors that can either enhance proliferation and lower the threshold for activation (e.g. the CD19/CD21/CD81 complex) or, conversely, can suppress these things (for example CD22 or FcγRIIB1). When the B cell comes into contact with antigen, then, it will have input from these and possibly other regulatory molecules such as the PIRs, allowing it to integrate multiple signals as it determines whether and how to respond to antigen. Thus the B cell acquires information about its microenvironment by the relative position in the membrane of CD22; it receives information about the potential microbial origin of the antigen through co-localisation of CD19/CD81 by CD21; and it detects pre-existing antigen-specific IgG using FcγRIIB1. An improved understanding of the biology of B-cell co-receptors should lead to strategies for enhancing or suppressing immune responses in man.

*Acknowledgements.* This work was supported by the Wellcome Trust through a Principal Research Fellowship (to D.T. FEARON) and a Biomedical Research Collaboration Grant (to K.G.C. SMITH), and also by an MRC Career Establishment Grant (to K.G.C. SMITH). We thank K. Thompson for expert secretarial assistance, S. Secker and L. O'Rourke for helpful discussions and G. Doody for sharing unpublished results.

# References

Ahearn JM, Fischer MB, Croix D, Goerg S, Ma M, Xia J, Zhou X, Howard RG, Rothstein TL, Carroll MC (1996) Disruption of the Cr2 locus results in a reduction in B-1a cells and in an impaired B cell response to T-dependent antigen. Immunity 4:251–262

Bléry M, Kubagawa H, Chen CC, Vély F, Cooper MD, Vivier E (1998) The paired Ig-like receptor PIR-B is an inhibitory receptor that recruits the protein-tyrosine phosphatase SHP-1. Proc Natl Acad Sci USA 95:2446–2451

Bradbury LE, Kansas GS, Levy S, Evans S, Tedder TF (1992) The CD19/CD21 signal transducing complex of human B lymphocytes includes the target of antiproliferative antibody -1 and Leu-13 molecules. J Immunol 149:2841–2850

Bradbury LE, Goldmacher VS, Tedder TF (1993) The CD19 signal transduction complex of B lymphocytes. Deletion of the CD19 cytoplasmic domain alters signal transduction but not complex formation with TAPA-1 and Leu 13. J Immunol 151:2915–2927

Braesch-Andersen S, Stamenkovic I (1994) Sialylation of the B lymphocyte molecule CD22 by alpha2, 6-sialyltransferase is implicated in the regulation of CD22-mediated adhesion. J Biol Chem 269:11783–11786

Budman DR, Merchant EB, Steinberg AD, Doft B, Gershwin ME, Lizzio E, Reeves JP (1977) Increased spontaneous activity of antibody-forming cells in the peripheral blood of patients with active SLE. Arthritis Rheum 20:829–833

Burshtyn DN, Yang W, Yi Taolin, Long EO (1997) A novel phosphotyrosine motif with a critical amino acid at position 2 for the SH2 domain mediated activation of the tyrosine phosphatase SHP-1. J Biol Chem 272:13066–13072

Campana D, Janossy G, Bofill M, Trejdosiewicz LK, Hoffbrand MD, Mason AV, Lebacq AM, Forster HK (1985) Human B cell development. I. Phenotypic differences of B lymphocytes in the bone marrow and peripheral lymphoid tissue. J Immunol 134:1524–1530

Campbell M, Klinman NR (1995) Phosphotyrosine-dependent association between CD22 and protein tyrosine phosphatase 1C. Eur J Immunol 25:1573–1579

Carter RH, Tuveson DA, Park DJ, Rhee SG, Fearon DT (1991) The CD19 complex of B lymphocytes. Activation of phospholipase C by a protein tyrosine kinase-dependent pathway that can be enhanced by the membrane IgM complex. J Immunol 147:3663–3671

Carter RH and Fearon DT (1992) CD19: lowering the threshold for antigen receptor stimulation by B lymphocytes. Science 256:105–107

Casillas A, Hanekom C, Williams K, Katz R, Nel AE (1991) Stimulation of B-cells via the membrane immunoglobulin receptor or with phorbol myristate 13-acetate induces tyrosine phosphorylation and activation of a 42-kDa microtubule-associated protein-2 kinase. J Biol Chem 266:19088–19094

Chalupny NJ, Aruffo A, Esselstyn JM, Chan P-Y, Bajorath J, Blake J, Gilliland LK, Ledbetter JA, Tepper MA (1995) Specific binding of Fyn and phosphatidylinositol 3-kinase to the B cell surface glycoprotein CD19 through their src homology domains. Eur J Immunol 25:2978–2984

Chan VWF, Meng F, Soriano, P, DeFranco AL, Lowell CA (1997) Characterization of the B lymphocyte populations in lyn-deficient mice and the role of lyn in signal initiation and down-regulation. Immunity 7:69–81

Chan VWF, Lowell CA, DeFranco AL (1998) Defective negative regulation of antigen receptor signaling in lyn-deficient B lymphocytes. Current Biology 8:545–553

Cornall RJ, Cyster JG, Hibbs ML, Dunn AR, Otipoby KL, Clark EA, Goodnow CC (1998) Polygenic autoimmune traits: lyn, CD22 and SHP-1 are limiting elements of a biochemical pathway regulating BCR signaling and selection. Immunity 8:497–508

Croix DA, Ahearn JM, Rosengard AM, Han S, Kelsoe G, Ma M, Carroll MC (1996) Antibody response to a T-dependent antigen requires B cell expression of complement receptors. J Exp Med 183:1857–1864

Cyster JG and Goodnow CC (1995) Protein tyrosine phosphatase 1C negatively regulates antigen receptor signaling in B lymphocytes and determines thresholds for negative selection. Immunity 2:13–24

Cyster JG and Goodnow CC (1997) Tuning antigen receptor signaling by CD22: integrating cues from antigens and the microenvironment. Immunity 6:509–517

Delibrias CC, Floettmann JE, Rowe M, Fearon DT (1997) Downregulated expression of SHP-1 in Burkitt lymphomas and germinal center B lymphocytes. J Exp Med 186:1575–1583

Dempsey PW, Allison MED, Akkaraju S, Goodnow CC, Fearon DT (1996) C3d of complement as a molecular adjuvant: bridging innate and acquired immunity. Science 271:348–350

Doody GM, Justement LB, Delibrias CC, Matthews RJ, Lin J, Thomas ML, Fearon DT (1995) A role in B cell activation for CD22 and the protein tyrosine phosphatase SHP. Science 269:242–244

Dörken B, Moldenhauer G, Pezzutto A, Schwartz R, Feller A, Kiesel S, Nadler LM (1986) HD39 (B3), a B lineage-restricted antigen whose cell surface expression is limited to resting and activated human B lymphocytes. J Immunol 136:24470–24479

Engel P, Wagner N, Miller A, Tedder TF (1995a) Identification of the ligand binding domains of CD22, a member of the immunoglobulin superfamily that uniquely binds a sialic acid-dependent ligand. J Exp Med 181:21581–21586

Engel P, Zhou L-J, Ord D C, Sato S, Kioller B, Tedder TF (1995) Abnormal B lymphocyte development, activation and differentiation in mice that lack or overexpress the CD19 signal transduction molecule. Immunity 3:39–50

Erickson LD, Tygrett LT, Bhatia SK, Grabstein KH, Waldschmidt TJ (1996) Differential expression of CD22 (Lyb8) on murine B cells. International Immunology 8:1121–1129

Evans SS, Lee DB, Han T, Tomasi TB, Evans RL (1990) Monoclonal antibody to the interferon-inducible protein Leu-13 triggers aggregation and inhibits proliferation of leukemic B cells. Blood 76:2583–2593

Fauci AS, Montsopoulos HM (1980) Polyclonally triggered B cells in the peripheral blood and bone marrow of normal individuals and in patients with SLE and Sjögren's syndrome. Arthritis Rheum 24:577–584

Fehr T, Rickert RC, Odermatt B, Roes J, Rajewsky K, Hengartner H, Zinkernagel RM (1998) Antiviral protection and germinal center formation, but impaired B cell memory in the absence of CD19. J Exp Med 188:145–155

Fischer MB, Minghe M, Goerg S, Zhou X, Xia J, Finco O, Han S, Kelsoe G, Howard RG, Rothstein TL, Kremmer E, Rosen FS, Carroll MC (1996) Regulation of the B cell response to T-dependent antigens by classical pathway complement. J Immunol 157:549–556

Fischer MB, Goerg S, Shen L, Prodeus AP, Goodnow CC, Kelsoe G, Carroll MC (1998) Dependence of germinal center B cells on expression of CD21/CD35 for survival. Science 280:582–585

Gold MR, Sanghera JS, Stewart J, Pelech SJ (1992) Selective activation of p42 mitogen-activated protein (MAP) kinase in murine B lymphoma cell lines by membrane immunoglobulin cross-linking. Biochem J 287:269–276

Gray D, Skarvall H (1988) B cell memory is short-lived in the absence of antigen. Nature 336:70–73

Hanasaki K, Varki A, Stamenkovic I, Bevilacqua MP (1994) Cytokine-induced β-galactoside α-2,6-sialyltransferase in human endothelial cells mediates α2,6-sialylation of adhesion molecules and CD22 ligands. J Biol Chem 269:10637–10643

Hennet T, Chui D, Paulson JC, Marth JD (1998) Immune regulation by the ST6Gal sialyltransferase. Proc Natl Acad Sci U S A 95:4504–4509

Hibbs ML, Tarlinton DM, Armes J, Grail D, Hodgson G, Maglitto R, Stacker SA, Dunn AR (1995) Multiple defects in the immune system of Lyn-deficient mice, culminating in autoimmune disease. Cell 83:301–311

Hippen KL, Buhl AM, D'Ambrosio D, Nakamura K, Persin C, Cambier JC (1997) FcγRIIB1 inhibition of BCR-mediated phosphoinositide hydrolysis and $Ca^{2+}$ mobilization is integrated by CD19 dephosphorylation. Immunity 7:49–58

Jacob J, Kassir R, Kelsoe G (1991) In situ studies of the primary immune response to (4-hydroxy-3-nitrophenyl) acetyl. I. The architecture and dynamics of responding cell populations. J Exp Med 173:1165–1175

Kozlowski M, Mlinaric-Rascan I, Feng GS, Shen R, Pawson T, Siminovitch KA (1993) Expression and catalytic activity of the tyrosine phosphatase PTP1 C is severely impaired in motheaten and viable motheaten mice. J Exp Med 178:2157–2163

Krop I, de Fougerolles AR, Hardy RR, Allison M, Schlissel MS, Fearon DT (1996) Self-renewal of B1 lymphocytes is dependent on CD19. Eur J Immunol 26:238–242

Kubagawa H, Burrows PD, Cooper MD (1997) A novel pair of immunoglobulin-like receptors expressed by B cells and myeloid cells. Proc Natl Acad Sci USA 94:5261–5266

Law CL, Torres RM, Sundberg HA, Parkhouse RME, Brannan CI, Copeland NG, Jenkins NA, Clark EA (1993) Organization of the murine CD22 locus. J Immunol 151:175–187

Law CL, Aruffo A, Chandran KA, Doty RT, Clark EA (1995) Ig domains 1 and 2 of murine CD22 constitute the ligand-binding domain and bind multiple sialylated ligands expressed on B and T cells. J Immunol 155:3368–3376

Law CL, Sidorenko SP, Chandran KA, Zhao Z, Shen SH, Fischer EH, Clark EA (1996) CD22 associates with protein tyrosine phosphatase 1C, Syk, and phospholipase Cγ1 upon B cell activation. J Exp Med 183:547–560

Leprince C, Draves KE, Geahlen RL, Ledbetter JA, Clark EA (1993) CD22 associates with the human surface IgM-B-cell antigen receptor complex. Proc Natl Acad Sci USA 90:3236–3240

Levy A, Todd SC, Maecker HT (1998) CD81 (TAPA-1): A molecule involved in signal transduction and cell adhesion in the immune system. Ann Rev Immunol 89–109

Li X, Sandoval D, Freeberg L, Carter RH (1997) Role of CD19 tyrosine 391 in synergistic activation of B lymphocytes by coligation of CD19 and membrane Ig. J Immunol 158:5649–5657

Li Y-Y, Baccam M, Waters SB, Pessin JE, Bishop GA, Koretzky GA (1996) CD40 ligation results in protein kinase C-independent activation of ERK and JNK in resting murine splenic B cells. J Immunol 157:1440–1447

Ling NR, MacLennan ICM, Mason DY (1987) B-cell and plasma antigens: new and previously defined clusters. In *Leukocyte Typing.III. White Cell Differentiation Antigens,* ed. AJ McMichael, pp. 302–305. Oxford Univ. Press

Liossis SN, Kovacs B, Dinnis G, Kammer GM, Tsokos GC (1996) B cells from patients with systemic lupus erythematosus display abnormal antigen receptor-mediated signal transduction events. J Clin Invest 98:2549–2557

Maecker HT, Levy S (1997) Normal lymphocyte development but delayed humoral immune response in CD81-null mice. J Exp Med 185:1505–1510

Maloney MD, Lingwood CA (1994) CD19 has a potential CD77 (globotriaosyl ceramide)- binding site with sequence similarity to verotoxin B-subunits: implications of molecular mimicry for B cell adhesion and enterohemorrhagic Escherichia coli pathogenesis. J Exp Med 180:191–201

Matsumoto AK, Kopicky-Burd J, Carter RH, Tuveson DA, Tedder TF, Fearon DT (1991) Intersection of the complement and immune systems: a signal transduction complex of the B lymphocyte-containing complement receptor type 2 and CD19. J Exp Med 173:55–64

Matsumoto AK, Martin DR, Carter RH, Klickstein LB, Ahearn JM, Fearon DT (1993) Functional dissection of the CD21/CD19/TAPA-1/Leu-13 complex of B lymphocytes. J Exp Med 178:1407–1417

Miyazaki T, Müller R, Campbell KS (1997) Normal development but differentially altered proliferative responses of lymphocytes in mice lacking CD81. EMBO 16:4217–4225

Mohan C, Morel L, Yang P, Wakeland EK (1997) Genetic dissection of systemic lupus erythematosus pathogenesis: Sle2 on murine chromosome 4 leads to B cell hyperactivity. J Immunol 159:454–465

Molina H, Holers VM, Li B, Fang Y-F, Mariathasan S, Goellner J, Strauss-Schoenberger J, Karr RW, Chaplin DD (1996) Markedly impaired humoral immune response in mice deficient in complement receptors 1 and 2. Proc Natl Acad Sci USA 93:3357–3361

Morel L, Rudofsky UH, Longmate JA, Schiffenbauer J, Wakeland EK (1994) Polygenic control of susceptibility to murine systemic lupus erythematosus. Immunity 1:219–229

Murakami M, Honjo T (1995) Involvement of B-1 cells in mucosal immunity and autoimmunity. Immunol Today 16:534–539

Nishizumi H, Taniuchi I, Yamanashi Y, Kitamura D, Ilic D, Mori S, Watanabe T, Yamamoto T (1995) Impaired proliferation of peripheral B cells and indication of autoimmune disease in lyn-deficient mice. Immunity 3:549–560

Nitschke L, Carsetti R, Ocker B, Kohler G, Lamers MC (1997) CD22 is a negative regulator of B cell receptor signaling. Curr Biol 7:133–143

O'Keefe TL, Williams GT, Davies SL, Neuberger MS (1996) Hyperresponsive B cells in CD22-deficient mice. Science 274:798–801

Ono M, Okada H, Bolland S, Yanagi S, Kurosaki T, Ravetch JV (1997) Deletion of SHIP or SHP-1 reveals two distinct pathways for inhibitory signaling. Cell 90:293–301

O'Rourke LM, Tooze R, Turner M, Sandoval DM, Carter RH, Tybulewicz VLJ, Fearon DT (1998) CD19 as a membrane-anchored adaptor protein of B lymphocytes: costimulation of lipid and protein kinases by recruitment of Vav. Immunity 8:365–645

Otipoby KL, Andersson KB, Draves KE, Klaus SJ, Farr AG, Kerner JD, Perlmutter RM, Law CL, Clark EA (1996) CD22 regulates thymus-independent responses and the lifespan of B cells. Nature 384:634–637

Pani G, Kozlowski M, Cambier JC, Mills GB, Siminovitch KA (1995) Identification of the tyrosine phosphatase PTP1C as a B cell antigen receptor-associated protein involved in the regulation of B cell signaling. J Exp Med 181:2077–2084

Peaker CJG, Neuberger MS (1993) Association of CD22 with the B cell antigen receptor. Eur J Immunol 23:1358–1363

Pepys MB (1974) Role of complement in induction of antibody production in vivo. Effect of cobra factor and other C3-reactive agents on thymus dependent and thymus-independent antibody responses. J Exp Med 140:126–145

Pezzutto A, Dörken B, Moldenhauer G, Clark EA (1987) Amplification of human B cell activation by a monoclonal antibody to the B cell-specific antigen CD22, Bp 130/140. J Immunol 138:98–103

Pezzutto A, Rabinovitch PS, Dörken B, Modenhauer G, Clark EA (1988) Role of the CD22 human B cell antigen in B cell triggering by anti-immunoglobulin. J Immunol 140:1791–1795

Razi N, Varki A (1998) Masking and unmasking of the sialic acid-binding lectin activity of CD2 (Siglec-2) on B lymphocytes. Proc Natl Acad Sci USA 95:7469–7474

Rickert RC, Rajewsky K, Roes J (1995) Impairment of T-cell-dependent B-cell responses and B-1 cell development in CD19 deficient mice. Nature 376:352–355

Rudofsky UK, Evans BD, Balaban SL, Mottironi VD, Gabrielsen AE (1993) Differences in expression of lupus nephritis in New Zealand Mixed H-$2^z$ homozygous inbred strains of mice derived from New Zealand Black and New Zealand White mice: origins and initial characterization. Lab Invest 68: 419–426

Sato S, Miller AS, Inaoki M, Bock CB, Jansen PJ, Tang MLK, Tedder TF (1996a) CD22 is both a positive and negative regulator of B lymphocyte antigen receptor signal transduction: altered signaling in CD22-deficient mice. Immunity 5:551–562

Sato S, Ono N, Steeber DA, Pisetsky DS, Tedder TF (1996b) CD19 regulates B lymphocyte signaling thresholds for the development of B-1 lineage cells. J Immunol 156:4371–4378

Sato S, Miller AS, Howard MC, Tedder TF (1997a) Regulation of B lymphocyte development and activation by the CD19/CD21/CD81/Leu 13 complex requires the cytoplasmic domain of CD19. J Immunol 159:3278–3287

Sato S, Jansen PJ, Tedder TF (1997b) CD19 and CD22 expression reciprocally regulates tyrosine phosphorylation of Vav protein during B lymphocyte signaling. Proc Natl Acad Sci USA 94:13158–13162

Scharenberg AM, Kinet JP (1996) The emerging field of receptor-mediated inhibitory signaling: SHP or SHIP? Cell 87:961–964

Schittek B, Rajewsky K (1990) Maintenance of B cell memory by long-lived cells generated from proliferating precursors. Nature 346:749–751

Schulte RJ, Campbell M-A, Fischer WH, Sefton BM (1992) Tyrosine phosphorylation of CD22 during B cell activation. Science 258:1001–1004

Schulz LD, Schweitzwer PA, Rajan TV, Yi T, Ihle JN, Mathews RJ, Thomas ML, Beier DR (1993) Mutations at the murine motheaten locus are within the hemopoietic cell protein tyrosine phosphatase (Heph) gene. Cell 73:1445–1454

Sgroi D, Koretzky GA, Stamenkovic I (1995) Regulation of CD45 engagement by the B-cell receptor CD22. Proc Natl Acad Sci USA 92:4026–4030

Smith KGC, Hewitson TD, Nossal GJV, Tarlinton DM (1996) The phenotype and fate of the antibody-forming cells of the splenic foci. Eur J Immunol 26:444–448

Smith KGC, Tarlinton DM, Doody GM, Hibbs ML, Fearon DT (1998) Inhibition of the B cell by CD22: A requirement for Lyn. J Exp Med 187:807–811

Stamenkovic I, Seed B (1990) The B cell antigen CD22 mediates monocyte and erythrocyte adhesion. Nature 344:74–77

Stephens L, Jackson TR, Hawkins PT (1993) Activation of phosphatidylinositol 4,5-bisphosphate supply agonists and non-hydrolysable GTP analogues. Biochem J 296:481–488

Stoddart A, Ray RJ, Paige CJ (1997) Analysis of murine CD22 during B cell development: CD22 is expressed on B cell progenitors prior to IgM. Int Immunol 9:1571–1579

Sutherland CL, Heath AW, Pelech SL, Young PR, Gold MR (1996) Differential activation of the ERK, JNK, and p38 mitogen-activated protein kinases by CD40 and the B cell antigen receptor. J Immunol 157:3381–3390

Symington FW, Subbarao B, Mosier DE, Sprent J (1992) Lyb-8.2: a new B cell antigen defined and characterized with a monoclonal antibody. Immunogenetics 16:381–391

Tedder TF, Zhou L-J, Engel P (1994) The CD19/CD21 signal transduction complex of B lymphocytes. Immunol Today 15:437–442

Tedder TF, Tuscano J, Sato S, Kehrl JH (1997a) CD22, a B lymphocyte-specific adhesion molecule that regulates antigen receptor signaling. Annu Rev Immunol 15:481–504

Tedder TF, Inaoki M, Sato S (1997b) The CD19-CD21 complex of B lymphocytes regulates signal transduction thresholds governing humoral immunity and autoimmunity. Immunity 6:107–118

Tooze RM, Doody GM, Fearon DT (1997b) Counterregulation by the coreceptors CD19 and CD22 of MAP kinase activation by membrane immunoglobulin. Immunity 7:59–67

Torres RM, Law CL, Santos-Argumedo L, Kirkham PA, Grabstein K, Parkhouse RM, Clark EA (1992) Identification and characterization of the murine homologue of CD22, a B lymphocyte-restricted adhesion molecule. J Immunol 149:2641–2649

Tsitsikov EN, Gutierrez-Ramos JC, Geha RS (1997) Impaired CD19 expression and signaling, enhanced antibody response to type II T independent antigen and reduction of B-1 cells in CD81-deficient mice. Proc Natl Acad Sci USA 94:10844–10849

Tsui HW, Siminovitch KA, de Souza L, Tsui FWL (1993) Motheaten and viable motheaten mice have mutations in the haematopoietic cell phosphatase gene. Nature Genetics 4:124–129

Tuscano JM, Engel P, Tedder TF, Kehrl JH (1996a) Engagement of the adhesion receptor CD22 triggers a potent stimulatory signal for B cells and blocking CD22/CD22L interactions impairs T-cell proliferation. Blood 87:4723–4730

Tuscano JM, Engel P, Tedder TF, Agarwal A, Kehrl JH (1996b) Involvement of p72syk kinase, p53/56lyn kinase and phosphatidyl inositol-3 kinase in signal transduction via the human B lymphocyte antigen CD22. Eur J Immunol 26:1246–1252

Tuveson DA, Carter RH, Soltoff SP, Fearon DT (1993) CD19 of B cells as a surrogate kinase insert region to bind phosphatidylinositol 3-kinase. Science 260:986–989

van Noesel CJM, Lankester AC, van Schijndel GMW, van Lier RAW (1993) The CR2/CD19 complex on human B cells contains the src-family kinase Lyn. Int Immunol 5:699–705

Wang J, Koizumi T, Watanabe T (1996) Altered antigen receptor signaling and impaired Fas-mediated apoptosis of B cells in Lyn-deficient mice. J Exp Med 184:831–838

Weng WK, Jarvis L, LeBien TW (1994) Signaling through CD19 activates Vav/mitogen-activated protein kinase pathway and induces formation of a CD19/Vav/phosphatidylinositol 3-kinase complex in human B cell precursors. J Biol Chem 269:32514–32521

Williams GT, Peaker CJ, Patel KJ, Neuberger MS (1994) The α/β sheath and its cytoplasmic tyrosines are required for signaling by the B-cell antigen receptor but not for capping or for serine/threonine-kinase recruitment. Proc Natl Acad Sci USA 91:474–478

Wilson GL, Najfeld V, Kozlow E, Menniger J, Ward D, Kehrl JH (1993) Genomic structure and chromosomal mapping of the human CD22 gene. J Immunol 150:5013–5024

Yamanashi Y, Terutaka K, Mizuguchi J, Yamamoto T, Toyoshima K (1991) Association of B cell antigen receptor with protein tyrosine kinase Lyn. Science 251:192–194

# Positive and Negative Signaling in B Lymphocytes

K.M. Coggeshall



## 1 Introduction

Stimulation of B lymphocytes through the antigen-receptor, surface immunoglobulin (sIg) activates B lymphocytes to proliferate and secrete soluble antigen-specific Ig. Numerous studies over the past decade have at least in part uncovered the complex biochemistry of the activation process through sIg; these studies will be

Department of Microbiology and the Comprehensive Cancer Center, Ohio State University, Columbus, OH 43210, USA
e-mail: coggeshall.1@osu.edu

reviewed below. Binding of antigen through sIg is the first step in the humoral response to foreign antigens. Antigen binding is specific and, so, initiates and is fundamentally responsible for the dominant feature of the immune system, i.e., antigen specificity. Thus, only B lymphocytes with an appropriate sIg surface receptor, formed by genomic rearrangement during B-cell development, are able to respond to antigen and later to helper T lymphocytes.

The B-lymphocyte activation process is regulated at several levels, including the interaction of B lymphocytes with helper T lymphocytes, which occurs through a variety of surface receptor proteins on both cells (reviewed in GRAY et al. 1996; FOY et al. 1996). B–T lymphocyte interaction is attended by formation and secretion of lymphokines by helper T lymphocytes to promote B-lymphocyte proliferation and differentiation into Ig-secreting cells or into memory B lymphocytes.

Likewise, secreted, antigen-specific Ig plays an important role in the regulation of B-cell activation. Many earlier experiments indicated that the presence of soluble Ig reduces or prevents antigen-triggered activation of naive B lymphocytes and subsequent secretion of nascent Ig. This suppressive effect of soluble Ig has been termed feedback suppression or "negative signaling" to indicate a regulatory loop in which secreted Ig prevents new, further or continued B-cell Ig production. Negative signaling in B lymphocytes was proposed as an immunoregulatory mechanism nearly 30 years ago (SINCLAIR et al. 1970; SINCLAIR and CHAN 1971). It occurs through antigen–antibody complexes that effect concomitant ligation of sIg and the B-cell IgG receptor, FcγRIIb, which produce a dominant negative signal and act in contrast to sIg stimulation alone, which induces a positive signal to promote B-cell activation. A model using F(ab′)2 fragments or intact anti-Ig antibodies, first developed by PHILLIPS and PARKER (1983), has been used to study the biochemistry and cell biology of positive and negative signaling. These activation arrangements are shown in Fig. 1. Recent studies have greatly advanced our knowledge of the negative signaling process, how the dominant-negative signal is generated, and the mechanisms by which B-lymphocyte activation is affected.

## 2 Historical Perspective

The ability of soluble antibody to negatively influence a humoral response has been known for over a century and was reviewed in 1968 (UHR and MOLLER 1968). Many prior experiments indicated that animals were able to mount impressive humoral responses upon injection of antigen alone. However, injection of antigen–antibody complexes or passive administration of antisera during antigen injections would reduce or eliminate the humoral response to that antigen (UHR and BAUMANN 1961a,b; NEIDERS et al. 1962; WIGZELL 1966; DIXON et al. 1967; GREENBURY and MOORE 1968; SINCLAIR 1969). Thus, the presence in the host of specific antibody prevented a new humoral response to the same antigen. Indeed, administration of specific antiserum many days after antigen injections would inhibit the humoral response (UHR and BAUMANN 1961a,b).

**a.** Positive signaling by antigen or F(ab')2 anti-Ig.

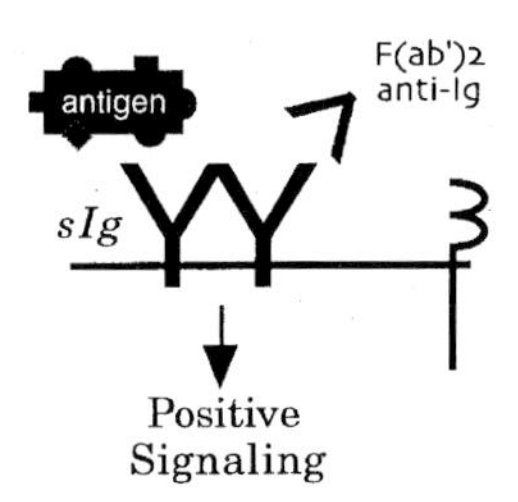


**b.** Negative signaling by antigen-antibody complexes.

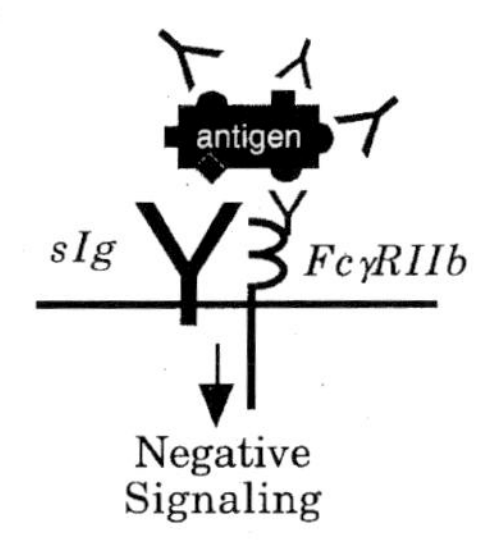


**c.** Negative signaling by intact anti-Ig.

Intact
anti-Ig
sIg
FcγRIIb
Negative
Signaling

**d.** Negative signaling by intact IgG as carrier for hapten.

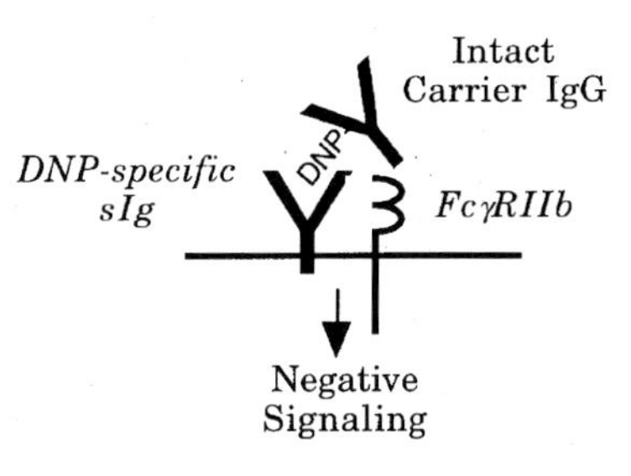


**Fig. 1a–d.** Modes of co-clustering sIg and FcγRIIb to induce negative signaling. **a** Positive signaling by antigen or F(ab')2 anti-Ig. Both antigen and the variable region of IgH and IgL clusters sIg without engaging FcγRIIb, thereby stimulating B lymphocytes. **b** Negative signaling antigen–antibody complexes. Free, unbound epitopes on an antigenic particle engage sIg while the Fc moiety of bound IgG engages FcγRIIb to induce co-clustering. **c** Negative signaling by intact anti-Ig. The variable region of IgH and IgL engage sIg while the Fc moiety of the antibody engages FcγRIIb to induce co-clustering. **d** Negative signaling by intact IgG as a carrier for hapten. The hapten engages the antigen-binding region of sIg while the Fc moiety of carrier IgG engages FcγRIIb to induce co-clustering

An important but unresolved issue at the time was whether the antigen–antibody complexes simply lacked a recognizable antigenic determinant, i.e., did the suppressive antibody shelter all available antigenic sites and so block B-cell recognition of antigen? In this regard, it was notable that some earlier studies used F(ab′)2 fragments of antibodies which suppressed the humoral response to the same extent as intact antibodies (Greenbury and Moore 1968; Cerottini et al. 1969). Suppression by F(ab′)2 antibodies is consistent with the notion that immune complexes blocked humoral responses simply by preventing immune recognition by B lymphocytes. However, others reported suppression only with intact antibodies and not with F(ab′)2 fragments (Sinclair et al. 1968; Chan and Sinclair 1971) consistent with the current interpretation of sIg-FcγR co-clustering as the suppressive event. Sinclair et al. (1968, 1970) speculated that F(ab′)2 preparations of antibodies used in earlier experiments were contaminated with intact antibodies due to incomplete digestion by pepsin in their preparation. Using rigorously purified, pepsin-digested anti-sheep red blood cell (RBC) antibodies, they observed that F(ab′)2 fragments failed to suppress the humoral response to sheep RBC antigens, while undigested, intact anti-sheep RBC antibodies or non-purified, pepsin-digested antibodies were able to suppress the response. These findings indicated that suppressive antibody required an Fc moiety and were instrumental in developing the model shown in Fig. 1 (SINCLAIR and CHAN 1971).

Further studies of humoral suppression to antigens upon passive administration of antisera indicated that B-lymphocyte tolerance was antigen-specific, such that humoral responses to unrelated antigens appeared as normal controls. Despite the antigen specificity, it was observed that the humoral response to other antigens on the same particle, such as RBCs coated with specific antisera, were similarly blocked (GREENBURY and MOORE 1968). For example, rabbits injected with sheep RBCs coated with anti-Forssman antisera were unable to respond to an unrelated sheep RBC "isophile" antigen (GREENBURY and MOORE 1968). The animals therefore displayed antibody-induced suppression toward the isophile antigen present on sheep RBC. Antibody suppression could be overcome by antigen excess, while antisera excess would eliminate humoral responses altogether. In light of our current knowledge, one would interpret these early experiments as indicating that epitopes of the isophile antigen were able to contact sIg while the Fc moiety of anti-Forssman IgG would bind FcγRIIb, thereby producing co-clustering of the two receptors and promoting the dominant-negative signal (Fig. 1b). Antibody-induced suppression by co-clustering sIg-FcγRIIb can therefore occur in B cells responding to an unrelated antigen on the same particle or protein. This finding led to the notion that signaling through FcγRIIb by itself was not suppressive; rather, suppression occurs when FcγRIIb is co-clustered with sIg.

These experimental findings might be explained by invoking clearance of antigen–antibody complexes through Fc receptor-dependent phagocytosis by monocytes. According to this possibility, injected antigen–antibody complexes, or passive administration of antisera to form such complexes in vivo, would be internalized by phagocytes and removed from circulation. Antigen removal by phagocytosis might then prevent any epitope from contacting the B lymphocyte compartment, thereby blocking humoral responses to injected antigens. This possibility is still a viable one and may at least be a contributing factor in antibody-mediated suppression. The recent advent of Fc receptor-deficient animals (TAKAI et al. 1994, 1996) may permit this issue to be better addressed.

Experiments on B-cell development and the effects of anti-IgM antibodies (anti-μ) likewise revealed a role for negative signaling. Prior studies had established that introduction of anti-Ig antibodies into chickens followed by bursectomy completely blocked B-lymphocyte development (KINCADE et al. 1970). Likewise, mice treated with intact anti-μ antibodies displayed a complete lack of circulating mature B lymphocytes and no humoral response to foreign antigen (LAWTON and COOPER 1974). However, pre-B lymphocytes could still be found since mature B lymphocytes would emerge upon in vitro culture of bone marrow. Subsequent experiments investigating the dose of anti-μ necessary to elicit suppression of B-lymphocyte maturation and the dose needed to affect B-lymphocyte function found that the two were not the same. Very low doses of anti-μ permitted normal B-lymphocyte maturation but completely suppressed responses to exogenous mitogens (PIKE et al. 1982). Thus, anti-μ concentrations of 0.1μg/ml or less failed to affect the appearance of $sIg^{+}$ lymphocytes from in vitro bone-marrow cultures, while the same concentrations completely blocked the proliferative response to bacterial lipopolysaccharide. These studies also reported a dramatic acceleration of the spontaneous death rate of B cells co-cultured with low dose anti-μ treatment, implying that virgin mature B lymphocytes that have recently emerged from immature cells are especially susceptible to cell death by intact anti-μ antibodies. Much of the effect of anti-Ig reagents on B-lymphocyte development is likely due to the well-established growth inhibition of immature B lymphocytes after triggering of sIg (BRINK et al. 1992; YELLEN-SHAW and MONROE 1992); reviewed in Rothstein (1996). However, the low-dose anti-μ experiments described above do not appear to be such an example since B-lymphocyte development proceeded normally in the in vitro cultures. In light of current information, it is likely that the anti-μ antibody induced a negative signal in the maturing B lymphocytes by co-clustering sIg through its anti-μ activity, and FcγRIIb through its Fc moiety (Fig. 1c).

Studies of tolerance induction also provided important insights into the phenomenon of negative signaling and its biological consequences; indeed, such studies were responsible for the term "negative signaling" (NOSSAL 1983). Borel and colleagues (GOLAN and BOREL 1971; BOREL and KILHAM 1974) speculated that humoral tolerance to a hapten might result when using autologous proteins as a carrier. To test this hypothesis, the hapten dinitrophenyl (DNP) was conjugated to normal mouse serum proteins, the DNP-conjugated serum proteins were injected into mice, and the animals were challenged with DNP conjugated to keyhole limpet hemocyanin (KLH). It was found that prior exposure to DNP-serum protein conjugates prevented a subsequent response to DNP–KLH, i.e., the animals were tolerant towards DNP. However, fractionation of the autologous serum carrier proteins revealed that IgG, but not other serum components, such as albumin or IgM, was responsible for induction of tolerance. Thus, the results revealed the unique ability of IgG as carrier to induce hapten tolerance. These experiments were followed by a number of other studies using intact IgG as a carrier, and which reported tolerance towards the haptens to which they had been conjugated (DOYLE et al. 1976a,b; PARKS et al. 1978; TITE and TAYLOR 1979; TITE et al. 1981). The induction of tolerance again required that the carrier IgG contained an Fc moiety, since F(ab′)2 fragments of normal mouse IgG as carriers for hapten induced potent anti-hapten responses, rather than hapten tolerance. In retrospect, it is likely that negative signaling was induced in these experiments and caused by co-clustering sIg of hapten-responsive B lymphocytes and FcγRIIb, recognizing the Fc portion of the carrier IgG (Fig. 1d).

To investigate the mechanism(s) of antibody-induced B-cell suppression, it was necessary to develop an in vitro model of antigen-induced positive signaling and antigen–antibody complex-induced negative signaling. In an earlier report (PARKER

1975), it was noted that murine B lymphocytes were unable to proliferate or secrete Ig when stimulated in vitro with soluble anti-Ig reagents. Scribner et al. (1978) and Phillips and Parker (1983, 1984, 1985) noted the considerable literature with regard to feedback suppression by antibody and speculated that both phenomena might be due to the Fc portion of anti-Ig (in the case of in vitro proliferation with anti-Ig reagents) or of antibody bound to antigens (in the cases of feedback suppression). They proposed that the Fc portion of anti-Ig or of antigen–antibody complexes would permit co-clustering of sIg and Fc receptors on B cells and provide a dominant-negative signal, as originally proposed by Sinclair and Chan (1971). To test this hypothesis, they examined the ability of intact and F(ab′)2 fragments of rabbit anti-mouse IgM or of polyclonal anti-Ig antibodies to induce proliferation and Ig secretion of murine B cells in in vitro cultures. They found that F(ab′)2 fragments but not intact anti-Ig antibodies induced vigorous B-cell proliferation (Scribner et al. 1978; Phillips and Parker 1983, 1984) and, in the presence of supernatants of activated T cells, induced Ig secretion (Phillips and Parker 1983, 1985). These findings confirmed that co-clustering of sIg and Fc receptor(s) by intact anti-Ig or by antigen–antibody complexes blocked B-lymphocyte activation.

It was later established that anti-Ig reagents of the IgG but not of IgM isotypes blocked B-lymphocyte activation (Phillips and Parker 1985), consistent with the interpretation that an IgG receptor, which we now know to be FcγRIIb, was responsible for the dominant-negative signal. Both proliferation and Ig secretion in the intact anti-Ig-stimulated B-cell populations were restored when the Fc portion of the stimulating antibody was bound to soluble protein A (Phillips and Parker 1983) or when the interaction of intact IgG anti-Ig with the Fc receptor of the responding B lymphocytes was inhibited by including a blocking anti-FcγRIIb-specific monoclonal antibody, 2.4G2 in the stimulation protocol (Phillips and Parker 1985). The paradigm shown in Fig. 1a and c emerged from these studies and provided a polyclonal in vitro model to study positive signaling events elicited by F(ab′)2 fragments of anti-Ig and negative signaling events elicited by intact anti-Ig.

In order to consider the mechanisms of negative signaling by FcγRIIb, it is useful to summarize the known signaling pathways in B lymphocytes. Any of the activating pathways described below are potential sites of action for the FcγRIIb-mediated dominant-negative signal.

## 3 B-Cell Activation: Positive Signaling

### 3.1 Receptor Components and the Immunoreceptor Tyrosine-Based Activation Motif

Immunoreceptors such as the B- and T-lymphocyte antigen receptors and the Ig receptors found on macrophages, monocytes, neutrophils and mast cells, have a similar, highly conserved composition and mechanism of action, although the exact individual components are variable. Each receptor is composed of three parts: a protein that recognizes antigen but otherwise does not participate in the signaling biochemistry; low-molecular-weight proteins associated with the antigen receptor proper which are responsible for surface expression of the receptor complex and carry out signaling functions; and protein tyrosine kinases (PTKs) of the Src family.

The B- and T-lymphocyte antigen receptors are the most variable of these three components. Both B- and T lymphocytes employ a disulfide-bonded heterodimeric protein that is the product of separate gene segments, rearranged in the genome during lymphocyte development and stoichastically expressed on individual cells. The antigen receptor of B lymphocytes is Ig, containing mu or delta heavy chains and kappa or lambda light chains and directed by variable, diversity, joining and constant gene segments. The antigen receptor of T lymphocytes consists of the α- and β-chain proteins, likewise directed by rearranged gene segments during T-lymphocyte development (reviewed in WILLERFORD et al. 1996). Both these lymphocyte antigen receptors contain very short cytoplasmic tails; three amino acid residues in the case of sIg and four to seven amino acid residues in the case of the T-cell α/β heterodimer. As such, there is little structural information in the cytoplasmic tails to promote signal transduction; this fact anticipated the discovery of other proteins that carry out intracellular signaling function.

Lymphocyte antigen receptors are associated with a series of differentially expressed low-molecular-weight proteins (16–34kDa) whose presence is obligatory for proper assembly and surface appearance of the antigen receptors (FRANK et al. 1990). The low molecular weight proteins in B lymphocytes, Igα and Igβ, and those in T lymphocytes, γ, δ, ε, and ζ, are each the product of separate genes and are expressed in invariant form from cell to cell. As the genes encoding these receptor-associated proteins were discovered and reported in the literature, it was noted (RETH 1989) that the cytoplasmic tails of each protein had a common feature within their primary sequence. The motif noted contained tandem tyrosine residues each followed after a two amino acid spacer by a leucine or isoleucine and separated by six to twelve residues. The motif of YxxI/L – x(6–12) – YxxI/L was later termed the immunoreceptor tyrosine-based activation motif (ITAM) (CAMBIER 1995).

The low-molecular-weight ITAM-containing proteins associated with immunoreceptors are listed in Table 1. The amino acid residues between the YxxI/L motifs were shown to be binding sites for members of the Src family of PTKs (CLARK et al. 1992, 1994; SALCEDO et al. 1993; GAUEN et al. 1994; JOUVIN et al. 1994; PLEIMAN et al. 1994) and explained earlier results showing that ITAM-containing proteins derived from resting lymphocytes were associated with Src-family PTKs (SAMELSON et al. 1990; SAROSI et al. 1992; TIMSON GAUEN et al. 1992; SAOUAF et al. 1995; BEWARDER et al. 1996; CASSARD et al. 1996; DUCHEMIN and ANDERSON 1997;). The importance of the ITAM and its association with Src-family PTKs was made clear in experiments of cells expressing chimeric proteins that were composed of one of the several cytoplasmic ITAMs and an irrelevant extracellular domain (IRVING and WEISS 1991; LETOURNEUR and KLAUSNER 1991; ROMEO and SEED 1991; KIM et al. 1992; LETOURNEUR and KLAUSNER 1992; ROMEO et al. 1992;

**Table 1.** Cytoplasmic tyrosines of immunoreceptor tyrosine-based activation motif (ITAM)-containing, receptor-associated proteins

| Receptor | Sequence | | | | | Reference |
|---|---|---|---|---|---|---|
| FcγRIIa | DGG | YMTL | NPRAPT DDDKNI | YLTL | PPN | STUART et al. 1987 |
| γ chain associated with FcεR/FcγRI/FcγIII | DGV | YTGL | STRNQET | YETL | KHE | KUSTER et al. 1990 |
| Fcε β chain | DRV | YEEL | NIYSAT | YSEL | EDP | KUSTER et al. 1992 |
| Ig α chain | ENL | YEGL | LNLDD CSM | YEDI | SRG | HA et al. 1992 |
| Ig β chain | DHT | YEGL | DIDQTAT | YEDI | VTL | WOOD et al. 1993 |
| CD3 γ chain | DQL | YQPL | KDREDDQ | YSHL | QQN | KRISSANSEN et al. 1986 |
| CD3 δ chain | DQV | YQPL | RDRDDAQ | YSHL | GGN | VAN DEN ELSEN et al. 1984 |
| CD3 ε chain | NPD | YEPI | RKGQRDL | YSGL | NQR | GOLD et al. 1986 |
| TCR ζ1 chain | NQL | YNEL | NLGRREE | YDVL | NQR | WEISSMAN et al. 1988 |
| TCR ζ2 chain | EGL | YNEL | QKDKM AEA | YSEI | GMK | WEISSMAN et al. 1988 |
| TCR ζ3 chain | DGL | YQGL | STATKDT | YDAL | HMQ | WEISSMAN et al. 1988 |

Cytoplasmic tyrosines of immunoreceptor tyrosine-based activation motif (ITAM)-containing, receptor-associated proteins. Sequences shown are of human origin. **Acidic residues in the +1 or +2 position relative to the tyrosine residues are underlined**

IRVING et al. 1993; SANCHEZ et al. 1993; BURKHARDT et al. 1994; IWASHIMA et al. 1994). Upon clustering the extracellular domain of the chimera using antibody or a physiological ligand, responding cells underwent activation that was indistinguishable from that induced by antigen-receptor triggering. These findings demonstrated that the ITAM was necessary and sufficient for lymphocyte activation and implied that the sole function of the antigen receptor proper was simply to provide an antigen-induced clustering function to promote ITAM aggregation which, in turn, activated the ITAM-associated Src-family PTKs.

Ig receptors are separated based on the class of Ig they bind. Thus, separate genes code for Ig receptors that bind IgA, IgD, IgE, IgG and IgM; these receptors are differentially expressed among hematopoietically derived cells. Like lymphocyte antigen receptors, all Ig receptors are members of the Ig-superfamily and have multiple genes with several mRNA splice variants. Only the IgG receptors are germane to this review; the structure and function of other Ig receptors have recently been reviewed (DAERON 1997). Three distinct groups of IgG receptors (FcγRI, II and III) have been identified; these are separated on a structural basis and, to some extent, by their affinity for IgG. There are three genes in mouse (FcγRI, II and III) while, in humans, each of these groups is encoded by three separate genes (A, B and C) and many have splice variants (reviewed in RAVETCH and KINET 1991). Like the antigen receptors, FcγRI and III are associated with a γ chain, a low-molecular-weight protein homologous with the ζ chain of the T-cell antigen receptor; the γ chain contains a canonical ITAM motif (KUSTER et al. 1990). Unlike the antigen receptors, the γ-chain subunit may not be required for FcγR expression, since a functional FcγR receptor can be expressed in lymphocytes or fibroblasts upon transfection with FcγR cDNA alone (ALLEN and SEED 1989;

LOWRY et al. 1998). However, the stability of Fc$\gamma$R expression and phagocytic function of the transfected cells is greatly improved by co-transfection of the $\gamma$ chain (VAN VUGT et al. 1996; LOWRY et al. 1998). Additionally, co-expression of the $\gamma$ subunit with Fc$\gamma$R increases the overall affinity of the receptor for IgG (MILLER et al. 1996).

The Fc$\gamma$RII family, which has the lowest affinity for IgG in humans, is comprised of three separate genes denoted by A, B or C. The single mouse Fc$\gamma$RII gene is structurally and functionally similar to the human Fc$\gamma$RIIB gene (LEWIS et al. 1986; RAVETCH et al. 1986). The human A and C gene products are unique among immunoreceptors in structure and function in that they contain a cytoplasmic ITAM as an intrinsic part of the receptor and so do not require an associated ITAM-bearing subunit. The Fc$\gamma$RIIB gene product is also unique in that it lacks an ITAM or an ITAM-containing associated protein, but contains a related sequence consisting of a single YxxL motif; this sequence motif is discussed in more detail below. Fc$\gamma$RIIb is highly expressed in B lymphocytes (BROOKS et al. 1989), mast cells (TIGELAAR et al. 1971; LOBELL et al. 1994), and a subfraction of peripheral T cells (MANTZIORIS et al. 1993); mRNA for Fc$\gamma$RIIb was reported to be expressed at low levels in monocytes and macrophages (BROOKS et al. 1989; CASSEL et al. 1993) but expression of the protein has not been demonstrated. Fc$\gamma$RIIA is abundantly expressed in human monocytes, neutrophils and other phagocytic cells, or tissue containing these cells (BROOKS et al. 1989; RAVETCH and KINET 1991), while Fc$\gamma$RIIC is expressed in natural killer (NK) cells (METES et al. 1998). The ITAM of Fc$\gamma$RIIA and C appears to confer signaling properties of these receptors (CHACKO et al. 1994; BEWARDER et al. 1996; Chacko et al. 1996a) and permits phagocytic function (INDIK et al. 1995; VAN DEN HERIK-OUDIJK et al. 1995).

## 3.2 Signaling Events Following ITAM Phosphorylation

Clustering of immunoreceptors by antigen or anti-receptor antibodies induces rapid phosphorylation of the tyrosines within the ITAM, most likely catalyzed by the ITAM-associated Src-family PTKs. The role of Src-family PTKs in this process is based on three findings. First, the linear sequence of the ITAM motif appears to be an optimal phosphorylation site of this PTK family (SONGYANG et al. 1994a, 1995). Second, experiments have shown that an ITAM-containing protein in fibroblasts is not phosphorylated unless the cells co-express a Src-family PTK (IWASHIMA et al. 1994). Third, B (TAKATA et al. 1994) or T (STRAUS and WEISS 1992; IWASHIMA et al. 1994) lymphocytes deficient in a Src-family PTK do not undergo receptor-triggered ITAM phosphorylation. The mechanism of activation of the Src PTK is not entirely clear, although it appears to involve dephosphorylation of a constitutively phosphorylated C-terminal tyrosine residue of the Src-family PTK by the CD45 phosphotyrosine phosphatase (MUSTELIN et al. 1989, 1992; MUSTELIN and ALTMAN 1990; HURLEY et al. 1993; XU and CHONG 1995), since CD45-deficient lymphocytes are unresponsive to receptor triggering (KORETZKY et al. 1990, 1991, 1992; JUSTEMENT et al. 1991; DESAI et al. 1994; KAWAUCHI et al. 1994; HOFFMEYER et al. 1995).

Once the tyrosines of the ITAM are phosphorylated, signaling proceeds by the concerted action of tyrosine kinases and protein–protein interactions, which result in plasma membrane re-localization of critical signaling enzymes. The tyrosine phosphorylated ITAM serves as the principal docking site for proteins containing Src homology 2 (SH2) domains. The SH2 domain (reviewed in SCHLESSINGER 1994) is a module of approximately 100 amino acids and is common to many proteins involved in growth-factor-induced cellular proliferation or other forms of activation. A functionally but not structurally related phosphotyrosine-binding (PTB) domain is found in the adapter protein Shc, among other proteins (reviewed in VAN DER GEER and PAWSON 1995; MARGOLIS 1996). The PTB domain associates with phosphotyrosines found in an asparagine-proline-x-tyrosine (NPxY) motif. All SH2 domains contain an essential and invariant arginine residue that intimately binds to the negatively charged oxygens of the phosphotyrosine (WAKSMAN et al. 1993). Binding of SH2 domains to tyrosine-phosphorylated proteins is ordered and specified by the primary sequence of the SH2 domain and by the amino acids surrounding the phosphotyrosine residue in the protein (SONGYANG et al. 1993; SONGYANG et al. 1994b). The tyrosines within the ITAM often contain acidic amino acids (D or E) in the +1 and/or +2 positions and a small hydrophobic amino acid (I or L) in the +3 position relative to the phosphotyrosine. As such, the ITAM is an optimal recognition site for the catalytic region of Src-family PTKs as well as an excellent high-affinity binding site for the SH2 domain of these kinases (SONGYANG et al. 1993, 1994b).

### 3.2.1 ITAM-Syk/ZAP70 Binding

Binding to the tyrosine-phosphorylated ITAM of the PTKs Syk in sIg (CAMBIER and JOHNSON 1995; ROWLEY et al. 1995; WEINLANDS et al. 1995) or FcR signaling (GHAZIZADEH et al. 1995; PARK and SCHREIBER 1995; SHIUE et al. 1995; CHACKO et al. 1996a; IBARROLA et al. 1997), or the Syk homolog ZAP70 in T-cell antigen receptor signaling (CHAN et al. 1991; STRAUS and WEISS 1993; IWASHIMA et al. 1994; ISAKOV et al. 1995; NEUMEISTER et al. 1995; OSMAN et al. 1995; LABADIA et al. 1996), is the next essential step. Syk/ZAP70 appears to carry out the bulk of receptor-triggered tyrosine phosphorylation, based on studies of Syk/ZAP70-deficient cells (IWASHIMA et al. 1994; TAKATA et al. 1994) or animals (CHAN et al. 1994, 1995; CROWLEY et al. 1997) and the use of receptor/Syk chimeras (KOLANUS et al. 1993; RIVERA and BRUGGE 1995; GREENBERG et al. 1996). SH2 engagement of Syk/ZAP70 by binding to phosphorylated ITAM and their subsequent phosphorylation by the Src-family PTK, activates the kinase activity of these PTKs (KUROSAKI et al. 1995; ROWLEY et al. 1995). Syk/ZAP70-mediated protein tyrosine phosphorylation promotes numerous additional SH2 interactions (DEFRANCO 1997), and alterations in enzymatic activity by phosphorylation of at least Vav (CRESPO et al. 1997; HAN et al. 1997; TERAMOTO et al. 1997) and phospholipase C (PLC)γ (NISHIBE et al. 1990; KIM et al. 1991; WAHL et al. 1992). Signaling events emanating from clustering of immunoreceptors are summarized in Fig. 2.

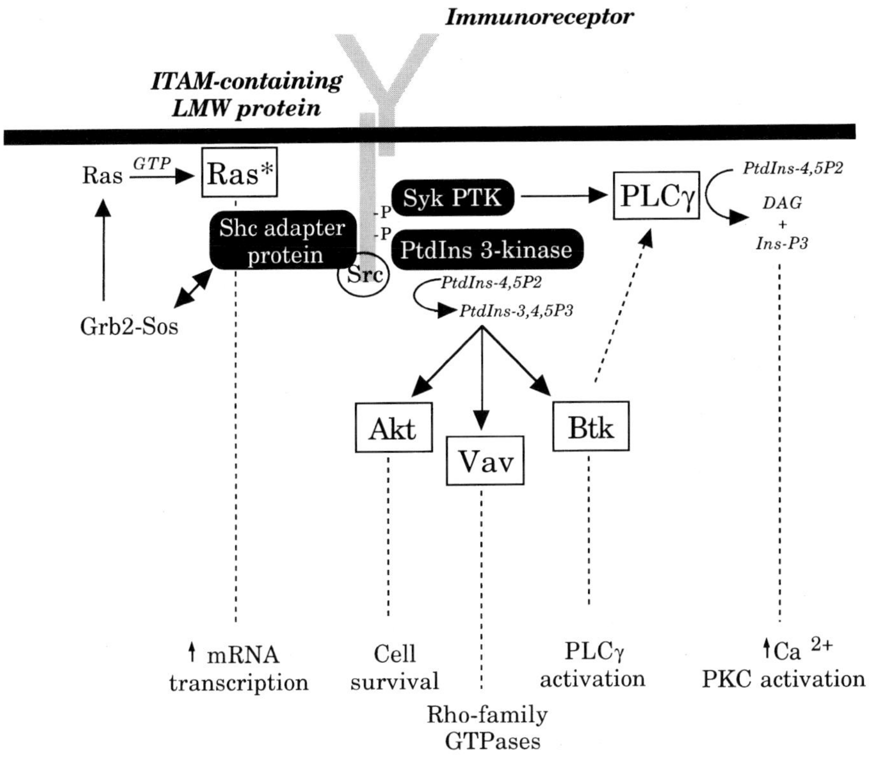

**Fig. 2.** Signaling events elicited by immunoreceptors. Following Src-family kinase activation and phosphorylation of the immunoreceptor tyrosine-based activation motif (ITAM) tyrosines, the SH2 domains of Syk, PtdIns 3-kinase and Shc are recruited. Syk phosphorylates a number of intracellular substrates, including phospholipase C (PLC)γ to stimulate PLCγ enzymatic activity. Shc binding to ITAM promotes Shc tyrosine phosphorylation and interaction with Grb2–Sos. Sos membrane translocation promotes Ras activation by catalyzing the exchange of GDP for GTP on Ras. PtdIns 3-kinase generates PtdIns-3,4,5P3, which acts as a mediator for several intracellular enzymes. PtdIns-3,4,5P3 activates the kinase activity of Btk, which supports PLCγ tyrosine phosphorylation, the GDP/GTP exchange activity of Vav towards Rho-family GTPases, involved in phagocytosis, and the kinase activity of Akt, which supports cell viability

### 3.2.2 ITAM-p85 Binding and Formation of 3-Phosphoinositides

Besides Syk/ZAP70, other proteins engage tyrosine-phosphorylated ITAMs. First, the p85 subunit of phosphatidylinositol 3-kinase (PtdIns 3-kinase) has been shown to bind synthetic ITAM phosphopeptides (Exley et al. 1994; Osman et al. 1996; Zenner et al. 1996; Ibarrola et al. 1997) or co-immunoprecipitate with ITAM-containing proteins after receptor stimulation (Clark et al. 1992; Exley et al. 1994; De Aos et al. 1997). PtdIns 3-kinase phosphorylates phosphoinositides on the D3 position of the inositol ring (reviewed in Hawkins et al. 1997). PtdIns 3-kinase activity and the attendant formation of 3-phosphorylated phosphoinositides [predominantly, phosphatidylinositol 3,4,5-trisphosphate (PtdIns-3,4,5P3) and phosphatidylinositol 3,4-bisphosphate (PtdIns-3,4P2); (Fukui et al. 1991;

Norgauer et al. 1992; Ptasznik et al. 1996)] is essential for a number of biological processes in many cell types, including mitogenesis, phagocytosis, and cell viability (reviewed in Hawkins et al. 1997; Toker and Cantley 1997). Earlier experiments established that membrane-targeting of the 110-kDa catalytic subunit of PtdIns 3-kinase or the 85-kDa regulatory subunit, is sufficient to elicit downstream biological events (Klippel et al. 1996). These findings imply that the function of ITAM engagement by the p85 subunit is to effect plasma-membrane recruitment of the enzyme, perhaps promoting enzyme access to phosphoinositide substrates.

The PtdIns 3-kinase products, PtdIns-3,4,5P3 and PtdIns-3,4P2, act as intracellular mediators to stimulate an impressive variety of intracellular enzymes. These include Vav (Han et al. 1998), a guanine nucleotide exchange factor for the Rho-family of GTPases (Crespo et al. 1996; Crespo et al. 1997; Han et al. 1997), the serine threonine kinase Akt (Burgering and Coffer 1995; Datta et al. 1996; Franke et al. 1997; Klippel et al. 1997), involved in protecting cells from apoptosis (Kennedy et al. 1997; Marte and Downward 1997), and Bruton's tyrosine kinase (Btk) (Salim et al. 1996; Hyvonen and Saraste 1997; Li et al. 1997), involved in B-lymphocyte development and activation (Tsukada et al. 1994; Desiderio 1997). Furthermore, recent experiments indicate a requirement for PtdIns 3-kinase and/or its 3-phosphoinositide products in stimulation of enzymes distal to Ras (Yamauchi et al. 1993; Urich et al. 1995; King et al. 1997), suggesting that the Ras pathway is similarly influenced by PtdIns 3-kinase. Binding and activation effects of 3-phosphoinositides to these proteins is likely mediated by their common pleckstrin homology (PH) domain (Musacchio et al. 1993), an approximate 100-amino acid motif found in a variety of proteins involved in signaling, that binds PtdIns-3,4,5P3 and PtdIns-3,4P2 (Frech et al. 1997; Rameh et al. 1997) to affect subcellular localization of enzymes (Buchsbaum et al. 1996; Lemmon et al. 1996; Andjelkovic et al. 1997; Falasca et al. 1998) or directly affect activity (Franke et al. 1997; Frech et al. 1997). Although PtdIns 3-kinase products are inefficient substrates for hydrolysis by PLCγ (Serunian et al. 1989), other studies indicate that 3-phosphoinositides contribute to PLCγ activation by direct binding to the SH2 domain of PLCγ (Rhee and Bae 1997; Bae et al. 1998). Thus, many studies have documented the central importance of receptor-mediated induction of PtdIns 3-kinase and/or its 3-phosphoinositide products acting proximal to stimulation of intracellular signaling enzymes and promoting biological effects.

### 3.2.3 ITAM-Shc Binding and Ras Activation

Experiments of the Shc adapter protein analyzed by synthetic phosphopeptide binding (Baumann et al. 1994; Zhou et al. 1995; Labadia et al. 1996; D'Ambrosio et al. 1996b; Marais et al. 1998) or co-immunoprecipitation of ITAM-bearing proteins (Ravichandran et al. 1993) have revealed findings similar to those of p85 described above. Upon ITAM recruitment through the Shc SH2 domain, Shc is phosphorylated at three tyrosine residues, Y239, Y240 (Van Der Geer et al. 1996) and Y317 (Pelicci et al. 1992; Salcini et al. 1994), whereupon it associates with the SH2 domain of Grb2, a second adapter protein. Grb2 contains an SH2 domain flanked by two Src homology 3 (SH3) domains; these associate to proline-rich sequences in proteins, albeit with a considerably lower affinity than that of SH2-protein interactions (reviewed in BAR-SAGI et al. 1993; YU et al. 1994). Engagement of the SH2 domain of Grb2 by tyrosine-phosphorylated Shc then promotes an interaction of the SH3 domain of Grb2 with proline-rich sequences of Sos (RAVICHANDRAN et al. 1995; HARMER and DEFRANCO 1997), a guanine nucleotide exchange protein for Ras. Formation of the Shc/Grb2/Sos complex at the site of receptor tyrosine phosphorylation results in plasma-membrane translocation of Sos from its cytosolic location and it is now positioned to stimulate GTP binding to the membrane-bound protein, Ras (LOWENSTEIN et al. 1992; CHARDIN et al. 1993; LI et al. 1993).

As with PtdIns 3-kinase, membrane-targeted versions of Sos promote constitutive Ras activation and downstream effects (ARONHEIM et al. 1994; HOLSINGER et al. 1995). Thus, ITAM association of Shc/Grb2/Sos would appear to function by re-localizing Sos to stimulate Ras-GTP binding. GTP-bound active Ras stimulates a series of serine/threonine kinases, Raf, MEK and Erk (MARAIS and MARSHALL 1996), which phosphorylates transcription factors of the ets family to promote new mRNA synthesis (reviewed in MARAIS and MARSHALL 1996). It should be noted that, although several studies have reported sIg-triggered induction of Shc tyrosine phosphorylation, interaction of Shc with Grb2, and formation of Grb2/Sos complexes in B lymphocytes (SAXTON et al. 1994; SMIT et al. 1994; KUMAR et al. 1995; CROWLEY et al. 1996; HARMER and DEFRANCO 1997), there is no genetic evidence that they are required for stimulation of the Ras pathway.

### 3.2.4 Activation of PLC

Besides induction of the PtdIns 3-kinase and of the Ras pathway, immunoreceptors stimulate the hydrolysis of phosphatidylinositol 4,5-bisphosphate (PtdIns-4,5P2) by PLCγ. Two isoforms of PLCγ, 1 and 2, have been described (OHTA et al. 1988; EMORI et al. 1989; BURGESS et al. 1990). T lymphocytes express predominantly PLCγ1, while B lymphocytes and phagocytic cells of myeloid origin express predominantly PLCγ2 (HEMPEL and DEFRANCO 1991; PARK et al. 1991; COGGESHALL et al. 1992; HEMPEL et al. 1992). Antigen-receptor triggering induces potent tyrosine phosphorylation of the predominant isoform. Although the primary sequence of both isoforms is ~50% identical, they are remarkably conserved in overall structure, including two SH2 domains, a PH domain, and a single SH3 domain separated by tyrosine phosphorylation sites. Tyrosine phosphorylation alone appears to be sufficient to activate the enzymatic activity, rather than SH2 engagement or enzyme recruitment to an ITAM (NISHIBE et al. 1990; KIM et al. 1991), but see report by WAHL et al. (1992). Indeed, although the SH2 domains of PLCγ associate with other phosphorylated growth-factor receptors (MARGOLIS et al. 1990; ROTIN et al. 1992a,b), the ITAM motif does not appear to be capable of associating with the SH2 domains of PLCγ isoforms (SONGYANG et al. 1993). Others have shown that PLCγ interacts with and is phosphorylated by Syk (LAW et al. 1996a), and that Syk-deficient B cells are incapable of activating PLCγ or inducing downstream biological events (Takata et al. 1994); thus, Syk may act as an adapter protein for PLCγ. However, genetic studies indicate an additional role for the PTKs of the Tec family in PLCγ tyrosine phosphorylation; specifically Btk expressed in B lymphocytes (Takata and Kurosaki 1996).

### 3.2.5 Inositol Metabolism and Changes in Cellular Calcium

Regardless of how the enzyme is activated, PLCγ hydrolyzes PtdIns-4,5P2 to generate two intracellular mediators: diacylglycerol and inositol-1,4,5-trisphosphate (Ins-P3) (reviewed in Rhee and Bae 1997). Ins-P3 induces the release of $Ca^{2+}$ from intracellular stores to raise cytoplasmic $Ca^{2+}$ levels while diacylglycerol allosterically activates protein kinase C isoforms (reviewed in Weinstein et al. 1997). The precise roles of these mediators in lymphocytes or phagocytes are enigmatic, although numerous pharmacological studies indicate they are important and potent stimulators of hematopoietic cell biology. Studies on $Ca^{2+}$ elevation have revealed two stages: an early phase, dependent on Ins-P3 formation catalyzed by PLCγ and insensitive to removal of extracellular $Ca^{2+}$ with ethyleneglycol tetraacetic acid (EGTA); and a late phase that is sensitive to EGTA. (Ransom et al. 1986, 1988; Dugas et al. 1987). Sensitivity of the phases to EGTA reveals the source of $Ca^{2+}$ since EGTA only binds extracellular $Ca^{2+}$ and does not affect release from intracellular stores. Thus, the later phase occurs by movement of $Ca^{2+}$ into the cell, although the mechanism permitting $Ca^{2+}$ entry is not clear.

One possible mechanism is $Ca^{2+}$-release activated current (CRAC) channels, non-voltage-dependent $Ca^{2+}$-specific membrane channels that permit $Ca^{2+}$ entry upon depletion of intracellular $Ca^{2+}$ stores (reviewed in Bennett et al. 1995; Berridge 1995; Friel 1996). CRAC channels are activated upon treatment of cells with thapsigargin, which blocks $Ca^{2+}$ uptake into intracellular storage sites. B lymphocytes respond to thapsigargin by increasing $Ca^{2+}$ uptake and thus appear to express CRAC channels (Lewis and Cahalan 1995; Sugawara et al. 1997). An alternate mechanism of extracellular $Ca^{2+}$ entry involves metabolism of Ins-P3 by phosphorylation to inositol-1,3,4,5P4 (Ins-P4). Ins-P4 has been shown to affect the permeability of various cellular membranes to $Ca^{2+}$ (Ivorra et al. 1991; Luckhoff and Clapham 1992). An Ins-P4-binding protein, $GAP^{IP4BP}$, was recently identified (Cullen et al. 1995) and platelet-membrane fractions containing $GAP^{IP4BP}$ increased their $Ca^{2+}$ permeability when Ins-P4 was added (O'Rourke et al. 1996). These findings suggest that extracellular $Ca^{2+}$ entry during the later phase of $Ca^{2+}$ influx can be mediated by Ins-P3 metabolism to Ins-P4, which acts on $GAP^{IP4BP}$ to permit entry of $Ca^{2+}$. Recent experiments of B lymphocytes defective in expression of Ins-P3 receptors showed no anti-Ig-induced $Ca^{2+}$ influx from either intracellular or extracellular sources. However, these cells were still able to generate Ins-P3 and were able to respond to thapsigargin by an increase in $Ca^{2+}$ (Sugawara et al. 1997), indicating that they were capable of capacitative $Ca^{2+}$ influx. Since Ins-P3 metabolism to Ins-P4 was presumably unaffected by the mutations, these findings suggest that the later influx of $Ca^{2+}$ is mediated by CRAC channels and not by $GAP^{IP4BP}$.

Thus, three signaling pathways are initiated by ITAM-containing immunoreceptors. PtdIns 3-kinase generates PtdIns-3,4,5P3 to stimulate enzymes dependent on this lipid; Shc/Grb2/Sos protein interaction induces GTP-binding of Ras; and PLCγ tyrosine phosphorylation generates intracellular mediators diacylglycerol and $Ca^{2+}$. All three pathways offer sites of inhibition by the dominant-negative signal emanating from FcγRIIb co-clustering with sIg. Thus, hydrolysis of 3-phosphoinositides would block the action of PtdIns 3-kinase and interrupt enzymes that are dependent on this lipid; disruption of Shc/Grb2/Sos protein interactions would inhibit the Ras pathway and recruitment of phosphotyrosine phosphatases would reverse the action of PTKs and inhibit the stimulation of PLCγ. Recent studies have documented a role for all three inhibitory strategies and these will be reviewed below.

# 4 B-Cell Desensitization Through FcγRIIb: Negative Signaling

## 4.1 The Immunoreceptor Tyrosine-Based Inhibitory Motif

As described above, all IgG Fc receptors either contain an ITAM as an intrinsic part of the receptor (FcγRIIA and C genes) or are associated with a low-molecular-weight, ITAM-containing γ chain (FcγRI and FcγRIII genes). The FcγRIIB gene is the single exception to this organization. First, it codes for a protein having a single cytoplasmic YxxI/L rather than the orthodox tandem YxxI/L found in an ITAM and, second, stimulation through FcγRIIb provokes inhibition of cell activation (negative signaling), rather than stimulation of cells.

Approaches sought to map the precise domain through which FcγRIIb mediated the negative signal used various deletion or point mutants of the receptor. These experiments (Muta et al. 1994) identified a region within FcγRIIb coding for a cytoplasmic 13-amino acid motif that promulgated the negative signal. The 13-amino acid motif contained the single YxxL primary sequence motif, mentioned above. Thus, B lymphocytes expressing deletion mutants lacking the entire 13 residues, or those expressing a mutation of the tyrosine within the YxxL to phenylalanine, responded similarly when stimulated by F(ab′)2 or intact forms of anti-Ig, i.e., the negative signal was ablated in B cells expressing these mutant versions of FcγRIIb. The presence of a tyrosine residue in this motif and the observation that tyrosine-to-phenyalanine FcγRIIb mutants lacked negative signaling function, prompted the investigators to examine phosphorylation of FcγRIIb at this tyrosine residue. B lymphocytes expressing the wild-type version of FcγRIIb displayed tyrosine phosphorylation of FcγRIIb, while those expressing the deletion or tyrosine-to-phenylalanine versions of FcγRIIb did not.

The killer cell inhibitory receptor (KIR) expressed on NK cells and a subpopulation of T cells appeared to act as an inhibitory receptor for NK-mediated killing of targets, similar to FcγRIIb in B lymphocytes (Binstadt et al. 1996, 1997;

Burshtyn et al. 1996; Fry et al. 1996; Blery et al. 1997; Cosman et al. 1997; Long et al. 1997). KIR recognizes major histocompatability complex (MHC) class I on potential targets; if the target expresses MHC class I and KIR is engaged, the cell is an inefficient NK cell target and is not lysed. The cytomegalovirus genome encodes a MHC class-I homolog such that infected cells are rendered NK-resistant, despite the fact that infected cells downregulate endogenous MHC class I (Farrell et al. 1997; Reyburn et al. 1997). Experiments to address the mechanisms through which KIR blocked NK-mediated killing revealed that the KIR is phosphorylated on a cytoplasmic tyrosine (Binstadt et al. 1996; Burshtyn et al. 1996; Fry et al. 1996; Cosman et al. 1997), which results in the recruitment of a tandem SH2 domain-containing phosphotyrosine phosphatase, SHP-1. These findings suggest that KIR blocks target lysis by recruiting SHP-1, which dephosphorylates PTK substrates and halts the signaling apparatus that promotes target killing. The linear sequence of the phosphorylated motif in KIR resembles the 13-amino acid motif of FcγRIIb, shown to be essential in inhibition of B-lymphocyte function (Muta et al. 1994). Accordingly, this region has been termed the immunoreceptor tyrosine-based inhibitory motif (ITIM), given its similarity to the ITAM in containing the YxxI/L sequence and the fact that it conferred an inhibitory phenotype. A list of ITIM-containing negative signaling receptors is provided in Table 2.

Since the definition of the ITIM, many other receptors were identified as conferring an inhibitory phenotype and likewise containing the ITIM (reviewed in Cambier 1997; Isakov 1997). In addition to KIR, these latter inhibitory receptors include the B-lymphocyte receptor CD22 (Leprince et al. 1993; Law et al. 1994; Campbell and Klinman 1995; Doody et al. 1995; Lankester et al. 1995; Law et al. 1996b), the T lymphocyte receptor CTLA-4 (Marengère et al. 1996), paired Ig-like receptors (PIR) A and B, expressed in B lymphocytes and monocytes (Kubagawa et al. 1997; Blery et al. 1998; Maeda et al. 1998), the mast cell B isoform of glycoprotein of 49kDa [gp49B, (Wang et al. 1997)], ILT3 (Cella et al. 1997), p91 (Hayami et al. 1997), expressed in macrophage/monocyte cells, and the signal-regulatory proteins (SIRPs), which blocked fibroblast proliferation elicited by various growth factors (Kharitonenkov et al. 1997). Besides a single YxxI/L, an ITIM generally, but not always, includes a small hydrophobic residue at the −2 position, relative to the tyrosine residue; the exception being CTLA4 which encodes a glycine residue at the −2 position.

Ectopic expression of FcγRIIb in various non-B-lymphocyte cell types, such as fibroblasts, T lymphocytes, or mast cells, along with an activating receptor such as the ζ chain of the T-cell antigen receptor (Daeron et al. 1995), or FcγRIIa (Hunter et al. 1998) was reported to block signal transduction events and biological events associated with the activating receptor. Thus, T-lymphocyte generation of interleukin (IL)2, mast cell secretion of serotonin, or FcγRIIa-mediated phagocytosis in transfected fibroblasts was inhibited by co-stimulation of FcγRIIb. The findings indicate that the ITIM of FcγRIIb can function in a variety of cellular contexts.

**Table 2.** Cytoplasmic tyrosines of negative signaling receptors present in an immunoreceptor tyrosine-based inhibitory motif (ITIM)

| Receptor | Sequence | | | Expressed | Reference |
|---|---|---|---|---|---|
| FcγRIIb | NTIT | **YSLL** | MHP | B lymphocytes and mast cells | Brooks et al. 1989 |
| KIR-1 (CL-11) | QEVT | **YAQL** | NHC | NK cells and T lymphocytes | Wagtmann et al. 1995a |
| KIR-2 (CL-11) | DIIV | **YTEL** | PNA | NK cells and T lymphocytes | |
| KIR-1 (CL-5) | QEVT | **YAQL** | DHC | NK cells and T lymphocytes | Wagtmann et al. 1995b |
| KIR-2 (CL-5) | DTSV | **YTEL** | PNA | NK cells and T lymphocytes | |
| KIR-1 (CL-2) | EEVT | **YAQL** | DHC | NK cells and T lymphocytes | |
| KIR-2 (CL-2) | DTIL | **YTEL** | PNA | NK cells and T lymphocytes | |
| NKG2-1 | QGVI | **YSDL** | NLP | NK cells and T lymphocytes | Houchins et al. 1991 |
| NKG2-2 | QEIT | **YAEL** | NLQ | NK cells and T lymphocytes | Yabe et al. 1993 |
| [1]CTLA4 | TTGV | **YVKM** | PPT | T lymphocytes | Harper et al. 1991; Dariavach et al. 1988 |
| gp49B-1 | GGIV | **YAGV** | KPS | Mast cells and NK cells | Castells et al. 1994 |
| gp49B-2 [2](murine) | QDVT | **YAQL** | CIR | | |
| PIR-B [2](murine) | QDVT | **YAGL** | ESR | B lymphocytes and monocytes | Kubagawa et al. 1997 |
| CD22-1 | DGIS | **YTTL** | RFP | B lymphocytes | Wilson et al. 1991 |
| CD22-2 | EGIH | **YSEL** | IQF | | |
| CD22-3 | ENVD | **YVIL** | KH- | | |
| p91-1 | EESL | **YASV** | EDM | Macrophages | Hayami et al. 1997 |
| p91-2 | QGET | **YAQV** | KPS | | |
| p91-3 | QDVT | **YAQL** | CSR | | |
| p91-4 [2](murine) | EPSV | **YATL** | AAA | | |
| ILT2-1 | CDVT | **YAQL** | HSF | Monocytes/ macrophages | Samaridis and Colonna 1997 |
| ILT2-2 | EPSV | **YATL** | AIH | | |
| ILT3-1 | CDVT | **YAQL** | HSL | Monocytes/ macrophages | Cella et al. 1997 |
| ILT3-2 | VPSI | **YATL** | AIH | | |

Cytoplasmic tyrosines of negative signaling receptors present in an immunoreceptor tyrosine-based inhibitory motif (ITIM) motif. Sequences shown are of human origin, with the exception of gp49B, PIR-B and p91. **Tyrosine residues in a cononical ITIM (I/L/VvYxxI/V/L) are shown in bold**

[1] Both human and murine CTLA4 lack the orthodox hydrophobic residue at the −2 position relative to the tyrosine

[2] A human homolog to these proteins has not been identified. The extracellular domain of p91 exhibited homology to gp49B, KIR and FcαR (Hayami et al. 1997). p91 protein migrated with a size of 110kDa and 130kDa in transfected COS cells and ex vivo macrophages (Hayami et al. 1997). The sequence shown for PIR-B was essential in confering a negative signal (Maeda et al. 1998) and recruits SHP-1 and SHP-2 phosphatases (Blery et al. 1998; Maeda et al. 1998)

## 4.2 Select Inhibition of Signaling Pathways

The model developed by Phillips and Parker (1983, 1984) using F(ab′)2 anti-Ig to cluster sIg alone and stimulate positive signaling or intact anti-Ig to co-cluster sIg and FcγRIIb to induce negative signaling was applied to B lymphocytes to understand the attendant signaling events. Early studies demonstrated that B lymphocytes stimulated under negative signaling conditions exhibited reduced formation of Ins-P3 (Bijsterbosch and Klaus 1985), indicating defective activation and/or tyrosine phosphorylation of PLCγ. Since Ins-P3 promotes increases in intracellular $Ca^{2+}$ (see above), these findings predicted that B lymphocytes stimulated by intact anti-Ig antibodies would display defects in $Ca^{2+}$ influx. Indeed, later experiments confirmed this prediction (Choquet et al. 1993; Diegel et al. 1994) but the results indicated that only the later phase and not the initial Ins-P3-dependent stage of $Ca^{2+}$ entry was affected by sIg-FcγRIIb co-clustering. In media containing EGTA, the early $Ca^{2+}$ influx was not affected in either B cells stimulated with F(ab′)2 anti-Ig or those stimulated with intact anti-Ig. When $Ca^{2+}$ was added back, those triggered under positive signaling conditions took up more $Ca^{2+}$, showing that a plasma-membrane $Ca^{2+}$ channel had been opened by sIg stimulation. In contrast, B lymphocytes triggered under negative signaling conditions were unable to take up $Ca^{2+}$, indicating that this channel was not opened or was actively closed by FcγRIIb co-clustering. Together, these experiments reveal that B lymphocytes stimulated under conditions of negative signaling exhibit defects in the entry of extracellular $Ca^{2+}$ through a plasma-membrane $Ca^{2+}$ channel.

The defect in $Ca^{2+}$ signaling and Ins-P3 formation following sIg-FcγRIIb co-clustering could be due to deficiencies in the activation of proximal PTKs, like the Src-family or Syk, associated with sIg and activated by stimulation (Burkhardt et al. 1991). Experiments assessing the activity of the Src-family PTK Lyn and of p72Syk derived from B lymphocytes stimulated under positive and negative signaling conditions indicated this was not the case (Sarkar et al. 1996). Likewise, the overall repertoire of PTK substrates appearing in B lymphocytes after positive or negative stimulation appeared similar (Sarkar et al. 1996; Kiener et al. 1997). These studies established that B cells stimulated under negative signaling conditions activate receptor-associated PTKs and that such cells carry out tyrosine phosphorylation to a similar extent as do B cells stimulated under positive signaling conditions. However, examination of particular PTK substrates revealed reduced tyrosine phosphorylation of PLCγ2 (Sarkar et al. 1996), the major PLCγ isoform expressed in B lymphocytes, and CD19 (Hippen et al. 1997; Kiener et al. 1997), an important sIg co-receptor (Tedder et al. 1994), in B lymphocytes stimulated under negative signaling conditions. Reduced PLCγ2 tyrosine phosphorylation may account for the reduced formation of Ins-P3 (Bijsterbosch and Klaus 1985) mentioned above. Reduced CD19 tyrosine phosphorylation was shown to result in decreased recruitment of the p85 subunit of PtdIns 3-kinase (Hippen et al. 1997; Kiener et al. 1997), thus potentially contributing to the reported block in entry of extracellular $Ca^{2+}$.

Other experiments revealed defects in stimulation of the Ras pathway in B cells stimulated by sIg-FcγRIIb co-clustering (Sarmay et al. 1996; Tridandapani et al. 1997a). To understand the cause of this defect, specific proximal events were analyzed, including ITAM tyrosine phosphorylation, ITAM-mediated Shc recruitment, Shc tyrosine phosphorylation and Shc–Grb2 interaction in B lymphocytes stimulated under positive or negative signaling conditions (Tridandapani et al. 1997a). Of these, only Shc–Grb2 interaction was blocked under negative signaling conditions. This observation was particularly intriguing since Shc was phosphorylated to an extent equal to that seen under positive signaling conditions and Shc interacted with the phosphorylated ITAM-containing α chain associated with sIg. Furthermore, tyrosine-phosphorylated Shc, obtained from B-cell lysates after stimulation under negative signaling conditions, was fully capable of binding to the SH2 domain of recombinant Grb2, when the latter was added to lysates in vitro. This finding argued that Shc was phosphorylated on tyrosine residues capable of interacting with the SH2 domain of Grb2; nevertheless, the two endogenous proteins did not associate in B cells. It was hypothesized that, rather than Grb2, Shc interacted with another tyrosine phosphorylated in B cells under negative signaling conditions. Indeed, Shc immunoprecipitates from B cells stimulated with intact anti-Ig contained very high levels of an unknown 145-kDa tyrosine-phosphorylated protein. The authors concluded that Shc selectively bound to p145 rather than Grb2 under negative signaling conditions, preventing the formation of the Shc/Grb2/Sos complex and leading to a block in the Ras pathway.

Although immune tolerance was invoked as the explanation for antibody-induced suppression, results of recent experiments (Ashman et al. 1996; Yamashita et al. 1996) suggested that apoptotic B-lymphocyte death is involved. In the former studies (Ashman et al. 1996), the rate of apoptotic cell death was compared in murine B lymphocytes treated with nothing or intact or F(ab′)2 fragments of anti-Ig. B lymphocytes displayed a spontaneous apoptotic rate which was accelerated by stimulation with intact but not with F(ab′)2 anti-Ig. Likewise, B lymphocytes treated with intact anti-Ig failed to enter the cell cycle, remaining in $G_o$ and very rapidly became hypo-diploid, characteristic of apoptotic cell death. The accelerated apoptosis induced by intact anti-Ig was reversed by co-stimulation with IL4, shown previously to overcome inhibition of proliferation by intact anti-Ig (Phillips et al. 1988). Likewise, the accelerated apoptosis by intact anti-Ig was reduced by preventing engagement of the FcγRIIb with the blocking monoclonal antibody 2.4G2. In the latter set of experiments (Yamashita et al. 1996), B-lymphocyte blasts were exposed to intact or F(ab′)2 fragments of anti-Ig. Both forms of anti-Ig promoted apoptosis; however, those exposed to intact anti-Ig reagents were arrested in $G_1$ and exhibited an accelerated rate of apoptotic cell death while those exposed to F(ab′)2 anti-Ig continued in their cell cycle transitions to S and $G_2$/M phases. The accelerated death induced by intact anti-Ig was shown to be independent of Fas expression. Together, these findings raise the interesting possibility that B cells undergoing negative signaling are not simply made tolerant to antigens but are actually removed from the repertoire of the host.

## 4.3 Recruitment of Negative Signaling Enzymes to the ITIM

### 4.3.1 Recruitment of the SHP-1 Phosphotyrosine Phosphatase

The observation that negative signaling receptors contained the common cytoplasmic ITIM, which included a phosphorylated tyrosine residue (Muta et al. 1994), suggested that these receptors recruit SH2-domain-containing enzymes to promote their inhibitory function. To address this issue, D'Ambrosio et al. (1995) used synthetic phosphopeptides corresponding to the 13-amino acid motif of FcγRIIb, described earlier (Muta et al. 1994), to identify proteins capable of binding. When incubated with lysates of B cells, the ITIM phosphopeptide bound three proteins, p65, p70 and p145 (D'Ambrosio et al. 1995). The primary amino acid sequence of p65 was identified as SHP-1, a phosphotyrosine phosphatase expressing two tandem N-terminal SH2 domains and a C-terminal catalytic domain. Later experiments identified p70 as the structurally related phosphotyrosine phosphatase, SHP-2 (Olcese et al. 1996).

Endogenous FcγRIIb, derived from B lymphocytes stimulated under negative signaling conditions, was shown to co-immunoprecipitate with SHP-1. B lymphocytes lacking FcγRIIb expression displayed neither defects in $Ca^{2+}$ influx, described above, nor association of FcγRIIb and SHP-1. Furthermore, engagement of the SH2 domain of SHP-1 by the ITIM phosphopeptide of FcγRIIb, but not ITAM phosphopeptides of Ig α chain or FcεRI β chain, induced an approximately sixfold increase in phosphatase activity towards a synthetic substrate. Lastly, proliferation of B cells derived from the motheaten strain of mice, genetically deficient in SHP-1, was induced by both F(ab′)2 and intact anti-Ig antibodies, arguing that SHP-1 was required for negative signaling as measured by proliferation. Later experiments, described below, indicated that protein dephosphorylation by SHP-1 is probably not the primary mechanism that procures negative signaling through the FcγRIIb in B lymphocytes. Nevertheless, these important results provided a testable model to explore the mechanisms of negative signaling: phosphorylation of ITIM tyrosines creates a docking site for SH2 domain-containing inhibitory enzymes that block cell activation.

### 4.3.2 Phosphorylation of the SH2-Domain-Containing Inositol 5-Phosphatase, SHIP

The p145 protein, observed as binding to the ITIM of FcγRIIb (D'Ambrosio et al. 1995), was later identified as the SH2-domain-containing inositol phosphatase, SHIP (Chacko et al. 1996b; Ono et al. 1996). SHIP was highly expressed in B lymphocytes but only phosphorylated under negative signaling conditions by stimulating with intact anti-Ig and co-clustering sIg and FcγRIIb (Chacko et al. 1996b). Only SHIP and the tyrosine within the ITIM of FcγRIIb share this unusual property. SHIP tyrosine phosphorylation was fully reversible when co-stimulating B lymphocytes with intact anti-Ig and 2.4G2 (Chacko et al. 1996b), the anti-FcγRII blocking antibody used to restore proliferation induced by the same stimulus (Phillips and Parker 1985). Additional experiments indicated that SHIP interacted with Shc, the adapter protein involved in Ras stimulation, and again the interaction was only observed upon sIg-FcγRIIb co-clustering. Other approaches (Ono et al. 1996) found that SHIP associated with a synthetic phosphopeptide corresponding to the 13-amino acid motif identified earlier (Muta et al. 1994) and that SHIP co-immunoprecipitated with FcγRIIb derived from mast cells or B lymphocytes stimulated under negative signaling conditions. SHIP was therefore the unidentified, tyrosine-phosphorylated protein previously observed interacting with Shc under conditions of negative signaling (Tridandapani et al. 1997a). These findings regarding the exclusive SHIP tyrosine phosphorylation under negative signaling conditions in B lymphocytes were later independently confirmed (D'Ambrosio et al. 1996a).

A number of earlier studies reported that cytokines induced tyrosine phosphorylation and Shc-association of a 145-kDa protein (Damen et al. 1993; Lioubin et al. 1994; Liu et al. 1994); and similar findings had been observed in B lymphocytes stimulated through sIg (Saxton et al. 1994; Smit et al. 1994; Crowley et al. 1996). Consequently, SHIP cDNA was cloned and described by three separate laboratories studying cytokine-induced protein phosphorylation (Damen et al. 1996; Kavanaugh et al. 1996; Lioubin et al. 1996). The enzyme was shown to hydrolyze the D5 phosphate of 3-phosphoinositides and 3-phosphorylated inositol phosphates; thus, SHIP hydrolyzes PtdIns-3,4,5P3 to PtdIns-3,4P2 and Ins-1,3,4,5P4 to Ins-1,3,4P3. SHIP encodes an N-terminal SH2 domain, a central catalytic core similar to other known inositol 5-phosphatases, and two NPxY motifs optimal for engaging the PTB domain of Shc upon phosphorylation of the tyrosine residue(s), and a proline-rich C-terminus that associates with the SH3 domain of Grb2. It is highly likely that the earlier reports of the Shc-associated p145 in B lymphocytes (Saxton et al. 1994; Smit et al. 1994; Crowley et al. 1996) was indeed SHIP and perhaps these earlier studies used intact anti-Ig antibodies in stimulation protocols which promoted sIg-FcγRIIb co-clustering to induce SHIP phosphorylation (Chacko et al. 1996b).

### 4.3.3 ITIM Engagement is the Mechanism of SHIP Tyrosine Phosphorylation

Other experiments sought to address the curiously efficient SHIP tyrosine phosphorylation upon sIg-FcγRIIb co-clustering. It was known that co-clustering induced phosphorylation of the tyrosine within the ITIM motif of FcγRIIb (Muta et al. 1994); thus, it was proposed (Tridandapani et al. 1997c) that the phosphorylated ITIM formed a docking site for the SH2 domain of SHIP, thereby bringing it within the range of sIg-associated PTKs. In support of this hypothesis, the recombinant SH2 domain of SHIP, or the intact endogenous protein, bound to synthetic phosphopeptides corresponding to the ITIM motif but not to other control phosphopeptides present in an ITAM motif. Likewise, an ITIM phosphopeptide corresponding to a cytoplasmic tyrosine of CD22 interacted with the SH2 domain of SHIP. Furthermore, B lymphocytes deficient in FcγRIIb expression were unable to phosphorylate SHIP upon stimulation with intact anti-Ig, revealing a genetic requirement for expression of the ITIM of FcγRIIb to promote

SHIP tyrosine phosphorylation. Lastly, the ability of SHIP to interact with Shc was examined and reported to be entirely dependent on SHIP tyrosine phosphorylation. Thus, B lymphocytes lacking FcγRIIb displayed efficient Shc tyrosine phosphorylation but not SHIP–Shc interaction. These observations (Tridandapani et al. 1997c) directly implicated the phosphorylated ITIM motif in promoting SHIP tyrosine phosphorylation and likely account for earlier findings of SHIP–FcγRIIb co-immunoprecipitation (Ono et al. 1996).

### 4.3.4 Is SHIP Engaged to ITAMs of Activating Receptors?

The above model (Tridandapani et al. 1997c) invoked a requirement of a phosphorylated ITIM in promoting SHIP tyrosine phosphorylation. If so, it is reasonable to ask how SHIP is tyrosine phosphorylated as a consequence of cytokine stimulation, in which there is no apparent "negative" signaling and thus no need for an ITIM motif. Oddly, many cytokine receptors do indeed contain a canonical ITIM motif and others possess a single YxxI/L without the small hydrophobic amino acid residue at the −2 position, similar to CTLA-4 (Table 3). Other experiments investigating the interaction of the SH2 domain of SHIP to various ITIM motifs revealed that the small hydrophobic residue at the −2 position relative to the tyrosine is dispensable (Vely et al. 1997) and thus the SH2 domain may simply engage a YxxI/L, rather than the orthodox I/L/VxYxxI/L primary sequence of the ITIM.

Mutational analysis of cytoplasmic tyrosines within cytokine receptors suggest that they do indeed confer an inhibitory function upon cytokine triggering, such that replacement of some tyrosines with a phenylalanine residue induces greater proliferation upon expression and stimulation of the mutant receptor in cells (Yi et al. 1993; Sokol et al. 1995; Prchal and Sokol 1996; Wang et al. 1996). Naturally occurring mutations have been identified in the granulocyte colony-stimulating factor (G-CSF) receptor (Dong et al. 1995; Touw and Dong 1996) that code for truncated forms lacking cytoplasmic tyrosine residues which promote negative signaling. Patients expressing these forms of the G-CSF receptor exhibit acute myelogenous leukemias, although the causal relationship between negative signaling and the leukemia is not established. However, recent studies indicate that the G-CSF (Hunter and Avalos 1998) and the IL4 receptors (P. Rothman, personal communication) recruit SHIP to the ITIMs in the cytoplasmic tail of the respective receptors, and that SHIP contributes to a reduction in cytokine-induced proliferation. The signaling strategy for these receptors may be truly a Yin/Yang condition (Isakov 1997) in which both positive and negative signaling elements combine to elicit a tempered proliferative response.

Other experiments employing a yeast three-hybrid system revealed that SHIP is capable of binding to an authentic ITAM motif (Osborne et al. 1996). In this study, yeast expressing Lck, a Src-family PTK, and the human sequences corresponding to the ITAM of FcεR β or γ chain, or the ζ chain of the T-cell antigen receptor were used as "bait" and were screened against a rat mast cell cDNA library to identify interacting proteins. SHIP cDNA was isolated from this screen.

**Table 3.** Cytoplasmic tyrosines of cytokine receptors present in motifs resembling the immunoreceptor tyrosine-based inhibitory motif (ITIM)

| Receptor (amino acid numbers) | Sequence | | | position in Cytoplasm | Reference |
|---|---|---|---|---|---|
| FcγRIIb (273–284) cononical ITIM motif | NTIT | **YSLL** | MHP | cyto Y 1/1 | BROOKS et al. 1989 |
| Epo-R (422–433) | ASFE | **YTIL** | DPS | cyto Y 3/9 | JONES et al. 1990 |
| Epo-R (450–461) | PHLK | **YLYL** | VVS | cyto Y 4/9 | |
| [1]G-CSF (748–759) | DQVL | **YGQL** | LGS | cyto Y 3/4 | LARSEN et al. 1990; FUKUNAGA et al. 1990 |
| [1]G-CSF (783–785) | SPKS | **YENL** | WFQ | cyto Y 4/4 | |
| gp130 (755–766) | STVQ | **YSTV** | VHS | cyto Y 2/6 | HIBI et al. 1990 |
| Kit-R (605–616) | EATA | **YGLI** | KSD | cyto Y 6/21 | YARDEN et al. 1987 |
| Kit-R (671–682) | EYCC | **YGDL** | LNF | cyto Y 9/21 | |
| Kit-R (699–710) | EAAL | **YKNL** | LHS | cyto Y 10/21 | |
| [2]beta/common (624–635) | GSLE | **YLCL** | PAG | cyto Y 4/8 | GORMAN et al. 1990 |
| [2]beta/common (818–829) | QVGD | YCFL | PGL | cyto Y 7/8 | |
| [3]IL4α (709–720) | SGIV | YSAL | TCH | cyto Y 5/6 | IDZERDA et al. 1990; GALIZZI et al. 1990 |
| IL7α (397–408) | GPHV | YQDL | LLS | cyto Y 1/3 | GOODWIN et al. 1990 |
| IL9α (298–309) | KRIF | YQNV | PSP | cyto Y 1/5 | RENAULD et al. 1992 |

Cytoplasmic tyrosines of cytokine receptors present in motifs resembling the immunoreceptor tyrosine-based inhibitory motif (ITIM). Sequences shown are of human origin; **tyrosine residues in a cononical ITIM (I/L/VvYxxI/V/L) are shown in bold**. The position of the tyrosine-containing motif relative to other vytoplasmic tyrosines is indicated in the third column.

[1] Patients with adult myelogenous leukemia harbor a G-CSF receptor truncated at residue 718 or 739 (DONG et al. 1995; TOUW and DONG 1996); these poteintial ITIM seqences are therefore deleted

[2] The beta/common subunit was reported to assocaite with SHP-1 phosphotyrosine phosphatase; the site was not determined (YI et al. 1993)

[3] This region was not included in an earlier study of tyrosine-to-phenylalanine mutants (WANG et al. 1996)

Further in vitro analysis showed that the recombinant SH2 domain of SHIP was capable of binding synthetic phosphopeptides corresponding to the above ITAMs. Oddly, the SH2 domain of SHIP bound to doubly phosphorylated CD3γ chain but not δ chain, despite the fact that the two sequences are nearly identical (Table 1). The concentrations of peptide (~0.8μM) and recombinant SH2 domains (35nM) used in this experiment were similar to those used in other studies. However, SHIP association with the Fcε β or γ chain did not increase upon receptor clustering in cells (OSBORNE et al. 1996), which induces ITAM phosphorylation. Likewise, SHIP did not co-immunoprecipitate with the ITAM-containing CD3 chains in T cells (LAMKIN et al. 1997). SHIP tyrosine phosphorylation was not induced upon stimulation of the FcεR in cultured mast cells (OSBORNE et al. 1996).

Some of these findings contrast with others using the identical cell line (RBL-2H3 rat basophilic cells) and identical stimulation protocols (Kimura et al. 1997). In this latter example, SHIP was able to bind the ITAM of both the FcεR β chain and the doubly phosphorylated FcεR γ chain; the sequences used corresponded to the rat proteins. The phosphotyrosine phosphatase SHP-1 did not bind either phosphopeptide. Additionally, SHIP tyrosine phosphorylation was stimulated by clustering of FcεR, whereupon SHIP bound to Shc in a manner similar to B lymphocytes stimulated under negative signaling conditions (Chacko et al. 1996b). The authors speculated that SHIP engaged the C-terminal tyrosine in the ITAM motif of FcεR β chain, which has the sequence SPIYSAL (human SATYSEL; Table 1). They noted the similarity of the rat C-terminal FcεR β chain motif to a region of the 13-amino acid ITIM of FcγRIIb; TITYSLL (Muta et al. 1994) (Table 2), while the N-terminal tyrosine residue of the rat FcεR β chain was present in a distinct motif of DRLYEEL (human, DRVYEEL; Table 1). The latter tyrosine is followed by acidic amino acids, an optimal motif for engagement by the SH2 domain of Src-family PTKs (Songyang et al. 1993), but perhaps excludes engagement by the SH2 domain of SHIP, as previously hypothesized (Tridandapani et al. 1997c).

Like signaling through FcεR, experiments exploring signaling mechanisms of FcγRIIa, which contains an endogenous ITAM, or of FcγRI, which is associated with the ITAM-containing γ chain, in U937 monocytes indicated potent and prolonged tyrosine phosphorylation of SHIP and interaction with Shc (Maresco et al., unpublished observation). Since the stimulation protocol employed antibodies to FcγRIIa, it was conceivable that FcγRIIb was co-clustered in these experiments, as it is with intact anti-Ig in the B lymphocyte paradigm. However, these workers carefully eliminated this possibility by showing that neither the primary anti-FcγRIIa nor the secondary clustering antibody was capable of stimulating FcγRIIb and, in any case, that expression of FcγRIIb in U937 monocytes was not detectable. Thus, it is probable that some ITAMs recruit SHIP and provoke its tyrosine phosphorylation, perhaps due to similarity in primary sequence with the ITIM motif of FcγRIIb, although that does not appear to be the case of B-lymphocyte antigen receptors (Chacko et al. 1996b). The role played by SHIP in these examples is not clear and is not necessarily negative. SHIP may enhance phagocytosis or degranulation by its enzymatic activity or protein interactions. Further study is necessary to resolve this intriguing issue.

# 5 Mechanisms of Action of SHP-1 and SHIP Phosphatase

## 5.1 SHP-1 Phosphotyrosine Phosphatase

The motheaten mutation arose at the Jackson Laboratories and was identified as the SHP-1 phosphotyrosine phosphatase gene (Kozlowski et al. 1993), resulting in severely reduced (motheaten-viable) or complete loss of (motheaten) expression of the phosphatase. Such animals exhibit impaired function of immune cells exhibited in in vitro assays of T and B lymphocytes, NK and cytotoxic T-cell function, systemic autoimmunity, and early death proximally due to inflammation of lung tissue but ultimately the result of accumulation of phagocytic cells (reviewed in SCHULTZ et al. 1997; SCHULTZ and SIDMAN 1987). The accumulation of these cells is likely due to the reported hyperproliferation of macrophages to multiple cytokines; ILs3 (YI et al. 1993), macrophage CSF (CHEN et al. 1996), G-CSF (TAPLEY et al. 1997), and granulocyte/macrophage CSF (JIAO et al. 1997). In vivo responses to these cytokines of macrophage precursors lacking the inhibitory phosphatase is associated with greatly increased numbers of cells (TAPLEY et al. 1997). Numbers of lymphocytes are reduced, apparently due to defective lymphocyte development, but the precise mechanism is unclear (reviewed in SCHULTZ et al. 1997). Nevertheless, the hyper-proliferative responses to cytokines and increased cell number in SHP-1-deficient animals argues that SHP-1 is involved in negative signaling by cytokines to keep cell numbers and activities in check.

SHP-1 phosphatase associates with the tyrosine-phosphorylated ITIM motif within KIR of human (BINSTADT et al. 1996; FRY et al. 1996; VELY et al. 1996; COSMAN et al. 1997; CARRETERO et al. 1998) or mouse (OLCESE et al. 1996) origin. Experiments have revealed that SHP-1 dephosphorylates a wide variety of PTK substrates, as shown in anti-phosphotyrosine immunoblots of whole-cell lysates of NK cells engaging targets that express the KIR ligand, MHC class I (NAKAMURA et al. 1997). Specific SHP-1 substrates identified include the PTK Syk (BRUMBAUGH et al. 1997), shown to be essential for target lysis by NK cells (BRUMBAUGH et al. 1997), the Janus-family kinase Jak2, involved in proliferation induced by the erythropoietin receptor (JIAO et al. 1996), and pp36, a possible adapter protein involved in PLCγ tyrosine phosphorylation (VALIANTE et al. 1996).

Defects in changes of cytoplasmic $Ca^{2+}$ have been exploited as an indicator of negative signaling by both KIR and FcγRIIb. However, the particular defect in $Ca^{2+}$ influx is quite distinct when triggered by KIR to that of FcγRIIb (BLERY et al. 1997) and has been used as a diagnostic tool to distinguish particular mechanisms (ONO et al. 1997). KIR engagement appears to block both release of $Ca^{2+}$ from intracellular stores as well as increased uptake of extracellular $Ca^{2+}$ (BLERY et al. 1997). FcγRIIb co-clustering appears to block the latter event in B lymphocytes but does not affect release from intracellular stores (CHOQUET et al. 1993; Diegel et al. 1994). The block in release of $Ca^{2+}$ from intracellular stores is consistent with the reported inhibition in Ins-P3 formation in NK cells recognizing targets expressing the KIR ligand (VALIANTE et al. 1996). However, while these findings argue that SHP-1 is a negative regulator of several signaling pathways, other studies showed that SHP-1 is actually a positive regulator of Src-family PTKs, due to its ability to dephosphorylate these PTKs at the inhibitory C-terminal tyrosine, similar to CD45 (SOMANI et al. 1997). Since activation of Src-family PTKs are the most proximal events elicited by immunoreceptor stimulation, receptor recruitment of SHP-1 may promote rather than inhibit signal transduction events by this mechanism.

## 5.2 SHIP Inositol Phosphatase

Since both SHIP and SHP-1 were reported to associate with FcγRIIb upon stimulation of B lymphocytes under negative signaling conditions, it was unclear whether these were redundant enzymes, whether each made a unique contribution, or whether one was more important than the other in generating the dominant negative signal. B lymphocytes from motheaten mice still display reduced CD19 tyrosine phosphorylation under negative signaling conditions (NADLER et al. 1997), indicating that SHP-1 does not play a role in CD19 dephosphorylation. In any case, it is unclear to what extent CD19 dephosphorylation contributes to the overall negative signaling process. CD19 is clearly important in B-lymphocyte development and signaling, as indicated in sIg-CD19 co-clustering experiments and in CD19-deficient animals (TEDDER et al. 1994; FEARON and CARTER 1995). Furthermore, CD19 contains cytoplasmic tyrosines that recruit SH2-domain-containing proteins and enzymes that support the signaling process initiated by sIg (TEDDER et al. 1994; FEARON and CARTER 1995). Since the primary purpose of CD19 is to render sIg more sensitive to antigen triggering, CD19 dephosphorylation may provoke an antigen-unresponsive state in B lymphocytes.

Animals deficient in SHP-1 (motheaten or motheaten-viable) exhibit numerous defects in hematopoietic cells, hyperresponses to cytokines and hematopoietic cell growth factors and early death due to infiltration of granulocytes into lung tissue (reviewed in SHULTZ et al. 1997). SHIP-deficient animals exhibit a very similar phenotype (HELGASON et al. 1998), including reduced B-lymphocyte development. Unfortunately, no additional information is available at this time.

### 5.2.1 SHIP Enzymatic Activity

More information regarding the precise role of SHP-1 and SHIP in signal transduction has been obtained from in vitro experiments. B-lymphocyte cell lines have been rendered deficient in expression of SHP-1 and in SHIP (ONO et al. 1997) and changes in cytoplasmic $Ca^{2+}$ upon co-clustering sIg-FcγRIIb or sIg-KIR was measured. The results indicated that B lymphocytes deficient in SHP-1 showed a $Ca^{2+}$-influx pattern consistent with negative signaling upon sIg-FcγRIIb co-clustering. In contrast, B lymphocytes deficient in SHIP displayed a normal $Ca^{2+}$ influx pattern, inconsistent with negative signaling. These findings suggest that, at least in regard to the changes in cytoplasmic $Ca^{2+}$, SHP-1 is dispensable while SHIP is essential for FcγRIIb-mediated negative signaling. However, B lymphocytes deficient in SHP-1 showed a $Ca^{2+}$ influx pattern inconsistent with negative signaling upon sIg-KIR co-clustering while SHIP-deficient B lymphocytes showed an aborted, negative signaling pattern of $Ca^{2+}$ influx. Thus, negative signaling by KIR, which likewise terminates the influx of $Ca^{2+}$, requires expression of the SHP-1 phosphatase. Similar findings were made regarding activation of nuclear factor (NF)-B (ONO et al. 1997), a transcription factor which acts distal to the Ras pathway, and to increased cytosolic $Ca^{2+}$ (reviewed in STANCOVSKI and BALTIMORE 1997; SHA 1998).

Other studies using B lymphocytes derived from SHP-1-deficient motheaten mice arrived at a similar conclusion (NADLER et al. 1997). These results indicated that SHP-1-deficient B lymphocytes displayed a reduced influx of extracellular $Ca^{2+}$ under negative signaling conditions, identical to normal B lymphocytes. Thus, B cells either expressing SHP-1 or not displayed a blunted influx of extracellular $Ca^{2+}$, the usual pattern of $Ca^{2+}$ influx under negative signaling conditions. The global panel of tyrosine-phosphorylated proteins arising in the SHP-1-expressing and -deficient B lymphocytes upon stimulation with intact anti-Ig reagents was indistinguishable. The phosphorylation level of CD19 was similarly reduced in SHP-1-deficient B lymphocytes, indicating that SHP-1 does not affect the CD19 tyrosine phosphorylation status. Since earlier studies indicated reduced tyrosine phosphorylation of PLCγ2 in B cells stimulated under negative signaling conditions (SARKAR et al. 1996), other experiments using the same B-lymphocyte cell lines examined the status of PLCγ2 phosphotyrosine. The data revealed that PLCγ2 tyrosine phosphorylation was equivalently reduced in SHP-1-expressing and -deficient lines (Kelley and Coggeshall, unpublished observations), suggesting that PLCγ2 is not a direct SHP-1 substrate.

Another approach to address the role of SHIP and SHP-1 in negative signaling by FcγRIIB and KIR used chimeric receptors expressed in NK cells recognizing targets expressing the KIR ligand, MHC class I (GUPTA et al. 1997). The results demonstrated that target lysis was inhibited when NK cells expressed a KIR chimera encoding the intracellular domain of either FcγRIIb or of wild-type KIR. Thus, both versions of the ITIM motif could promote negative signaling in this context. Co-expression of a catalytic-deficient version of SHP-1 blocked the negative signal delivered by the cytoplasmic tail of wild-type KIR but not that of FcγRIIb. However, over-expression of the SH2 domain of SHIP blocked the negative signal delivered by the cytoplasmic tail of FcγRIIb but not KIR. These findings indicated an important feature of negative signaling. As concluded in the SHIP/SHP-1 knock-out studies described above (ONO et al. 1997), the negative signaling mechanism used by KIR and FcγRIIb is not identical. Each receptor or, more precisely, each ITIM motif recruits a different phosphatase; SHIP in FcγRIIb-mediated negative signaling and SHP-1 in KIR-mediated negative signaling. This dichotomy occurred in the same cell context (NK cells) and acted on the same cellular process (target lysis). Thus, despite the expression of both SHIP and SHP-1 in these cells, and despite the reported ability of both phosphatases to engage the phosphorylated ITIM motif of both receptors, KIR selected SHP-1 and FcγRIIb selected SHIP from the milieu of SH2-domain-containing proteins, as determined in a functional assay. The data further revealed that over-expression of the SH2 domain of SHIP can block SHIP function. While it is difficult to infer the mechanism of action in this case, it is likely that the over-expressed SH2 domain in isolation bound the phosphorylated ITIM of FcγRIIb and prevented membrane recruitment of endogenous SHIP, where it would have access to 3-phosphoinositide substrates. This ability of the isolated SH2 domain of SHIP to act as a dominant-negative version suggests that plasma membrane localization and subsequent lipid hydrolysis may be one means by which SHIP acts to influence B lymphocytes.

Studies of *Xenopus* oocyte maturation likewise revealed potent inhibitory effects of SHIP through its ability to hydrolyze PtdIns-3,4,5P3 (Deuter-Reinhard et al. 1997). Previous experiments indicated that microinjection of the SH2 domain of p85 into *Xenopus* oocytes blocked their maturation as well as their induction of Erk kinase activity stimulated by insulin (Chuang et al. 1994), indicating a role for PtdIns 3-kinase in these processes. Consistent with this hypothesis, microinjection of a constitutively active PtdIns 3-kinase, p110*, was able to induce oocyte maturation (Deuter-Reinhard et al. 1997). Furthermore, microinjection of the SHIP protein but not a catalytic-deficient form of SHIP was shown to block inhibit insulin-stimulated maturation and activation of Erk. Inactivating mutations in the SH2 domain of SHIP did not affect its inhibitory influence, suggesting that SH2 engagement by SHIP was not required. The ability of SHIP to block maturation correlated with a reduction in levels of PtdIns-3,4,5P3 in oocytes. These experiments suggest that in *Xenopus* oocytes the induction of Erk kinase by insulin requires PtdIns-3,4,5P3 formation by PtdIns 3-kinase and that Erk induction can be negatively influenced by SHIP-mediated hydrolysis of PtdIns-3,4,5P3. The SH2 domain of SHIP had no effect in this cellular context and in contrast to findings in NK cells described above. This fact suggests that SHIP carries out its functions without recruitment to an ITIM motif and may therefore have a distinct mechanism of action.

The influence of SHIP on 3-phosphoinositide metabolism in mammalian cells has been recently reported. Earlier experiments showed that sIg-activated B lymphocytes transiently accumulate both PtdIns-3,4,5P3 and PtdIns-3,4P2 through the action of PtdIns 3-kinase (Gold and Aebersold 1994). Other studies in fibroblasts indicated that expression of a constitutively active version of PtdIns 3-kinase promoted the activation of Btk (Li et al. 1997). Furthermore, Btk synergized with activated PtdIns 3-kinase and a partially activated version of the PTK Src in promoting transformation of fibroblasts (Li et al. 1997). These findings suggest that Btk acts distal to both PtdIns 3-kinase and Src-family PTKs. To investigate the influence of PtdIns 3-kinase and Btk on B-lymphocyte activation, the constitutively active version of PtdIns 3-kinase, p110*, was stably expressed in the murine B-cell lymphoma A20 (Scharenberg et al. 1998). Cells co-expressing both p110* and Btk displayed enhanced levels of PtdIns-3,4,5P3 as well as products of PLCγ activation, including Ins-P3 and greatly increased influx of intracellular $Ca^{2+}$. These findings suggest that PtdIns 3-kinase and Btk acted proximal to PLCγ in $Ca^{2+}$ signaling. Previous genetic experiments using Btk-deficient B lymphocytes had come to a similar conclusion (Takata and Kurosaki 1996).

Consistent with this hypothesis, co-expression of p110* and Btk enhanced PLCγ2 tyrosine phosphorylation and the enhancement was sensitive to the PtdIns 3-kinase inhibitor, wortmannin (Scharenberg et al. 1998). The investigators thus established a B-lymphocyte paradigm, in which a series of signaling events were demonstrated to be mediated by the SHIP substrate, PtdIns-3,4,5P3, acting on the Tec-family kinase, Btk. Similar results have been reported for Itk, the T-lymphocyte version of Tec-family kinases, as a link to PLCγ1 tyrosine phosphorylation and regulating $Ca^{2+}$ entry (Liu et al. 1998). Again exploiting the A20 B-cell lymphoma, the authors revealed (SCHARENBERG et al. 1998) no detectable PtdIns-3, 4,5P3 in B lymphocytes stimulated under negative signaling conditions, accompanied by reduced induction of Btk activity and reduced or absent PLCγ2 tyrosine phosphorylation, as reported previously (SARKAR et al. 1996). The influence of SHIP in this process was demonstrated in B lymphocytes expressing p110*; cells co-expressing wild-type SHIP but not catalytically inactive SHIP displayed reduced Btk activity.

These findings are consistent with the hypothesis that SHIP hydrolyzes PtdIns-3,4,5P3, which prevents induction of the Btk kinase and which, in turn, blocks PLCγ2 tyrosine phosphorylation. However, these results are likewise consistent with the possibility that stimulation of PtdIns 3-kinase activity is reduced under conditions of negative signaling and are thus not necessarily due to increased SHIP consumption but rather lowered production of PtdIns-3,4,5P3. This possibility is especially attractive, given the documented reduction in CD19 tyrosine phosphorylation and decreased association of PtdIns 3-kinase to CD19 seen under conditions of negative signaling. Similar findings regarding the regulation of $Ca^{2+}$ entry by Btk in B lymphocytes were reported by others (BOLLAND et al. 1998); however, these experiments employed cells expressing a chimeric receptor composed of FcγRIII extracellular domain linked to Btk. The sIg-FcγRIIb co-clustering protocol using intact anti-Ig likely stimulated the FcγRIII chimera, making difficult the precise interpretation of the results.

### 5.2.2 SHIP Protein Interactions

Considerably less is known about SHIP protein interactions. SHIP appears to associate through its C-terminal proline-rich domain to the SH3 domain of Grb2 and was originally cloned by virtue of this feature (DAMEN et al. 1996). SHIP inducibly associates with tyrosine-phosphorylated Shc (LIU et al. 1994, 1997a; LAMKIN et al. 1997; PRADHAN and COGGESHALL 1997), insulin receptor substrate-1 (VERDIER et al. 1997), and SHP-2 (LIU et al. 1997b), a phosphotyrosine phosphatase structurally but not functionally related to SHP-1 (reviewed in FREARSON and ALEXANDER 1997). Besides the C-terminal proline-rich region that can associate with SH3 domains, SHIP contains an N-terminal SH2 domain and two NPxY phosphorylation sites optimal for association with a PTB domain.

Previous experiments, described in more detail above, showed that tyrosine-phosphorylated Shc did not associate with the SH2 domain of Grb2 under conditions of negative signaling but rather associated with p145/SHIP (TRIDANDAPANI et al. 1997a). Shc did not bind Grb2, despite the fact that Shc was phosphorylated on appropriate tyrosines able to interact with recombinant forms of the SH2 domain of Grb2. Based on these results, it was proposed (TRIDANDAPANI et al. 1997b) that the SH2 domains of SHIP and Grb2 compete for binding to phosphorylated sites of Shc. SHIP is successful in this competition under negative signaling conditions while Grb2 is successful under positive signaling conditions. Experiments using recombinant interaction domains of SHIP and Shc or synthetic phosphopeptides of both proteins (PRADHAN and COGGESHALL 1997) revealed that SHIP–

Shc interaction is bidentate. Thus, the PTB domain of Shc binds to either of the two NPxY sites on SHIP. At the same time, the SH2 domain of SHIP binds to Y317 or doubly phosphorylated Y239, Y240 of Shc. The SH2 domain of Grb2 has been shown to recognize these identical sites on phosphorylated Shc (van der Geer et al. 1996). Other independent experiments have observed an identical bidentate interaction of SHIP and Shc (Liu et al. 1997a); however, studies in T lymphocytes detected only the Shc–PTB interaction and not the SHIP–SH2 interaction (Lamkin et al. 1997). Thus, the mode of SHIP–Shc interaction may be influenced by the cellular context in which it is studied.

Perhaps a more fundamental issue is how the single SH2 domain of SHIP can concomitantly engage both the phosphorylated ITIM motif of FcγRIIb and the Grb2-binding site on phospho-Shc. Immunoprecipitation analysis revealed that the SHIP protein was present in two distinct and mutually exclusive pools: one bound to FcγRIIb and one bound to Shc (Tridandapani et al. 1999a,b). There was no evidence of a trimeric complex of all three proteins; FcγRIIb immunoprecipitates contained only FcγRIIb and SHIP, while Shc immunoprecipitates contained only Shc and SHIP. These findings suggested that SHIP initially engaged FcγRIIb, became phosphorylated and disengaged the receptor, and bound to Shc. This proposal was consistent with the kinetics of SHIP's association with FcγRIIb, which was maximal at 30s of stimulation, and with Shc, which required several minutes of stimulation.

This model raises another issue – why would the SH2 domain of SHIP disengage FcγRIIb? A possible explanation is that the intrinsic affinity of the SH2 domain of SHIP is greater for tyrosine-phosphorylated Shc than for the ITIM motif of FcγRIIb. To test this hypothesis, the SH2 domain of SHIP was applied in direct affinity measurements to biotinylated, immobilized phosphopeptides corresponding to the ITIM motif of FcγRIIb or of Y317 in Shc. The results revealed approximately equal association rate constants for SHIP–SH2 binding either peptide but a 10-fold greater dissociation rate for FcγRIIb compared with Shc (Tridandapani et al. 1999b). The overall affinity of the SH2 domain of SHIP for the ITIM motif of FcγRIIb was 2.1μM while the affinity for Y317 of Shc was 0.26μM. The SH2 domain of Grb2 binds Y317 of Shc with a reported affinity of 0.03μM (Liu et al. 1997a). Based on these observations, a model can be formulated that describes these protein interactions of SHIP. Early in the process, the SH2 domain of SHIP is engaged to the tyrosine-phosphorylated ITIM motif of FcγRIIb, which is necessary for SHIP tyrosine phosphorylation (Tridandapani et al. 1997c). Once SHIP is phosphorylated, Shc first engages SHIP through its PTB binding either or both of the NPxpY motifs within SHIP. This interaction presents the phosphorylated tyrosine residues within Shc to the SH2 domain of SHIP. Since Shc phosphotyrosines have an intrinsically higher affinity for the SH2 domain of SHIP than FcγRIIb phosphotyrosines ($K_d$ 0.26μM vs 2.1μM, respectively), the initial engagement of Shc promotes the SH2 domain of SHIP to release FcγRIIb and bind Shc.

This model accounts for several previous and unusual observations regarding SHIP in B lymphocytes. First, SHIP is only tyrosine phosphorylated under conditions of negative signaling (CHACKO et al. 1996b). This fact appears to be due to the relatively weak affinity of the SH2 domain of SHIP for the 13- amino acid ITIM motif of FcγRIIb. Second, SHIP associates with Shc only under negative signaling conditions, i.e., there is minimal interaction between these two proteins under positive signaling conditions, despite the fact that Shc is highly phosphorylated and that the SH2 domain of SHIP displays affinity for phospho-Shc (CHACKO et al. 1996b; LIU et al. 1997a; PRADHAN and COGGESHALL 1997). The deficiency of SHIP–Shc interaction under positive signaling conditions is likely due to the relatively higher affinity of the Grb2 SH2 domain for tyrosine-phosphorylated Shc than that of the SHIP SH2 domain ($K_d$ 0.03μM vs 0.26μM) (LIU et al. 1997a). Thus, Shc phosphorylation in the absence of SHIP phosphorylation favors Shc–Grb2 interaction rather than Shc–SHIP interaction. Third, Shc and Grb2 do not associate under negative signaling conditions, despite the high affinity of Grb2 for phospho-Shc (TRIDANDAPANI et al. 1997a). However, SHIP–Shc complexes are readily apparent under negative but not positive signaling (CHACKO et al. 1996b; TRIDANDAPANI et al. 1997a). These previous findings in conjunction with the affinity and kinetic data mentioned above indicate that the formation of a stable SHIP–Shc interaction complex requires SHIP tyrosine phosphorylation and interaction through the PTB domain of Shc. The added contribution of the PTB domain of Shc binding to phospho-SHIP, along with the SH2 domain of SHIP engaging phospho-Shc, generates a stable, bidentate complex and one in which the SH2 domain of SHIP can successfully compete with that of Grb2 for binding to Shc (TRIDANDAPANI et al. 1997a,b).

## 6 Conclusion

Antibody-induced suppression of the humoral immune response has been documented and studied for many years. Progress regarding the mechanism of humoral suppression required several seminal findings. First, the role of FcγRIIb was indicated through in vivo experiments studying the effect of Ig on the humoral response, followed by later in vitro experiments comparing B-lymphocyte activation with stimulation by F(ab′)2 and intact anti-Ig antibodies. Second, information regarding positive signaling mechanisms emerged over the past decade and provided potential sites of inhibition by the dominant negative signal of FcγRIIb. These latter studies revealed the central role of PtdIns 3-kinase and the SHIP substrate PtdIns-3,4,5P3 in B-lymphocyte activation. Perhaps most telling and useful were the findings of MUTA et al. (1994), who initiated the concept of a common motif mediating negative signaling, and the paradigm established by D'AMBROSIO et al. (1995), who sought SH2-domain-containing inhibitory enzymes that engaged the phosphorylated ITIM. These findings led to the discovery of SHIP inositol phosphatase (ONO et al. 1996; D'AMBROSIO et al. 1996a; CHACKO et al. 1996b), and the most recent findings regarding its mechanism of action (Fig. 3).

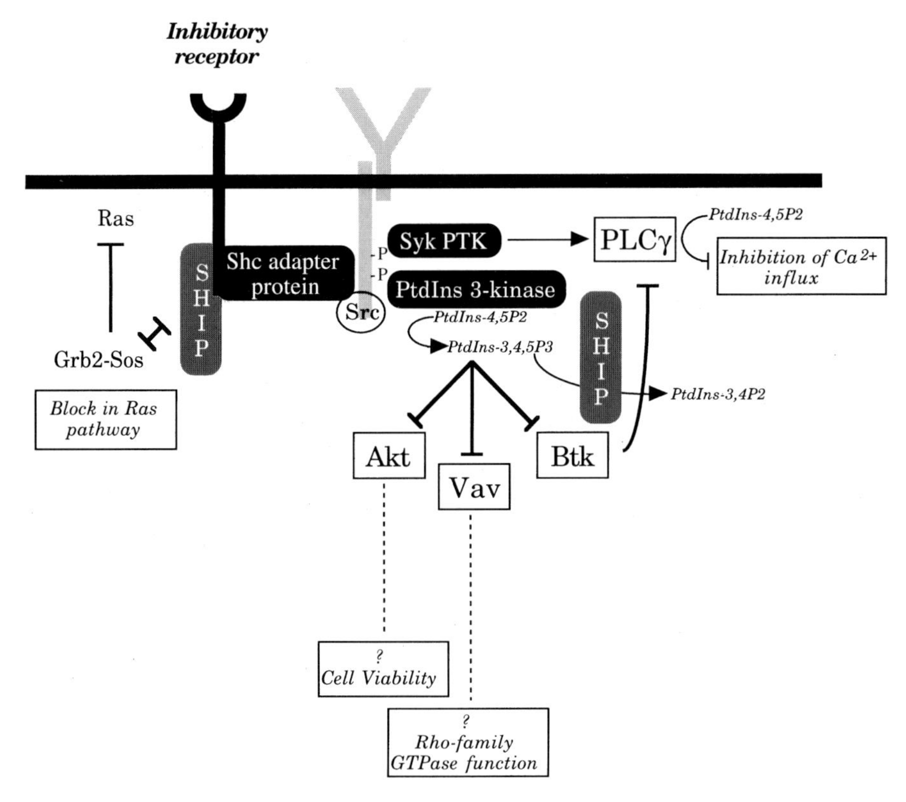


**Fig. 3.** Negative signaling by SH2-domain-containing inositol 5-phosphatase (SHIP). SHIP is recruited to an immunoreceptor tyrosine-based inhibitory motif (ITIM)-containing inhibitory receptor and undergoes tyrosine phosphorylation and interaction with Shc. SHIP–Shc interaction is associated with deficient Shc–Grb2 interaction and can lead to a block in the Ras signaling pathway. Through its enzymatic activity, SHIP hydrolyzes PtdIns-3,4,5P3 to prevent activation of Btk, resulting in inhibition of phospholipase C (PLC)γ tyrosine phosphorylation and activation. The deficient PLCγ activity results in insufficient Ins-P3 to stimulate $Ca^{2+}$ influx and open CRAC channels; thus, influx of $Ca^{2+}$ is blocked. SHIP-mediated consumption of PtdIns-3,4,5P3 may also block the activation of Vav and Akt and thus block the activation of Rho-family GTPases or promote apoptotic cell death

SHIP inositol phosphatase activity appears to block accumulation of PtdIns-3,4,5P3 in B lymphocytes (Scharenberg et al. 1998), which leads to inhibition of enzymes dependent on this lipid mediator. These enzymes include Tec-family kinases Btk (Bolland et al. 1998; Scharenberg et al. 1998), and Itk (Liu et al. 1998), involved in PLCγ tyrosine phosphorylation. Inhibition of Tec-family kinases and the resulting block in PLCγ activation prevent increased cytoplasmic $Ca^{2+}$ through CRAC channels. The activation of Vav (Han et al. 1998) and Akt (Alessi et al. 1997; Franke et al. 1997) are also influenced by PtdIns-3,4P2 and PtdIns-3,4,5P3; future experiments will be required to determine whether SHIP can influence the activity of these. Lastly, the SH2 domain of SHIP binds phosphotyrosines on Shc, which appears to exclude the SH2 domain of Grb2 (Tridandapani et al. 1997a). This SH2 domain competition of SHIP and Grb2 has been proposed to lead to a block in activation of the Ras pathway (Tridandapani et al. 1997a,b).

Despite the tremendous efforts in recent years that have provided a mechanistic basis for B-lymphocyte suppression produced by sIg-FcγRIIb co-clustering, there are a number of unresolved issues. sIg-FcγRIIb co-clustering produces dramatic effects in in vitro cultures of B lymphocytes and yet it is not clear to what extent negative signaling by this means contributes to the overall humoral response. Similarly, the contribution of FcγR-mediated clearance of antigen–antibody complexes to the humoral suppression is not understood. While studies of FcγRII-deficient mice (Takai et al. 1996) indicated an elevated IgG level in response to injected antigen, the elevation was very slight and not as dramatic as in vitro experimental results would suggest. The slight elevation in IgG is likely due to deficient feedback suppression, as earlier hypothesized (Sinclair and Chan 1971), since these animals genetically lack the FcγRII-mediated negative signal. The reason for the discrepancy between in vivo and in vitro findings is not clear. Perhaps other control mechanisms of the humoral response, such as features of B-T interaction, have a greater contribution than FcγRIIb-mediated negative signaling. Long-term studies in which hyper-IgG may have unpredictable pathologies have not been performed on these animals and the B-lymphocyte phenotype of SHIP-deficient mice (Helgason et al. 1998) will be telling.

However, given the recent studies reported here, it is clear that the inositol polyphosphate 5-phosphatase SHIP is an important mediator in the negative signaling process mediated by sIg-FγRIIb co-clustering. Nevertheless, some details in the mechanism of action of SHIP are not understood. First, the deficient PLCγ2 tyrosine phosphorylation and activation by Btk leads to deficient capacitative $Ca^{2+}$ entry but not $Ca^{2+}$ release from intracellular stores. Since $Ca^{2+}$ entry by CRAC channels is dependent on $Ca^{2+}$ release from intracellular stores, it is not clear why the CRAC channel entry is inhibited but not that released from intracellular stores. Second, the affect of negative signaling on activation of PtdIns 3-kinase has not been studied and it is unclear whether the reported lack of PtdIns-3,4,5P3 accumulation (Scharenberg et al. 1998) is due to SHIP-mediated hydrolysis or due to the reduced recruitment of PtdIns 3-kinase to CD19 (Hippen et al. 1997; Kiener et al. 1997). Likewise, other enzymes dependent on PtdIns-3,4,5P3, such as Akt and Vav, have not been studied in this context. Third, the mechanism through which SHIP blocks the Ras signaling pathway is not known and compelling arguments can be made for SHIP protein interactions or enzymatic activity in this regard. Last, the role of SHIP tyrosine phosphorylation in signaling by cytokine receptors or other positive signaling receptors has not been resolved. In these latter models, we may find ways in which SHIP can promote rather than inhibit receptor-mediated biological events.

*Acknowledgements.* This work was supported in part by grants from the National Institutes of Health CA64268, AI41447 and P30 CA16058. Dr. Coggeshall is a Scholar of the Leukemia Society of America. The author is grateful for many helpful discussions with and critical reading of the manuscript by Dr. C.L. Anderson at the Ohio State University. The author is also grateful for many illuminating discussions of antibody-mediated suppression and its long history with Dr. Nicholas Sinclair, University of Western Ontario. The author is especially thankful for the efforts and dedication of the members of his laboratory, particulary George Chacko and Susheela Tridandapani.

# References

Alessi DR, James SR, Downes CP, Holmes AB, Gaffney PR, Reese CB, Cohen P (1997) Characterization of a 3-phosphoinositide-dependent protein kinase which phosphorylates and activates protein kinase Balpha. Curr Biol 7:261–269

Allen JM, Seed B (1989) Isolation and expression of functional high-affinity Fc receptor complementary DNAs. Science 243:378–381

Andjelkovic M, Alessi DR, Meier R, Fernandez A, Lamb NJ, Frech M, Cron P, Cohen P, Lucocq JM, Hemmings BA (1997) Role of translocation in the activation and function of protein kinase. B J Biol Chem 272:31515–31524

Aronheim A, Engelberg D, Li N, al-Alawi N, Schlessinger J, Karin M (1994) Membrane targeting of the nucleotide exchange factor Sos is sufficient for activating the Ras signaling pathway. Cell 78:949–961

Ashman RF, Peckham D, Stunz LL (1996) Fc receptor off-signal in the B cell involves apoptosis. J Immunol 157:5–11

Bae YS, Cantley L-C, Chen CS, Kim SR, Kwon K-S, Rhee, S (1998) Activation of phospholipase Cgamma by phosphatidylinositol 3,4,5-trisphosphate. J Biol Chem 273:4465–4469

Bar-Sagi D, Rotin D, Batzer A, Mandiyan V, Schlessinger, J (1993) SH3 domains direct cellular localization of signaling molecules. Cell 74:83–91

Baumann G, Maier D, Freuler F, Tschopp C, Baudisch K, Weinands J (1994) In vitro characterization of major ligands for Src homology 2 domains derived from protein tyrosine kinases, from the adapter protein Shc and from GTPase-activating protein in Ramos B cells Eur. J Immunol 24:1799–1807

Bennett DL, Petersen CC, Cheek TR (1995) Calcium signaling racking ICRAC in the eye Curr Biol 5:1225–1228

Berridge MJ (1995) Capacitative calcium entry Biochem. J 312:1–11

Bewarder N, Weinrich V, Budde P, Hartmann D, Flaswinkel H, Reth M, Frey J (1996) In vivo and in vitro specificity of protein tyrosine kinases for immunoglobulin G receptor (FcgammaRII) phosphorylation. Mol Cell Biol 6:4735–4743

Bijsterbosch MK, Klaus GG (1985) Crosslinking of surface immunoglobulin and Fc receptors on B lymphocytes inhibits stimulation of inositol phospholipid breakdown via the antigen receptors. J Exp Med 162:1825–1836

Binstadt BA, Brumbaugh KM, Dick CJ, Scharenberg AM, Williams BL, Colonna M, Lanier LL, Kinet JP, Abraham RT, Leibson PJ (1996) Sequential involvement of Lck and SHP-1 with MHC-recognizing receptors on NK cells inhibits FcR-initiated tyrosine kinase activation. Immunity 5: 629–638

Binstadt BA, Brumbaugh KM, Leibson PJ (1997) Signal transduction by human NK cell MHC-recognizing receptors. Immunol Rev 155:197–203

Blery M, Delon J, Trautmann A, Cambiaggi A, Olcese L, Biassoni R, Moretta L, Chavrier P, Moretta A, Daeron M, Vivier E (1997) Reconstituted killer cell inhibitory receptors for major histocompatibility complex class I molecules control mast cell activation induced via immunoreceptor tyrosine-based activation motifs. J Biol Chem 272:8989–8996

Blery M, Kubagawa H, Chen CC, Vely F, Cooper MD, Vivier E (1998) The paired Ig-like receptor PIR-B is an inhibitory receptor that recruits the protein-tyrosine phosphatase SHP-1. Proc Nat Acad Sci (USA) 95:2446–2451

Bolland S, Pearse RN, Kurosaki T, Ravetch JV (1998) SHIP modulates immune receptor responses by regulating membrane association of Btk. Immunity 8:509–516

Borel Y, Kilham L (1974) Carrier-determined tolerance in various strains of mice: The role of isogenic IgG in the induction of hapten-specific tolerance. Proc Soc Exp Biol Med 145:470–474

Brink R, Goodnow CC, Crosbie J, Adams E, Eris J Mason DY, Hartley SB, Basten A (1992) Immunoglobulin M and D antigen receptors are both capable of mediating B lymphocyte activation, deletion, or anergy after interaction with specific antigen. J Exp Med 176:991–1005

Brooks DG, Qiu WQ, Luster AD, Ravetch JV (1989) Structure and expression of human IgG FcRII(CD32) Functional heterogeneity is encoded by the alternatively spliced products of multiple genes. J Exp Med 170:1369–1385

Brumbaugh KM, Binstadt BA, Billadeau DD, Schoon RA, Dick CJ, Ten RM, Leibson PJ (1997) Functional role for Syk tyrosine kinase in natural killer cell-mediated natural cytotoxicity. J Exp Med 186:1965–1974

Buchsbaum R, Telliez JB, Goonesekera S, Feig LA (1996) The N-terminal pleckstrin, coiled-coil, IQ domains of the exchange factor Ras-GRF act cooperatively to facilitate activation by calcium. Mol Cell Biol 16:4888–4896

Burgering BM, Coffer PJ (1995) Protein kinase B (c-Akt) in phosphatidylinositol-3-OH kinase signal transduction Nature 376:599–602

Burgess WH, Dionne CA, Kaplow J Mudd R, Friesel R, Zilberstein A, Schlessinger J, Jaye M (1990) Characterization and cDNA cloning of phospholipase C-gamma, a major substrate for heparin-binding growth factor 1 (acidic fibroblast growth factor)-activated tyrosine kinase. Mol Cell Biol 10:4770–4777

Burkhardt AL, Brunswick M, Bolen JB, Mond JJ (1991) Anti-immunoglobulin stimulation of B lymphocytes activates src-related protein-tyrosine kinases. Proc Nat Acad Sci (USA) 88:7410–7414

Burkhardt AL, Costa T, Misulovin Z, Stealy B, Bolen JB, Nussenzweig MC (1994) Ig alpha and Ig beta are functionally homologous to the signaling proteins of the T-cell receptor. Mol Cell Biol 14: 1095–1103

Burshtyn DN, Scharenberg AM, Wagtmann N, Rajagopalan S, Berrada K, Yi T, Kinet JP, Long EO (1996) Recruitment of tyrosine phosphatase HCP by the killer cell inhibitor receptor. Immunity 4:77–85

Cambier JC (1995) New nomenclature for the Reth motif (or ARH1/TAM/ARAM/YXXL). Immunol Today 16:110

Cambier JC (1997) Inhibitory receptors abound? Proc Nat Acad Sci (USA) 94:5993–5995

Cambier JC, Johnson SA (1995) Differential binding activity of ARH1/TAM motifs. Immunol Letts 44:77–80

Campbell MA, Klinman NR, (1995) Phosphotyrosine-dependent association between CD22 and protein tyrosine phosphatase 1C. Eur J Immunol 25:1573–1579

Carretero M, Palmieri G, Llano M, Tullio V, Santoni A, Geraghty DE, Lopez-Botet M (1998) Specific engagement of the CD94/NKG2-A killer inhibitory receptor by the HLA-E class Ib molecule induces SHP-1 phosphatase recruitment to tyrosine-phosphorylated NKG2-A: evidence for receptor function in heterologous transfectants. Eur J Immunol 28:1280–1291

Cassard S, Choquet D, Fridman WH, Bonnerot C (1996) Regulation of ITAM signaling by specific sequences in Ig-beta B cell antigen receptor subunit. J Biol Chem 271:23786–23791

Cassel DL, Keller MA, Surrey S, Schwartz E, Schreiber AD, Rappaport EF, McKenzie SE (1993) Differential expression of Fc gamma RIIA Fc gamma RIIB and Fc gamma RIIC in hematopoietic cells: analysis of transcripts. Mol Immunol 30:451–460

Castells MC, Wu X, Arm JP, Austen KF, Katz HR (1994) Cloning of the gp49B gene of the immunoglobulin superfamily and demonstration that one of its two products is an early-expressed mast cell surface protein originally described as gp49. J Biol Chem 269:8393–8401

Cella M, Dohring C, Samaridis J, Dessing M, Brockhaus M, Lanzavecchia A, Colonna M (1997) A novel inhibitory receptor (ILT3) expressed on monocytes Macrophages and dendritic cells involved in antigen processing. J Exp Med 185

Cerottini J-C, McConahey PJ, Dixon FJ (1969) Studies of the immunosuppression caused by passive administration of antibody. J Immunol 103:268–275

Chacko GW, Brandt JT, Coggeshall KM, Anderson CL (1996a) Phosphoinositide 3-kinase and p72syk noncovalently associate with the low affinity Fc gamma receptor on human platelets through an immunoreceptor tyrosine-based activation motif. Reconstitution with synthetic phosphopeptides. J Biol Chem 271:10775–10781

Chacko GW, Duchemin AM, Coggeshall KM, Osborne JM, Brandt JT, Anderson CL (1994) Clustering of the platelet Fc gamma receptor induces noncovalent association with the tyrosine kinase p72syk. J Biol Chem 269:32435–32440

Chacko GW, Tridandapani S, Damen J, Liu L, Krystal G, Coggeshall KM (1996b) Negative signaling in B-lymphocytes induces tyrosine phosphorylation of the 145kDa inositol polyphosphate 5-phosphatase SHIP. J Immunol 157:2234–2238

Chan AC, Irving BA, Fraser JD, Weiss A (1991) The zeta chain is associated with a tyrosine kinase and upon T-cell antigen receptor stimulation associates with ZAP-70 a 70-kDa tyrosine phosphoprotein. Proc Natl Acad Sci (USA) 88:9166–9170

Chan AC, Kadlecek TA, Elder ME, Filipovich AH, Kuo WL, Iwashima M, Parslow TG, Weiss A (1994) ZAP-70 deficiency in an autosomal recessive form of severe combined immunodeficiency. Science 264:1599–1601

Chan PL, Sinclair NRS (1971) Regulation of the immune response. V. An analysis of the function of the Fc portion of antibody in suppression of an immune response with respect to interaction with components of the immune system. Immunol 21:967–981

Chardin P, Camonis JH, Gale NW, van Aelst L, Schlessinger J, Wigler MH, Bar-Sagi D (1993) Human Sos1: a guanine nucleotide exchange factor for Ras that binds to GRB2. Science 260:1338–1343

Chen HE, Chang S, Trub T, Neel BG (1996) Regulation of colony-stimulating factor-1 receptor signaling by SH2 domain-containing tyrosine phosphatase SHPTP1. Mol Cell Biol 16:3685–3697

Cheng AM, Rowley B, Pao W, Hayday A, Bolen JB, Pawson T (1995) Syk tyrosine kinase required for mouse viability and B-cell development. Nature 378

Choquet D, Partiseti M, Amigorena S, Bonnerot C, Fridman WH, Korn H (1993) Cross-linking of IgG receptors inhibits membrane immunoglobulin-stimulated calcium influx in B lymphocytes. J Cell Biol 121:355–363

Chuang L-M, Hausdorff SF, Myers MG, White MF, Birnbaum MJ, Kahn RC, (1994) Interactive roles of Ras, insulin receptor substrate-1, and proteins with Src homology-2 domains in insulin signaling in Xenopus oocytes. J Biol Chem 269:27645–27649

Clark MR, Campbell KS, Kazlauskas A, Johnson SA, Hertz M, Potter TA, Pleiman C, Cambier JC (1992) The B cell antigen receptor complex: association of Ig-alpha and Ig-beta with distinct cytoplasmic effectors. Science 258:123–126

Clark MR, Johnson SA, Cambier JC (1994) Analysis of Ig-alpha-tyrosine kinase interaction reveals two levels of binding specificity and tyrosine phosphorylated Ig-alpha stimulation of Fyn activity. EMBO J 13:1911–1919

Coggeshall KM, McHugh JC, Altman A (1992) Predominant expression and activation-induced tyrosine phosphorylation of phospholipase C-gamma 2 in B lymphocytes. Proc Natl Acad Sci (USA) 89: 5660–5664

Cosman D, Fanger N, Borges L, Kubin M, Chin W, Peterson L, Hsu ML (1997) A novel immunoglobulin superfamily receptor for cellular and viral MHC class I molecules. Immunity 7:273–282

Crespo P, Bustelo XR, Aaronson DS, Coso OA, Lopez-Barahona M, Barbacid M, Gutkind JS (1996) Rac-1 dependent stimulation of the JNK/SAPK signaling pathway by Vav. Oncogene 13: 455–460

Crespo P, Schuebel KE, Ostrom AA, Gutkind JS, Bustelo XR (1997) Phosphotyrosine-dependent activation of Rac-1 GDP/GTP exchange by the vav proto-oncogene product. Nature 385:169–172

Crowley MT, Costello PS, Fitzer-Attas CJ, Turner M, Meng F, Lowell C, Tybulewicz VL, DeFranco AL (1997) A critical role for Syk in signal transduction and phagocytosis mediated by Fcgamma receptors on macrophages. J Exp Med 186:1027–1039

Crowley MT, Harmer SL, DeFranco AL (1996) Activation-induced association of a 145-kDa tyrosine-phosphorylated protein with Shc and Syk in B lymphocytes and macrophages. J Biol Chem 271: 1145–1152

Cullen PJ, Hsuan JJ, Truong O, Letcher AJ, Jackson TR, Dawson AP, Irvine RF (1995) Identification of a specific Ins(1,3,4,5)P4-binding protein as a member of the GAP1 family. Nature 376:527–530

D'Ambrosio D, Fong DC, Cambier JC (1996a) The SHIP phosphatase becomes associated with Fc gammaRIIB1 and is tyrosine phosphorylated during 'negative' signaling. Immunol Lett 54;77–82

D'Ambrosio D, Hippen KL, Cambier JC (1996b) Distinct mechanisms mediate Shc association with the activated and resting B cell antigen receptor. Eur J Immunol 26:1960–1965

D'Ambrosio D, Hippen KL, Minskoff SA, Mellman I, Pani G, Siminovitch KA, Cambier JC (1995) Recruitment and activation of PTP1 C in negative regulation of antigen receptor signaling by Fc gamma RIIB1. Science 268:293–297

Daeron M (1997) Fc receptor biology. Ann Rev Immunol 15:203–234

Daeron M, Latour S, Malbec O, Espinosa E, Pina P, Pasmans S, Fridman WH (1995) The same tyrosine-based inhibition motif, in the intracytoplasmic domain of Fc gamma RIIB, regulates negatively BCR-, TCR-, and FcR-dependent cell activation. Immunity 3:635–646

Damen JE, Liu L, Cutler RL, Krystal G (1993) Erythropoietin stimulates the tyrosine phosphorylation of Shc and its association with Grb2 and a 145-Kd tyrosine phosphorylated protein. Blood 82: 2296–2303

Damen JE, Liu L, Rosten P, Humphries RK, Jefferson AB, Majerus PW, Krystal G (1996) The 145-kDa protein induced to associate with Shc by multiple cytokines is an inositol tetraphosphate and phosphatidylinositol 3,4,5-trisphosphate 5-phosphatase. Proc Natl Acad Sci (USA) 93:1689–1693

Dariavach P, Mattei MG, Golstein P, Lefranc MP (1988) Human Ig superfamily CTLA-4 gene: chromosomal localization and identity of protein sequence between murine and human CTLA-4 cytoplasmic domains. Eur J Immunol 18:1901–1905
Datta K, Bellacosa A, Chan TO, Tsichlis PN (1996) Akt is a direct target of the phosphatidylinositol 3-kinase. Activation by growth factors, v-src and v-Ha-ras, in Sf9 and mammalian cells. J Biol Chem 271:30835–30839
de Aos I, Metzger MH, Exley M, Dahl CE, Misra S, Zheng D, Varticovski L, Terhorst C, Sancho J (1997) Tyrosine phosphorylation of the CD3-epsilon subunit of the T cell antigen receptor mediates enhanced association with phosphatidylinositol 3-kinase in Jurkat T cells. J Biol Chem 272: 25310–25318
DeFranco AL (1997) The complexity of signaling pathways activated by the BCR. Curr Op Immunol 9:296–308
Desai DM, Sap J, Silvennoinen O, Schlessinger J, Weiss A (1994) The catalytic activity of the CD45 membrane-proximal phosphatase domain iSRequired for TCR signaling anDRegulation. EMBO J 13:4002–4010
Desiderio S (1997) Role of Btk in B cell development and signaling. Curr Op Immunol 9:534–540
Deuter-Reinhard M, Apell G, Pot D, Klippel A, Williams LT, Kavanaugh WM (1997) SIP/SHIP inhibits Xenopus oocyte maturation induced by insulin and phosphatidylinositol 3-kinase. Mol Cell Biol 17:2559–2565
Diegel ML, Rankin BM, Bolen JB, Dubois PM, Kiener PA (1994) Cross-linking of Fc gamma receptor to surface immunoglobulin on B cells provides an inhibitory signal that closes the plasma membrane calcium channel. J Biol Chem 269:11409–11416
Dixon FJ, Jacot-Guillarmod H, McConahey PJ (1967) The effect of passively administered antibody on antibody synthesis. J Exp Med 125:1119–1136
Dong F, Brynes RK, Tidow N, Welte K, Lowenberg B, Touw IP (1995) Mutations in the gene for the granulocyte colony stimulating factor receptor in patients with acute myeloid leukemia preceded by severe congenital neutropenia. New Eng J Med 333:487–493
Doody GM, Justement LB, Delibrias CC, Matthews RJ, Lin J, Thomas ML, Fearon DT (1995) A role in B cell activation for CD22 and the protein tyrosine phosphatase SHP. Science 269:242–244
Doyle MV, Parks DE, Weigle WO (1976a) Specific suppression of the immune response by HGG-tolerant spleen cells I Parameters affecting the level of suppression. J Immunol 16:1640–1645
Doyle MV, Parks DE, Weigle WO (1976b) Specific, transient suppression of the immune response by HGG-tolerant spleen cells II Effector cells and target cells. J Immunol 117:1152–1158
Duchemin AM, Anderson CL (1997) Association of non-receptor protein tyrosine kinases with the Fc gamma RI/gamma-chain complex in monocytic cells. J Immunol 158:865–871
Dugas B, Calenda A, Delfraissy JF, Vazquez A, Bach JF, Galanaud P (1987) The cytosolic free calcium in anti-mu-stimulated human B cells is derived partly from extracellular medium and partly from intracellular stores. Eur J Immunol 17:1323–1328
Emori Y, Homma Y, Sorimachi H, Kawasaki H, Nakanishi O, Suzuki K, Takenawa T (1989) A second type of rat phosphoinositide-specific phospholipase C containing a src-related sequence not essential for phosphoinositide-hydrolyzing activity. J Biol Chem 264:21885–21890
Exley M, Varticovski L, Peter M, Sancho J, Terhorst C (1994) Association of phosphatidylinositol 3-kinase with a specific sequence of the T cell receptor zeta chain is dependent on T cell activation. J Biol Chem 269:15140–15146
Falasca M, Logan SK, Lehto VP, Baccante G, Lemmon MA, Schlessinger J (1998) Activation of phospholipase C gamma by PI 3-kinase-induced PH domain-mediated membrane targeting. EMBO J 17:414–422
Farrell HE, Vally H, Lynch DM, Fleming P, Shellman GR, Scalzo AA, Davis-Poyneter NJ (1997) Inhibition of natural killer cells by a cytomegalovirus MHC class I homolog in vivo. Nature 386: 510–514
Fearon DT, Carter RH (1995) The CD19/CR2/TAPA-1 complex of B lymphocytes: linking natural to acquired immunity. Ann Rev Immunol 13:127–149
Foy TM, Aruffo A, Bajorath J, Buhlmann JE, Noelle RJ (1996) Immune regulation by CD40 and its ligand GP39. Ann Rev Immunol 14:591–617
Frank SJ, Samelson LE, Klausner RD (1990) The structure and signaling functions of the invariant T cell receptor components. Semin Immunol 2:89–97
Franke TF, Kaplan DR, Cantley LC, Toker A (1997) Direct regulation of the Akt proto-oncogene product by phosphatidylinositol-3,4-bisphosphate. Science 275:665–668

Frearson JA, Alexander DR (1997) The role of phosphotyrosine phosphatases in haematopoietic cell signal transduction. Bioessays 19:417–427

Frech M, And jelkovic M, Ingley E, Reddy KK, Falck JR, Hemmings BA (1997) High affinity binding of inositol phosphates and phosphoinositides to the pleckstrin homology domain of RAC/protein kinase B and their influence on kinase activity. J Biol Chem 272:8474–8481

Friel DD (1996) TRP: Its role in phototransduction and store-operated Ca2+ entry. Cell 85:617–619

Fry AM, Lanier LL, Weiss A (1996) Phosphotyrosines in the killer cell inhibitory receptor motif of NKB1 are required for negative signaling and for association with protein tyrosine phosphatase 1 C. J Exp Med 184:295–300

Fukui Y, Saltiel AR, Hanafusa H (1991) Phosphatidylinositol-3 kinase is activated in v-src, v-yes, and v-fps transformed chicken embryo fibroblasts. Oncogene 6:407–411

Fukunaga R, Seto Y, Mizushima S, Nagata S (1990) Three different mRNAs encoding human granulocyte colony-stimulating factor receptor. Proc Nat Acad Sci (USA) 87:8702–8706

Galizzi JP, Zuber CE, Harada N, Gorman DM, Djossou O, Kastelein R, Banchereau J, Howard M, Miyajima A (1990) Molecular cloning of a cDNA encoding the human interleukin 4 receptor. Int Immunol 2:669–675

Gauen LK, Zhu Y, Letourneur F, Hu Q, Bolen JB, Matis LA, Klausner RD, Shaw AS (1994) Interactions of p59fyn and ZAP-70 with T-cell receptor activation motifs: defining the nature of a signaling motif. Mol Cell Biol 14:3729–3741

Ghazizadeh S, Bolen JB, Fleit HB (1995) Tyrosine phosphorylation and association of Syk with Fc gamma RII in monocytic THP-1 cells. Biochem J 305:669–674

Golan DT, Borel Y (1971) Nonantigenicity and immunologic tolerance: The role of the carrier in the induction of tolerance to hapten. J Exp Med 134:1046–1061

Gold DP, Puck JM, Pettey CL, Cho M, Coligan J, Woody JN, Terhorst C (1986) isolation of cDNA clones encoding the 20 K non-glycosylated polypeptide chain of the human T cell receptor/T3 complex. Nature 321:431–434

Gold MR, Aebersold R (1994) Both phosphatidylinositol 3-kinase and phosphatidylinositol 4-kinase products are increased by antigen receptor signaling in B cells. J Immunol 152:42–50

Goodwin RG, Friend D, Ziegler SF, Jerzy R ,Falk BA, Gimpel S, Cosman D, Dower SK, March CJ, Namen AE et al (1990) Cloning of the human and murine interleukin-7 receptors: demonstration of a soluble form and homology to a new receptor superfamily. Cel 60:941–951

Gorman DM, Itoh N, Kitamura T, Schreurs J, Yonehara S, Yahara I, Arai K, Miyajima A (1990) Cloning and expression of a gene encoding an interleukin 3 receptor-like protein: identification of another member of the cytokine receptor gene family. Proc Nat Acad Sci (USA) 87:5459–5463

Gray D, Siepmann K, van Essen D, Poudrier J, Wykes M, Jainandunsing S, Bergthorsdottir S, Dullforce P (1996) B-T lymphocyte interactions in the generation and survival of memory cells. Immunol Rev 150:45–61

Greenberg S, Chang P, Wang DC, Xavier R, Seed B (1996) Clustered syk tyrosine kinase domains trigger phagocytosis. Proc Nat Acad Sci (USA) 93:1103–1107

Greenbury CL, Moore DH (1968) Non-specific antibody-induced suppression of the immune response. Nature 219:526–527

Gupta N, Scharenberg AM, Burshtyn DN, Wagtmann N, Lioubin MN, Rohrschneider LR, Kinet JP, Long EO (1997) Negative signaling pathways of the killer cell inhibitory receptor and Fc gamma RIIb1 require distinct phosphatases. J Exp Med 186:473–478

Ha HJ, Kubagawa H, Burrows PD (1992) Molecular cloning and expression pattern of a human gene homologous to the murine mb-1 gene. J Immunol 148:1526–1531

Han J, Das B, Wei W, Van Aelst L, Mosteller RD, Khosravi-Far R, Westwick JK, Der CJ, Broek D (1997) Lck regulates Vav activation of members of the Rho family of GTPases. Mol Cell Biol 17:1346–1353

Han J, Luby-Phelps K, Das B, Shu X, Xia Y, Mosteller RD, Krishna UM, Falck JR, White MA, Broek D (1998) Role of substrates and products of PI 3-kinase in regulating activation of Rac-related guanosine triphosphatases by Vav. Science 279:558–560

Harmer SL, DeFranco AL (1997) Shc contains two Grb2 binding sites needed for efficient formation of complexes with SOS in B lymphocytes. Mol Cell Biol 17:4087–4095

Harper K, Balzano C, Rouvier E, Mattei MG, Luciani MF, Golstein P (1991) CTLA-4 and CD28 activated lymphocyte molecules are closely related in both mouse and human as to sequence message expression, gene structure, and chromosomal location. J Immunol 147:1037–1044

Hawkins PT, Welch H, McGregor A, Eguinoa A, Gobert S, Krugmann S, Anderson K, Stokoe D, Stephens L (1997) Signaling via phosphoinositide 3OH kinases. Biochem Soc Trans 25:1147–1151

Hayami K, Fukuta D, Nishikawa Y, Yamashita Y, Inui M, Ohyama Y, Hikida M, Ohmori H, Takai T (1997) Molecular cloning of a novel murine cell-surface glycoprotein homologous to killer cell inhibitory receptors. J Biol Chem 272:7320–7327
Helgason CD, Damen JE, Rosten P, Grewal R, Sorensen P, Chappel SM, Borowski A, Jirik F, Krystal G, Humphries RK (1998) Targeted disruption of SHIP leads to hemopoietic perturbations, lung pathology and a shortened life span. Genes Dev 12:1610–1620
Hempel WM, DeFranco AL (1991) Expression of phospholipase C isozymes by murine B lymphocytes. J Immunol 146:3713–3720
Hempel WM, Schatzman RC, DeFranco AL (1992) Tyrosine phosphorylation of phospholipase C-gamma 2 upon cross-linking of membrane Ig on murine B lymphocytes. J Immunol 148:3021–7
Hibi M, Murakami M, Saito M, Hirano T, Taga T, Kishimoto T (1990) Molecular cloning and expression of an IL-6 signal transducer, gp130. Cell 63:1149–1157
Hippen KL, Buhl AM, D'Ambrosio D, Nakamura K, Persin C, Cambier JC (1997) FcgammaRIIB1 inhibition of BCR-mediated phosphoinositide hydrolysis and Ca2+ mobilization is integrated by CD19 dephosphorylation. Immunity 7:49–58
Hoffmeyer F, Witte K, Gebhardt U, Schmidt RE (1995) The low affinity FcgammaRIIa and FcgammaRIIIb on polymorphonuclear neutrophils are differentially regulated by CD45 phosphatase. J Immunol 155:4016–4023
Holsinger LJ, Spencer DM, Austin DJ, Schreiber SL, Crabtree GR (1995) Signal transduction in T lymphocytes using a conditional allele of Sos. Proc Natl Acad Sci USA 92:9810–9814
Houchins JP, Yabe T, McSherry C, Bach FH (1991) DNA sequence analysis of NKG2, a family of related cDNA clones encoding type II integral membrane proteins on human natural killer cells. J Exp Med 173:1017–1020
Hunter MG, Avalos BR (1998) Phosphatidylinositol 3-kinase and SH2-containing inositol phosphatase (SHIP) are recruited by distinct positive and negative growth regulatory domains in the granulocyte colony-stimulating factor receptor. J Immunol 160:4979–4987
Hunter S, Indik ZK, Kim M-K, Cauley MD, Park J-G, Schreiber AD (1998) Inhibition of Fcgamma receptor-mediated phagocytosis by a nonphagocytic Fcgamma receptor. Blood 91:1762–1768
Hurley TR, Hyman R, Sefton BM (1993) Differential effects of expression of the CD45 tyrosine protein phosphatase on the tyrosine phosphorylation of the lck, fyn, and c-src tyrosine protein kinases. Mol Cell Biol 13:1651–1656
Hyvonen M, Saraste M (1997) Structure of the PH domain and Btk motif from Bruton's tyrosine kinase: molecular explanations for X-linked agammaglobulinaemia. EMBO J 16:3396–3404
Ibarrola I, Vossebeld PJ, Homburg CH, Thelen M, Roos D, Verhoeven AJ (1997) Influence of tyrosine phosphorylation on protein interaction with FcgammaRIIa. Biochem Biophys Acta 1357:348–358
Idzerda RL, March CJ, Mosley B, Lyman SD, Vanden Bos T, Gimpel SD, Din WS, Grabstein KH, Widmer MB, Park LS et al (1990) Human interleukin 4 receptor confers biological responsiveness and defines a novel receptor superfamily. J Exp Med 171:861–873
Indik ZK, Park JG, Hunter S, Schreiber AD (1995) The molecular dissection of Fc gamma receptor mediated phagocytosis. Blood 86:4389–4399
Irving BA, Chan AC, Weiss A (1993) Functional characterization of a signal transducing motif present in the T cell antigen receptor zeta chain. J Exp Med 177:1093–1103
Irving BA, Weiss A (1991) The cytoplasmic domain of the T cell receptor zeta chain is sufficient to couple to receptor-associated signal transduction pathways. Cell 64:891–901
Isakov N (1997) ITIMs and ITAMs. The Yin and Yang of antigen and Fc receptor-linked signaling machinery. Immunol Res 16:85–100
Isakov N, Wange RL, Burgess WH, Watts JD, Aebersold R, Samelson LE (1995) ZAP-70 binding specificity to T cell receptor tyrosine-based activation motifs: the tandem SH2 domains of ZAP-70 bind distinct tyrosine-based activation motifs with varying affinity. J Exp Med 181:375–380
Ivorra I, Gigg R, Irvine RF, Parker I (1991) Inositol 1,3,4,6-tetrakisphosphate mobilizes calcium in Xenopus oocytes with high potency. Biochem J 273:317–321
Iwashima M, Irving BA, van Oers NS, Chan AC, Weiss A (1994) Sequential interactions of the TCR with two distinct cytoplasmic tyrosine kinases. Science 263:1136–1139
Jiao H, Berrada K, Yang W, Tabrizi M, Platanias LC, Yi T (1996) Direct association with and dephosphorylation of Jak2 kinase by the SH2-domain-containing protein tyrosine phosphatase. SHP-1 Mol Cell Biol 16:6985–6992
Jiao H, Yang W, Berrada K, Tabrizi M, Shultz L, Yi T (1997) Macrophages from motheaten and viable motheaten mutant mice show increased proliferative responses to GM-CSF: detection of potential HCP substrates in GM-CSF signal transduction. Exp Hematol 25:592–600

Jones SS, D'Andrea AD, Haines LL, Wong GG (1990) Human erythropoietin receptor: cloning, expression, and biologic characterization Blood. 76:31–35

Jouvin MH, Adamczewski M, Numerof R, Letourneur O, Valle A, Kinet JP (1994) Differential control of the tyrosine kinases Lyn and Syk by the two signaling chains of the high affinity immunoglobulin E receptor. J Biol Chem 269:5918–5925

Justement LB, Campbell KS, Chien NC, Cambier JC (1991) Regulation of B cell antigen receptor signal transduction and phosphorylation by CD45. Science, 252:1839–1842

Kavanaugh WM, Pot DA, Chin SM, Deuter-Reinhard M, Jefferson AB, Norris FA, Masiarz FR, Cousens LS, Majerus PW, Williams LT (1996) Multiple forms of an inositol polyphosphate 5-phosphatase form signaling complexes with Shc and Grb2. Curr Biol 6:438–445

Kawauchi K, Lazarus AH, Rapoport MJ, Harwood A, Cambier JC, Delovitch TL (1994) Tyrosine kinase and CD45 tyrosine phosphatase activity mediate p21ras activation in B cells stimulated through the antigen receptor. J Immunol 152:3306–3316

Kennedy SG, Wagner AJ, Conzen SD, Jordan J, Bellacosa A, Tsichlis PN, Hay N (1997) The PI 3-kinase/Akt signaling pathway delivers an anti-apoptotic signal. Genes Dev 11:701–713

Kharitonenkov A, Chen Z, Sures I, Wang H, Schilling J, Ullrich A (1997) A family of proteins that inhibit signaling through tyrosine kinase receptors. Nature 386:181–186

Kiener PA, Lioubin MN, Rohrschneider LR, Ledbetter JA, Nadler SG, Diegel ML (1997) Co-ligation of the antigen and Fc receptors gives rise to the selective modulation of intracellular signaling in B cells. Regulation of the association of phosphatidylinositol 3-kinase and inositol 5′-phosphatase with the antigen receptor complex. J Biol Chem 272:3838–3844

Kim HK, Kim JW, Zilberstein A, Margolis B, Kim JG, Schlessinger J, Rhee SG (1991) PDGF stimulation of inositol phospholipid hydrolysis requires PLC-gamma 1 phosphorylation on tyrosine residues 783 and 1254. Cell 65:435–441

Kim KM, Alber G, Weiser P, Reth M (1992) Differential signaling through the Ig-alpha and Ig-beta components of the B cell antigen receptor. Eur J Immunol 23:911–916

Kimura T, Sakamoto H, Appella E, Siraganian RP (1997) The negative signaling molecule SH2 domain-containing inositol polyphosphate 5-phosphatase (SHIP) binds to the tyrosine-phosphorylated beta subunit of the high affinity IgE receptor. J Biol Chem 272

Kincade PW, Lawton AR, Bockman DE, Cooper MD (1970) Suppression of immunoglobulin G synthesis as a result of antibody-mediated suppression of immunoglobulin M synthesis in chickens. Proc Nat Acad Sci (USA) 67:1918–1925

King WG, Mattaliano MD, Chan TO, Tsichlis PN, Brugge JS (1997) Phosphatidylinositol 3-kinase is required for integrin-stimulated AKT and Raf-1/mitogen-activated protein kinase pathway activation. Mol Cell Biol 17:4406–4418

Klippel A, Kavanaugh WM, Pot D, Williams LT (1997) A specific product of phosphatidylinositol 3-kinase directly activates the protein kinase Akt through its pleckstrin homology domain. Mol Cell Biol 17:338–344

Klippel A, Reinhard C, Kavanaugh WM, Apell G, Escobedo MA, Williams LT (1996) Membrane localization of phosphatidylinositol 3-kinase is sufficient to activate multiple signal-transducing kinase pathways. Mol Cell Biol 16:4117–4127

Kolanus W, Romeo C, Seed B (1993) T cell activation by clustered tyrosine kinases. Cell 74:171–183

Koretzky GA, Kohmetscher MA, Kadleck T, Weiss A (1992) Restoration of T cell receptor-mediated signal transduction by transfection of CD45 cDNA into a CD45-deficient variant of the Jurkat T cell line. J Immunol 149:1138–1142

Koretzky GA, Picus J, Schultz T, Weiss A (1991) Tyrosine phosphatase CD45 is required for T-cell antigen receptor and CD2-mediated activation of a protein tyrosine kinase and interleukin 2 production. Proc Nat Acad Sci (USA) 88:2037–2041

Koretzky GA, Picus J, Thomas ML, Weiss A (1990) Tyrosine phosphatase CD45 is essential for coupling T-cell antigen receptor to the phosphatidyl inositol pathway. Nature 346:66–68

Kozlowski M, Mlinaric-Rascan I, Feng GS, Shen R, Pawson T, Siminovitch KA (1993) Expression and catalytic activity of the tyrosine phosphatase PTP1C is severely impaired in motheaten and viable motheaten mice. J Exp Med 178:2157–2163

Krissansen GW, Owen MJ, Verbi W, Crumpton MJ (1986) Primary structure of the T3 gamma subunit of the T3/T cell antigen receptor complex deduced from cDNA sequences: evolution of the T3 gamma and delta subunits. EMBO J 5:1799–1808

Kubagawa H, Burrows PD, Cooper MD (1997) A novel pair of immunoglobulin-like receptors expressed by B cells and myeloid cells. Proc Nat Acad Sci (U S A) 94:5261–5266

Kumar G, Wang S, Gupta S, Nel A (1995) The membrane immunoglobulin receptor utilizes a Shc/Grb2/hSOS complex for activation of the mitogen-activated protein kinase cascade in a B-cell line. Biochem J 307:215–223

Kurosaki T, Johnson SA, Pao L, Sada K, Yamamura H, Cambier JC (1995) Role of the Syk autophosphorylation site and SH2 domains in B cell antigen receptor signaling. J Exp Med 182: 1815–1823

Kuster H, Thompson H, Kinet J-P (1990) Characterization and expression of the gene for the human Fc receptor gamma subunit: definition of a new gene family. J Biol Chem 265:6448–6452

Kuster H, Zhang L, Brini AT, MacGlashan DW, Kinet J-P (1992) The gene and cDNA for the high affinity immunoglobulin E receptor beta chain and expression of the complete human receptor. J Biol Chem 267:12782–12787

Labadia ME, Ingraham RH, Schembri-King J, Morelock MM, Jakes S (1996) Binding affinities of the SH2 domains of ZAP-70 p56lck and Shc to the zeta chain ITAMs of the T-cell receptor determined by surface plasmon resonance. J Leukocyte Biol 59:740–746

Lamkin TD, Walk SF, Liu L, Damen JE, Krystal G, Ravichandran KS (1997) Shc interaction with Src homology 2 domain containing inositol phosphatase (SHIP) in vivo requires the Shc-phosphotyrosine binding domain and two specific phosphotyrosines on SHIP. J Biol Chem 272:10396–10401

Lankester AC, van Schijndel GM, van Lier RA (1995) Hematopoietic cell phosphatase is recruited to CD22 following B cell antigen receptor ligation. J Biol Chem 270:20305–20308

Larsen A, Davis T, Curtis BM, Gimpel S, Sims JE, Cosman D, Park L, Sorensen E, March CJ, Smith CA (1990) Expression cloning of a human granulocyte colony-stimulating factor receptor a structural mosaic of hematopoietin receptor, immunoglobulin, fibronectin domains. J Exp Med 172: 1559–1570

Law CL, Chandran KA, Sidorenko SP, Clark EA (1996a) Phospholipase C-gamma1 interacts with conserved phosphotyrosyl residues in the linker region of Syk and is a substrate for Syk. Mol Cell Biol 16:1305–1315

Law CL, Sidorenko SP, Chandran KA, Zhao Z, Shen SH, Fischer EH, Clark EA (1996b) CD22 associates with protein tyrosine phosphatase 1C, Syk, and phospholipase C-gamma(1) upon B cell activation. J Exp Med 183:547–560

Law CL, Sidorenko SP, Clark EA (1994) Regulation of lymphocyte activation by the cell-surface molecule CD22. Immunol Today 15:442–549

Lawton AR, Cooper MD (1974) Modification of B lymphocyte differentiation by anti-immunoglobulins. Contemp Topics Immunobiol 3:193–225

Lemmon MA, Ferguson KM, Schlessinger J (1996) PH domains: Diverse sequences with a common fold recruit signaling molecules to the cell surface. Cell 85:621–624

Leprince C, Draves KE, Geahlen RL, Ledbetter JA, Clark EA (1993) CD22 associates with the human surface IgM-B-cell antigen receptor complex. Proc Nat Acad Sci (USA) 90:3236–3240

Letourneur F, Klausner RD (1991) T-cell and basophil activation through the cytoplasmic tail of T cell receptor zeta family proteins. Proc Nat Acad Sci (USA) 88:8905–8909

Letourneur F, Klausner RD (1992) Activation of T cells by a tyrosine kinase activation domain in the cytoplasmic tail of CD3 epsilon. Science 255:79–82

Lewis RS, Cahalan MD (1995) Potassium and calcium channels in lymphocytes. Ann Rev Immunol 13:623–653

Lewis VA, Koch T, Plutner H, Mellman I (1986) A complementary DNA clone for a macrophage-lymphocyte Fc receptor. Nature 324:372–375

Li N, Batzer A, Daly R, Yajnik V, Skolnik E, Chardin P, Bar-Sagi D, Margolis B, Schlessinger J (1993) Guanine-nucleotide-releasing factor hSos1 binds to Grb2 and links receptor tyrosine kinases to Ras signaling. Nature 363:85–88

Li Z, Wahl MI, Eguinoa A, Stephens LR, Hawkins PT, Witte ON (1997) Phosphatidylinositol 3-kinase-gamma activates Bruton's tyrosine kinase in concert with Src family kinases. Proc Nat Acad Sci (USA) 94:13820–13825

Lioubin MN, Algate PAST, Carlberg K, Aebersold R, Rohrschneider LR (1996) p150Ship, a signal transduction molecule with inositol polyphosphate-5-phosphatase activity. Genes Dev 10: 1084–1095

Lioubin MN, Myles GM, Carlberg K, Bowtell D, Rohrschneider LR (1994) Shc, Grb2, Sos1, and a 150-kilodalton tyrosine-phosphorylated protein form complexes with Fms in hematopoietic cells. Mol Cell Biol 14:5682–5691

Liu K-Q, Bunnell SC, Gurniak CB, Berg LJ (1998) T cell receptor initiated calcium release is uncoupled from capacitative calcium entry in Itk-deficient T cells. J Exp Med 187:1721–1727

Liu L, Damen JE, Cutler RL, Krystal G (1994) Multiple cytokines stimulate the binding of a common 145-kilodalton protein to Shc at the Grb2 recognition site of Shc. Mol Cell Biol 14:6926–6935

Liu L, Damen JE, Hughes MR, Babic I, Jirik FR, Krystal G (1997a) The Src homology 2 (SH2) domain of SH2-containing inositol phosphatase (SHIP) is essential for tyrosine phosphorylation of SHIP, its association with Shc and its induction of apoptosis. J Biol Chem 272:8983–8988

Liu L, Damen JE, Ware MD, Krystal G (1997b) Interleukin-3 induces the association of the inositol 5-phosphatase SHIP with SHP2. J Biol Chem 272:10998–11001

Lobell RB, Austen KF, Katz HR (1994) FcgammaR-mediated endocytosis and expression of cell surface FcgammaRIIb1 and FcgammaRIIb2 by mouse bone marrow culture-derived progenitor mast cells. J Immunol 152:811–818

Long EO, Burshtyn DN, Clark WP, Peruzzi M, Rajagopalan S, Rojo S, Wagtmann N, Winter CC (1997) Killer cell inhibitory receptors: diversity, specificity, and function. Immunol Rev 155:135–144

Lowenstein EJ, Daly RJ, Batzer AG, Li W, Margolis B, Lammers R, Ullrich A, Skolnik EY, Bar-Sagi D, Schlessinger J (1992) The SH2 and SH3 domain-containing protein GRB2 linkSReceptor tyrosine kinases to ras signaling. Cell 70:431–442

Lowry MB, Duchemin AM, Robinson JM, Anderson CL (1998) Functional separation of pseudopod extension and particle internalization during Fc gamma receptor-mediated phagocytosis. J Exp Med 187:161–176

Luckhoff A, Clapham DE (1992) Inositol 1,3,4,5-tetrakisphosphate activates an endothelial Ca(2+)-permeable channel. Nature 355:356–358

Maeda A, Kurosaki M, Ono M, Takai T, Kurosaki T (1998) Requirement of SH2-containing protein tyrosine phosphatases SHP-1 and SHP-2 for paired immunoglobulin-like receptor B (PIR-B)-mediated inhibitory signal. J Exp Med 187:1355–1360

Mantzioris BX, Berger MF, Sewell W, Zola H (1993) Expression of the Fc receptor for IgG (FcgammaRII/CDw32) by human circulating T and B lymphocytes. J Immunol 150:5175–5184

Marais R, Light Y, Mason C, Paterson H, Olson MF, Marshall CJ (1998) Requirement of Ras-GTP-Raf complexes for activation of Raf-1 by protein kinase. C Science 280:109–112

Marais R, Marshall CJ (1996) Control of the ERK MAP kinase cascade by Ras and Raf. Cancer Surveys 27:101–125

Marengère LEM, Waterhouse P, Duncan GS, Mittrücker HW, Feng GS, Mak TW (1996) Regulation of T cell receptor signaling by tyrosine phosphatase SYP association with CTLA-4. Science 272: 1170–1173

Margolis B (1996) The PI/PTB domain: a new protein interaction domain involved in growth factor receptor signaling. J Lab Clin Med 128:235–241

Margolis B, Bellot F, Honegger AM, Ullrich A, Schlessinger J, Zilberstein A (1990) Tyrosine kinase activity is essential for the association of phospholipase C-gamma with the epidermal growth factor receptor. Mol Cell Biol 10:435–441

Marte BM, Downward J (1997) PKB/Akt: connecting phosphoinositide 3-kinase to cell survival and beyond Trends. Biochem Sci 22:355–358

Metes D, Ernst LK, Chambers WH, Sulica A, Herberman RB, Morel PA (1998) Expression of functional CD32 molecules on human NK cells is determined by an allelic polymorphism of the FcgammaRIIC gene. Blood 91:2369–2380

Miller KL, Duchemin AM, Anderson CL (1996) A novel role for the Fc receptor gamma subunit: enhancement of Fc gamma R ligand affinity. J Exp Med 183:2227–2233

Musacchio A, Gibson T, Rice P, Thompson J, Saraste M (1993) The PH domain: a common piece in the structural patchwork of signaling proteins. Trends Biochem Sci 18:343–348

Mustelin T, Altman A (1990) Dephosphorylation and activation of the T cell tyrosine kinase pp56lck by the leukocyte common antigen (CD45). Oncogene 5:809–813

Mustelin T, Coggeshall KM, Altman A (1989) Rapid activation of the T-cell tyrosine protein kinase pp56lck by the CD45 phosphotyrosine phosphatase. Proc Nat Acad Sci (USA) 86:6302–6306

Mustelin T, Pessa-Morikawa T, Autero M, Gassmann M, ersson LC, Gahmberg CG, Burn P (1992) Regulation of the p59fyn protein tyrosine kinase by the CD45 phosphotyrosine phosphatase. Eur J Immunol 22:1173–1178

Muta T, Kurosaki T, Misulovin Z, Sanchez M, Nussenzweig MC, Ravetch JV (1994) A 13-amino-acid motif in the cytoplasmic domain of Fc gamma RIIB modulates B-cell receptor signaling. Nature 368:70–73

Nadler MJS, Chen B, Anderson JS, Wortis HH, Neel BG (1997) Protein-tyrosine phosphatase SHP-1 is dispensable for FcgammaRIIB-mediated inhibition of B cell antigen receptor activation. J Biol Chem 272:20038–20043

Nakamura MC, Niemi EC, Fisher MJ, Shultz LD, Seaman WE, Ryan JC (1997) Mouse Ly-49A interrupts early signaling events in natural killer cell cytotoxicity and functionally associates with the SHP-1 tyrosine phosphatase. J Exp Med 185:673–684

Neiders ME, Rowley DA, Fitch FW (1962) The sustained suppression of hemolysin response in passively immunizeDRats. J Immunol 88:718–724

Neumeister EN, Zhu Y, Richard S, Terhorst C, Chan AC, Shaw AS (1995) Binding of ZAP-70 to phosphorylated T-cell receptor zeta and eta enhances its autophosphorylation and generates specific binding sites for SH2 domain-containing proteins. Mol Cell Biol 15:3171–3178

Nishibe S, Wahl MI, Hernandez-Sotomayor SM, Tonks NK, Rhee SG, Carpenter G (1990) Increase of the catalytic activity of phospholipase C-gamma 1 by tyrosine phosphorylation. Science 250: 1253–1256

Norgauer J, Eberle M, Lemke HD, Aktories K (1992) Activation of human neutrophils by mastoparan Reorganization of the cytoskeleton, formation of phosphatidylinositol 3,4,5-trisphosphate. secretion up-regulation of complement receptor type 3 and superoxide anion production are stimulated by mastoparan. Biochem J 282:393–397

Nossal GJV (1983) Cellular mechanisms of immunologic tolerance. Ann Rev Immunol 1:33–62

O'Rourke F, Matthews E, Feinstein MB (1996) Isolation of InsP4 and InsP6 binding proteins from human platelets: InsP4 promotes Ca2+ efflux from inside-out plasma membrane vesicles containing 104kDa GAP1IP4BP protein. Biochem J 315:1027–1034

Ohta S, Matsui A, Nazawa Y, Kagawa Y (1988) Complete cDNA encoding a putative phospholipase C from transformed human lymphocytes. FEBS Lett 242:31–35

Olcese L, Lang P, Vely F, Cambiaggi A, Marguet D, Blery M, Hippen KL, Biassoni R, Moretta A, Moretta L, Cambier JC, Vivier E (1996) Human and mouse killer-cell inhibitory receptor recruit PTP1C and PTP1D protein tyrosine phosphatases. J Immunol 156:4531–4534

Ono M, Bolland S, Tempst P, Ravetch JV (1996) Role of the inositol phosphatase SHIP in negative regulation of the immune system by the receptor Fc(gamma)RIIB. Nature 383:263–266

Ono M, Okada H, Bolland S, Yanagi S, Kurosaki T, Ravetch JV (1997) Deletion of SHIP or SHP-1 reveals two distinct pathways for inhibitory signaling. Cell 90:293–301

Osborne MA, Zenner G, Lubinus M, Zhang X, Songyang Z, Cantley LC, Majerus P, Burn P, Kochan JP (1996) The inositol 5′-phosphatase SHIP binds to immunoreceptor signaling motifs and responds to high affinity IgE receptor aggregation. J Biol Chem 271:29271–29278

Osman N, Lucas SC, Turner H, Cantrell D (1995) A comparison of the interaction of Shc and the tyrosine kinase ZAP-70 with the T cell antigen receptor zeta chain tyrosine-based activation motif. J Biol Chem 270:13981–13986

Osman N, Turner H, Lucas S, Reif K, Cantrell DA (1996) The protein interactions of the immunoglobulin receptor family tyrosine-based activation motifs present in the T cell receptor zeta subunits and the CD3 gamma, delta and epsilon chains. Eur J Immunol 26:1063–1068

Park DJ, Rho HW, Rhee SG (1991) CD3 stimulation causes phosphorylation of phospholipase C-gamma 1 on serine and tyrosine residues in a human T-cell line. Proc Nat Acad Sci (USA) 88:5453–5456

Park JG, Schreiber AD (1995) Determinants of the phagocytic signal mediated by the type IIIA Fc gamma receptor, Fc gamma RIIIA: sequence requirements and interaction with protein-tyrosine kinases. Proc Nat Acad Sci (USA) 92:7381–7385

Parker DC (1975) Stimulation of mouse lymphocytes by insoluble anti-mouse immunoglobulin. Nature 258:361–363

Parks DE, Doyle MV, Weigle WO (1978) Induction and mode of action of suppressor cells generated against human gamma globulin. I. An immunologic unresponsive state devoid of demonstrable suppressor cells. J Exp Med 148:625–638

Pelicci G, Lanfrancone L, Grignani F, McGlade J, Cavallo F, Forni G, Nicoletti I, Grignani F, Pawson T, Pelicci PG (1992) A novel transforming protein (SHC) with an SH2 domain is implicated in mitogenic signal transduction. Cell 70:93–104

Phillips NE, Gravel KA, Tumas K, Parker DC (1988) IL-4 (B cell stimulatory factor 1) overcomes Fc gamma receptor-mediated inhibition of mouse B lymphocyte proliferation without affecting inhibition of c-myc mRNA induction. J Immunol 141:4243–4249

Phillips NE, Parker DC (1983) Fc-dependent inhibition of mouse B cell activation by whole anti-mu antibodies. J Immunol 130:602–606

Phillips NE, Parker DC (1984) Cross-linking of B lymphocyte Fc gamma receptors and membrane immunoglobulin inhibits anti-immunoglobulin-induced blastogenesis. J Immunol 132:627–632

Phillips NE, Parker DC (1985) Subclass specificity of Fc gamma receptor-mediated inhibition of mouse B cell activation. J Immunol 134:2835–2838

Pike BL, Boyd AW, Nossal GJV (1982) Clonal anergy: The universally anergic B lymphocyte. Proc Nat Acad Sci (USA) 79:2013–2017

Pleiman CM, Abrams C, Gauen LT, Bedzyk W, Jongstra J, Shaw AS, Cambier JC (1994) Distinct p53/56lyn and p59fyn domains associate with nonphosphorylated and phosphorylated Ig-alpha. Proc Nat Acad Sci (USA) 91:4268–4272

Pradhan M, Coggeshall KM (1997) Activation-induced bi-dentate SHIP and Shc interaction in B lymphocytes. J Cell Biochem 67:32–42

Prchal JT, Sokol L (1996) "Benign erythrocytosis" and other familial and congenital polycythemias. Eur J Haematol 57:263–268

Ptasznik A, Prossnitz ER, Yoshikawa D, Smrcka A, Traynor-Kaplan AE, Bokoch GM (1996) A tyrosine kinase signaling pathway accounts for the majority of phosphatidylinositol 3,4,5-trisphosphate formation in chemoattractant-stimulated human neutrophils. J Biol Chem 271:25204–25207

Rameh LE, Arvidsson A, Carraway KLR, Couvillon AD, Rathbun G, Crompton A, VanRenterghem B, Czech MP, Ravichandran KS, Burakoff SJ, Wang DS, Chen CS, Cantley LC (1997) A comparative analysis of the phosphoinositide binding specificity of pleckstrin homology domains. J Biol Chem 272:22059–22066

Ransom JT, Chen M, Sandoval VM, Pasternak JA, Digiusto D, Cambier JC (1988) Increased plasma membrane permeability to Ca2+ in anti-Ig-stimulated B lymphocytes is dependent on activation of phosphoinositide hydrolysis. J Immunol 140:3150–3155

Ransom JT, Harris LK, Cambier JC (1986) Anti-Ig induceSRelease of inositol 1,4,5-trisphosphate which mediates mobilization of intracellular Ca++ stores in B lymphocytes. J Immunol 137:708–714

Ravetch JV, Kinet JP (1991) Fc receptors. Ann Rev Immunol 9:457–492

Ravetch JV, Luster AD, Weinshank R, Kochan J, Pavlovec A, Portnoy DA, Hulmes J, Pan YC, Unkeless JC (1986) Structural heterogeneity and functional domains of murine immunoglobulin G Fc receptors. Science 234:718–725

Ravichandran KS, Lee KK, Songyang Z, Cantley LC, Burn P, Burakoff SJ (1993) Interaction of Shc with the zeta chain of the T cell receptor upon T cell activation. Science 262:902–905

Ravichandran KS, Lorenz U, Shoelson SE, Burakoff SJ (1995) Interaction of Shc with Grb2 regulates association of Grb2 with mSOS. Mol Cell Biol 15:593–600

Renauld JC, Druez C, Kermouni A, Houssiau F, Uyttenhove C, Van Roost E, Van Snick J (1992) Expression cloning of the murine and human interleukin 9 receptor cDNAs. Proc Nat Acad Sci (USA) 89:5690–5694

Reth M (1989) Antigen receptor tail clue. Nature 338:383–384

Reyburn HT, Mandelboim O, Vales-Gomez M, Davis DM, Pazmany L, Strominger JL (1997) The class I MHC homolog of human cytomegalovirus inhibits attack by natural killer cells. Nature 386:514–516

Rhee SG, Bae YS (1997) Regulation of phosphoinositide-specific phospholipase C isozymes. J Biol Chem 272:15045–15048

Rivera VM, Brugge JS (1995) Clustering of Syk is sufficient to induce tyrosine phosphorylation and release of allergic mediators from rat basophilic leukemia cells. Mol Cell Biol 15:1582–1590

Romeo C, Amiot M, Seed B (1992) Sequence requirements for induction of cytolysis by the T cell antigen/Fc receptor zeta chain. Cell 68:889–897

Romeo C, Seed B (1991) Cellular immunity to HIV activated by CD4 fused to T cell or Fc receptor polypeptides. Cell 64:1037–1046

Rothstein TL (1996) Signals and susceptibility to programmed death in B cells. Curr Op Immunol 8: 362–371

Rotin D, Honegger AM, Margolis BL, Ullrich A, Schlessinger J (1992a) Presence of SH2 domains of phospholipase C gamma 1 enhances substrate phosphorylation by increasing the affinity toward the epidermal growth factor receptor. J Biol Chem 267:9678–9683

Rotin D, Margolis B, Mohammadi M, Daly RJ, Daum G, Li N, Fischer EH, Burgess WH, Ullrich A, Schlessinger J (1992b) SH2 domains prevent tyrosine dephosphorylation of the EGF receptor: identification of Tyr992 as the high-affinity binding site for SH2 domains of phospholipase C gamma EMBO J 11:559–567

Rowley RB, Burkhardt AL, Chao HG, Matsueda GR, Bolen JB, (1995) Syk protein-tyrosine kinase is regulated by tyrosine-phosphorylated Ig alpha/Ig beta immunoreceptor tyrosine activation motif binding and autophosphorylation. J Biol Chem 270:11590–11594

Salcedo TW, Kurosaki T, Kanakaraj P, Ravetch JV, Perussia B (1993) Physical and functional association of p56lck with Fc gamma RIIIA (CD16) in natural killer cells. J Exp Med 177:1475–1480

Salcini AE, McGlade J, Pelicci G, Nicoletti I, Pawson T, Pelicci PG (1994) Formation of Shc-Grb2 complexes is necessary to induce neoplastic transformation by overexpression of Shc proteins. Oncogene 9:2827–2836
Salim K, Bottomley MJ, Querfurth E, Zvelebil MJ, Gout I, Scaife R, Margolis RL, Gigg R, Smith CI, Driscoll PC, Waterfield MD, Panayotou G (1996) Distinct specificity in the recognition of phosphoinositides by the pleckstrin homology domains of dynamin and Bruton's tyrosine kinase. EMBO J 15:6241–6450
Samaridis J, Colonna M (1997) Cloning of novel immunoglobulin superfamily receptors expressed on human myeloid and lymphoid cells: structural evidence for new stimulatory and inhibitory pathways. Eur J Immunol 27:660–665
Samelson LE, Phillips AF, Luong ET, Klausner RD (1990) Association of the fyn protein-tyrosine kinase with the T-cell antigen receptor. Proc Nat Acad Sci (USA) 87:4358–4362
Sanchez M, Misulovin Z, Burkhardt AL, Mahajan S, Costa T, Franke R, Bolen JB, Nussenzweig M (1993) Signal transduction by immunoglobulin is mediated through Ig alpha and Ig beta. J Exp Med 178:1049–1055
Saouaf SJ, Kut SA, Fargnoli J, Rowley RB, Bolen JB, Mahajan S (1995) Reconstitution of the B cell antigen receptor signaling components in COS cells. J Biol Chem 270:27072–27078
Sarkar S, Schlottmann K, Cooney D, Coggeshall KM (1996) Negative signaling via FcgammaIIB1 in B cells blocks phospholipase Cgamma2 phosphorylation but not Syk or Lyn activation. J Biol Chem 271:20182–20186
Sarmay G, Koncz G, Gergely J (1996) Human type II Fcgamma receptors inhibit B cell activation by interacting with the p21(ras)-dependent pathway. J Biol Chem 271:30499–30504
Sarosi GA, Thomas PM, Egerton M, Phillips AF, Kim KW, Bonvini E, Samelson LE (1992) Characterization of the T cell antigen receptor-p60fyn protein tyrosine kinase association by chemical cross-linking. Int Immunol 4:1211–1217
Saxton TM, van Oostveen I, Bowtell D, Aebersold R, Gold MR (1994) B cell antigen receptor cross-linking induces phosphorylation of the p21ras oncoprotein activators SHC and mSOS1 as well as assembly of complexes containing SHC GRB-2 MSOS1 a 145-kDa tyrosine-phosphorylated protein. J Immunol 153:623–636
Scharenberg AM, El-Hillal O, Fruman DA, Beitz LO, Li Z, Lin S, Gout I, Cantley LC, Rawlings DJ, Kinet J-P (1998) Phosphatidylinositol-3,4,5-trisphosphate (PtdIns-3,4,5-P3)/Tec kinase-dependent calcium signaling pathways: a target for SHIP-mediated inhibitory signals. EMBO J 17:1961–1972
Schlessinger J (1994) SH2/SH3 signaling proteins. Curr Op Gen Dev 4:25–30
Schultz LD, Rajan TV, Greiner DL (1997) Severe defects in immunity and hematopoiesis caused by SHP-1 protein tyrosine phosphatase deficiency. Trends Biotechnol 15:302–307
Schultz LD, Sidman CL (1987) Gentically-determined murine models of immunodeficiency. Ann Rev Immunol 5:367–403
Scribner DJ, Weiner H, Moorhead JW (1978) Anti-immunoglobulin stimulation of murine lymphocytes. V. Age-related decline in Fc receptor-mediated immunoregulation. J Immunol 121:377–382
Serunian LA, Haber MT, Fukui T, Kim JW, Rhee SG, Lowenstein JM, Cantley LC, (1989) Polyphosphoinositides produced by phosphatidylinositol 3-kinase are poor substrates for phospholipases C from rat liver and bovine brain. J Biol Chem 264:17809–17815
Sha WC (1998) Regulation of immune responses by NF-kappa B/Rel transcription factors. J Exp Med 187:143–146
Shiue L, Green J, Green OM, Karas JL, Morgenstern JP, Ram MK, Taylor MK, Zoller MJ, Zydowsky LD, Bolen JB et al (1995) Interaction of p72syk with the gamma and beta subunits of the high-affinity receptor for immunoglobulin E, Fc epsilon. RI Mol Cell Biol 15:272–281
Shultz LD, Rajan TV, Greiner DL (1997) Severe defects in immunity and hematopoiesis caused by SHP-1 protein-tyrosine-phosphatase deficiency. Trens Biotechnol 15:302–307
Sinclair NRS, Lees RK, Chan PL, Khan RH (1970) Regulation of the immune response. II. Further studies on differences in ability of F(ab')2 and 7 S antibodies to inhibit an antibody response. Immunol 19:105–116
Sinclair NRS (1969) Regulation of the immune response. I. Reduction in ability of specific antibody to inhibit long-lasting IgG immunological priming following removal of the Fc-fragment. J Exp Med 129:1183–1201
Sinclair NRS, Chan PL (1971) Regulation of the immune response IV The role of Fc-fragment in feedback inhibition by antibody. Adv Exp Med Biol 12:609–615
Sinclair NRS, Lees R, Elliot EV (1968) Role of Fc fragment in the regulation of the primary immune response. Nature 220:1048–1049

Smit L, de Vries-Smits AM, Bos JL, Borst J (1994) B cell antigen receptor stimulation induces formation of a Shc-Grb2 complex containing multiple tyrosine-phosphorylated proteins. J Biol Chem 269:20209–20212
Sokol L, Luhovy M, Guan Y, Prchal JF, Semenza GL, Prchal JT (1995) Primary familial polycythemia: a frameshift mutation in the erythropoietin receptor gene and increased sensitivity of erythroid progenitors to erythropoietin. Blood 86:15–22
Somani AK, Bignon JS, Mills GB, Siminovitch KA, Branch DR (1997) Src kinase activity is regulated by the SHP-1 protein-tyrosine phosphatase. J Biol Chem 272:21113–21119
Songyang Z, Blechner S, Hoagland N, Hoekstra MF, Piwnica-Worms H, Cantley LC, (1994a) Use of an oriented peptide library to determine the optimal substrates of protein kinases. Curr Biol 4: 973–982
Songyang Z, Carraway KLr, Eck MJ, Harrison SC, Feldman RA, Mohammadi M, Schlessinger J, Hubbard SR, Smith DP, Eng C, Lorenzo MJ, Ponder BAJ, Mayer BJ, Cantley LC (1995) Catalytic specificity of protein-tyrosine kinases is critical for selective signaling. Nature 373:536–539
Songyang Z, Shoelson SE, Chaudhuri M, Gish G, Pawson T, Haser WG, King F, Roberts T, Ratnofsky S, Lechleider RJ et al (1993) SH2 domains recognize specific phosphopeptide sequences. Cell 72: 767–778
Songyang Z, Shoelson SE, McGlade J, Olivier P, Pawson T, Bustelo XR, Barbacid M, Sabe H, Hanafusa H, Yi T et al (1994b) Specific motifs recognized by the SH2 domains of Csk 3BP2 fps/fes GRB-2 HCP SHC Syk Vav Mol Cell Biol 14:2777–2785
Stancovski I, Baltimore D (1997) NF-kappaB activation: the I kappaB kinase revealed? Cell 91:299–302
Straus DB, Weiss A (1992) Genetic evidence for the involvement of the lck tyrosine kinase in signal transduction through the T cell antigen receptor. Cell 70:585–593
Straus DB, Weiss A (1993) The CD3 chains of the T cell antigen receptor associate with the ZAP-70 tyrosine kinase and are tyrosine phosphorylated after receptor stimulation. J Exp Med 178:1523–1530
Stuart SG, Trounstine ML, Vaux DJ, Koch T, Martens CL, Mellman I, Moore KW (1987) Isolation and expression of cDNA clones encoding a human receptor for IgG (Fcgamma RII). J Exp Med 166:1668–1684
Sugawara H, Kurosaki M, Takata M, Kurosaki T (1997) Genetic evidence for involvement of type 1, type 2, and type 3 inositol 1,4,5-trisphosphate receptors in signal transduction through the B-cell antigen receptor. EMBO J 16:3078–3088
Takai T, Li M, Sylvestre D, Clynes R, Ravetch JV (1994) FcR gamma chain deletion results in pleiotrophic effector cell defects. Cell 76:519–529
Takai T, Ono M, Hikida M, Ohmori H, Ravetch JV (1996) Augmented humoral and anaphylactic responses in Fc gamma RII-deficient mice. Nature 379:346–349
Takata M, Kurosaki T (1996) A role for Bruton's tyrosine kinase in B cell antigen receptor-mediated activation of phospholipase C-gamma 2. J Exp Med 184:31–40
Takata M, Sabe H, Hata A, Inazu T, Homma Y, Nukada T, Yamamura H, Kurosaki T (1994) Tyrosine kinases Lyn and Syk regulate B cell receptor-coupled Ca2+ mobilization through distinct pathways. EMBO J 13:1341–1349
Tapley P, Shevde NK, Schweitzer PA, Gallina M, Christianson SW, Lin IL, Stein RB, Shultz LD, Rosen J, Lamb P (1997) Increased G-CSF responsiveness of bone marrow cells from hematopoietic cell phosphatase deficient viable motheaten mice. Exp Hematol 25:122–131
Tedder TF, Zhou LJ, Engel P (1994) The CD19/CD21 signal transduction complex of B lymphocytes. Immunol Today 15:437–442
Teramoto H, Salem P, Robbins KC, Bustelo XR, Gutkind JS (1997) Tyrosine phosphorylation of the vav proto-oncogene product links FcepsilonRI to the Rac1-JNK pathway. J Biol Chem 272:10751–10755
Tigelaar RE, Vaz NM, Ovary Z (1971) Immunoglobulin receptors on mouse mast cells. J Immunol 106:661–669
Timson Gauen LK, Kong AN, Samelson LE, Shaw AS (1992) p59fyn tyrosine kinase associates with multiple T-cell receptor subunits through its unique amino-terminal domain. Mol Cell Biol 12: 5438–5446
Tite JP, Morrison CA, Taylor RB (1981) Immunoregulatory effects of covalent antigen-antibody complexes. II. Enhancement or suppression depending on the time of administration of complex relative to a T-independent antigen. Immunol 42:355–362
Tite JP, Taylor RB (1979) Immunoregulation by covalent antigen-antibody complexes II Suppression of a T cell independent anti-hapten response. Immunol 38:325–331
Toker A, Cantley LC (1997) Signaling through the lipid products of phosphoinositide-3-OH kinase. Nature 387:673–676

Touw IP, Dong F (1996) Severe congenital neutropenia terminating in acute myeloid leukemia: Disease progression associated with mutations in the granulocyte colony stimulating factor receptor gene. Leukemia Res 20:629–631

Tridandapani S, Chacko GW, Brocklyn JRv, Coggeshall KM (1997a) Negative signaling in B cells causes reduced as activity by reducing Shc-Grb2 interactions. J Immunol 158:1125–1132

Tridandapani S, Kelley T, Cooney D, Pradhan M, Coggeshall KM (1997b) Negative signaling in B cells: SHIP Grbs Shc. Immunol Today 18:424–427

Tridandapani S, Kelley T, Pradhan M, Cooney D, Justement LB, Coggeshall KM (1997c) Recruitment and phosphorylation of SHIP and Shc to the B cell Fcgamma ITIM peptide motif. Mol Cell Biol 17:4305–4311

Tridanapani S, Phee H, Shivakumar L, Kelley TW, Coggeshall KM (1999a) Role of SHIP in FcγRIIb-mediated inhibition of Ras activation in B cells. Mol Immunol 35:1135–1146

Tridanapani S, Pradhan M, LaDine JR, Garber S, Anderson CL, Coggeshall KM (1999b) Protein interactions of SHIP: association with Shc displaces SHIP from FCγRIIb in B cells. J Immunol 162:1408–1414

Tsukada S, Rawlings DJ, Witte ON (1994) Role of Bruton's tyrosine kinase in immunodeficiency. Curr Op Immunol 6:623–630

Uhr JD, Baumann JB (1961a) Antibody formation: I The suppression of antibody formation by passively administered antibody. J Exp Med 113:935–957

Uhr JW, Baumann JB (1961b) Antibody formation II The specific anamnestic antibody response. J Exp Med 113:959–970

Uhr JW, Moller G (1968) Regulatory effect of antibody on the immune response. Adv Immunol 8:81–127

Urich M, el Shemerly MY, Besser D, Nagamine Y, Ballmer-Hofer K (1995) Activation and nuclear translocation of mitogen-activated protein kinases by polyomavirus middle-T or serum depend on phosphatidylinositol 3-kinase. J Biol Chem 270:29286–29292

Valiante NM, Phillips JH, Lanier LL, Parham P (1996) Killer cell inhibitory receptor recognition of human leukocyte antigen (HLA) class I blocks formation of a pp36/PLC-gamma signaling complex in human natural killer (NK) cells. J Exp Med 184:2243–2250

van den Elsen P, Shepley BA, Borst J, Coligan JE, Markham AF, Orkin S, Terhorst C (1984) Isolation of cDNA clones encoding the 20 K T3 glycoprotein of human T-cell receptor complex. Nature 312: 413–418

van den Herik-Oudijk IE, Capel PJ, van der Bruggen T, Van de Winkel JG (1995) Identification of signaling motifs within human Fc gamma RIIa and Fc gamma RIIb isoforms. Blood 85: 2202–2211

van der Geer P, Pawson T (1995) The PTB domain: a new protein module implicated in signal transduction. Trends Biochem Sci 20:277–280

van der Geer P, Wiley S, Gish GG, Pawson T (1996) The Shc adaptor protein is highly phosphorylated at conserved, twin tyrosine residues (Y239/240) that mediate protein-protein interactions. Curr Biol 6:1435–1444

van Vugt MJ, Heijnen AF, Capel PJ, Park SY, Ra C, Saito T, Verbeek JS, van de Winkel JG (1996) FcR gamma-chain is essential for both surface expression and function of human Fc gamma RI (CD64) in vivo. Blood 87:3593–3599

Vely F, Olcese L, Blery M, Vivier E (1996) Function of killer cell inhibitory receptors for MHC class I molecules. Immunol Lett 54:145–150

Vely F, Olivero S, Olcese L, Moretta A, Damen JE, Liu L, Krystal G, Cambier JC, Daeron M, Vivier E (1997) Differential association of phosphatases with hematopoietic co-receptors bearing immunoreceptor tyrosine-based inhibition motifs. Eur J Immunol 27:1994–2000

Verdier F, Chretien S, Billat C, Gisselbrecht S, Lacombe C, Mayeux P (1997) Erythropoietin induces the tyrosine phosphorylation of insulin receptor substrate-2 An alternate pathway for erythropoietin-induced phosphatidylinositol 3-kinase activation. J Biol Chem 272:26173–26178

Wagtmann N, Biassoni R, Cantoni C, Verdiani S, Malnati MS, Vitale M, Bottino C, Moretta L, Moretta A, Long EO (1995a) Molecular clones of the p58 NK cell receptor reveal immunoglobulin-related molecules with diversity in both the extra- and intracellular domains. Immunity 2:439–449

Wagtmann N, Rajagopalan S, Winter CC, Peruzzi M, Long EO (1995b) Killer cell inhibitory receptors specific for HLA-C and HLA-B identified by direct binding and by functional transfer. Immunity 3:801–809

Wahl MI, Jones GA, Nishibe S, Rhee SG, Carpenter G (1992) Growth factor stimulation of phospholipase C-gamma 1 activity Comparative properties of control and activated enzymes. J Biol Chem 267:10447–10456

Waksman G, Shoelson SE, Pant N, Cowburn D, Kuriyan J (1993) Binding of a high affinity phosphotyrosyl peptide to the Src SH2 domain: crystal structures of the complexed and peptide-free forms. Cell 72:779–790

Wang HY, Paul WE, Keegan AD (1996) IL-4 function can be transferred to the IL-2 receptor by tyrosine-containing sequences found in the IL-4 receptor alpha chain. Immunity 4:113–121

Wang LL, Mehta IK, LeBlanc PA, Yokoyama WM (1997) Mouse natural killer cells express gp49B1, a structural homologue of human killer inhibitory receptors. J Immunol 158:13–17

Weinlands J, Freuler F, Bauman G (1995) Tyrosine-phosphorylated forms of Ig-beta, CD22, TCR-zeta and HOSS are major ligands for tandem SH2 domains of Syk. Int Immunol 7:1701–1708

Weinstein IB, Kahn SM, O'Driscoll K, Borner C, Bang D, Jiang W, Blackwood A, Nomoto K (1997) The role of protein kinase C in signal transduction, growth control, and lipid metabolism. Adv Exp Med Biol 400

Weissman AM, Hou D, Orloff DG, Modi WS, Seuanez H, O'Brien SJ, Kluasner RD (1988) Molecular cloning and chromosomal localization of the human T-cell receptor zeta chain: distinction from the CD3 complex. Proc Nat Acad Sci (USA) 85:9709–9713

Wigzell H (1966) Antibody synthesis at the cellular level. Antibody-induced suppression of 7S antibody synthesis. J Exp Med 124:953–969

Willerford DM, Swat W, Alt FW (1996) Developmental regulation of V(D)J recombination and lymphocyte differentiation. Curr Op Gen Dev 6:603–609

Wilson GL, Fox CH, Fauci AS, Kehrl JH (1991) cDNA cloning of the B cell membrane protein CD22: a mediator of B-B cell interactions. J Exp Med 173:137–146

Wood WJ, Thompson AA, Korenberg J, Chen XN, May W, Wall R, Denny CT (1993) Isolation and chromosomal mapping of the human immunoglobulin-associated B29 gene (IGB). Genomics 16: 187–192

Xu XL, Chong ASF (1995) Cross-linking of CD45 on NK cells stimulates p56lck-mediated tyrosine phosphorylation and IFN-gamma production. J Immunol 155:5241–5248

Yabe T, McSherry C, Bach FH, Fisch P, Schall RP, Sondel PM, Houchins JP (1993) A multigene family on human chromosome 12 encodes natural killer-cell lectins. Immunogen 37:455–460

Yamashita Y, Miyake K, Miura Y, Kaneko Y, Yagita H, Suda T, Nagata S, Nomura J, Sakaguchi N, Kimoto M (1996) Activation mediated by RP105 but not CD40 makes normal B cells susceptible to anti-IgM-induced apoptosis: a role for Fc receptor coligation. J Exp Med 184:113–120

Yamauchi K, Holt K, Pessin JE (1993) Phosphatidylinositol 3-kinase functions upstream of Ras and Raf in mediating insulin stimulation of c-fos transcription. J Biol Chem 268:14597–14600

Yarden Y, Kuang WJ, Yang-Feng T, Coussens L, Munemitsu S, Dull TJ, Chen E, Schlessinger J, Francke U, Ullrich A (1987) Human proto-oncogene c-kit: a new cell surface receptor tyrosine kinase for an unidentified ligand. EMBO J 6:3341–3351

Yellen-Shaw A, Monroe JG (1992) Differential responsiveness of immature- and mature-stage murine B cells to anti-IgM reflects both FcR-dependent and -independent mechanisms. Cell Immunol 145: 339–350

Yi T, Mui AL, Krystal G, Ihle JN (1993) Hematopoietic cell phosphatase associates with the interleukin-3 (IL-3) receptor beta chain and down-regulates IL-3-induced tyrosine phosphorylation and mitogenesis. Mol Cell Biol 13:7577–7586

Yu H, Chen JK, Feng S, Dalgarno DC, Brauer AW, Schreiber SL (1994) Structural basis for the binding of proline-rich peptides to SH3 domains. Cell 76:933–945

Zenner G, Vorherr T, Mustelin T, Burn P (1996) Differential and multiple binding of signal transducing molecules to the ITAMs of the TCR-zeta chain. J Cell Biochem 63:94–103

Zhou MM, Harlan JE, Wade WS, Crosby S, Ravichandran KS, Burakoff SJ, Fesik SW (1995) Binding affinities of tyrosine-phosphorylated peptides to the COOH-terminal SH2 and NH2-terminal phosphotyrosine binding domains of Shc. J Biol Chem 270:31119–31123

# Subject Index

# Current Topics in Microbiology and Immunology

## Volumes published since 1989 (and still available)

Vol. 204: **Saedler, Heinz; Gierl, Alfons (Eds.):** Transposable Elements. 1995. 42 figs. IX, 234 pp. ISBN 3-540-59342-X

Vol. 205: **Littman, Dan R. (Ed.):** The CD4 Molecule. 1995. 29 figs. XIII, 182 pp. ISBN 3-540-59344-6

Vol. 206: **Chisari, Francis V.; Oldstone, Michael B. A. (Eds.):** Transgenic Models of Human Viral and Immunological Disease. 1995. 53 figs. XI, 345 pp. ISBN 3-540-59341-1

Vol. 207: **Prusiner, Stanley B. (Ed.):** Prions Prions Prions. 1995. 42 figs. VII, 163 pp. ISBN 3-540-59343-8

Vol. 208: **Farnham, Peggy J. (Ed.):** Transcriptional Control of Cell Growth. 1995. 17 figs. IX, 141 pp. ISBN 3-540-60113-9

Vol. 209: **Miller, Virginia L. (Ed.):** Bacterial Invasiveness. 1996. 16 figs. IX, 115 pp. ISBN 3-540-60065-5

Vol. 210: **Potter, Michael; Rose, Noel R. (Eds.):** Immunology of Silicones. 1996. 136 figs. XX, 430 pp. ISBN 3-540-60272-0

Vol. 211: **Wolff, Linda; Perkins, Archibald S. (Eds.):** Molecular Aspects of Myeloid Stem Cell Development. 1996. 98 figs. XIV, 298 pp. ISBN 3-540-60414-6

Vol. 212: **Vainio, Olli; Imhof, Beat A. (Eds.):** Immunology and Developmental Biology of the Chicken. 1996. 43 figs. IX, 281 pp. ISBN 3-540-60585-1

Vol. 213/I: **Günthert, Ursula; Birchmeier, Walter (Eds.):** Attempts to Understand Metastasis Formation I. 1996. 35 figs. XV, 293 pp. ISBN 3-540-60680-7

Vol. 213/II: **Günthert, Ursula; Birchmeier, Walter (Eds.):** Attempts to Understand Metastasis Formation II. 1996. 33 figs. XV, 288 pp. ISBN 3-540-60681-5

Vol. 213/III: **Günthert, Ursula; Schlag, Peter M.; Birchmeier, Walter (Eds.):** Attempts to Understand Metastasis Formation III. 1996. 14 figs. XV, 262 pp. ISBN 3-540-60682-3

Vol. 214: **Kräusslich, Hans-Georg (Ed.):** Morphogenesis and Maturation of Retroviruses. 1996. 34 figs. XI, 344 pp. ISBN 3-540-60928-8

Vol. 215: **Shinnick, Thomas M. (Ed.):** Tuberculosis. 1996. 46 figs. XI, 307 pp. ISBN 3-540-60985-7

Vol. 216: **Rietschel, Ernst Th.; Wagner, Hermann (Eds.):** Pathology of Septic Shock. 1996. 34 figs. X, 321 pp. ISBN 3-540-61026-X

Vol. 217: **Jessberger, Rolf; Lieber, Michael R. (Eds.):** Molecular Analysis of DNA Rearrangements in the Immune System. 1996. 43 figs. IX, 224 pp. ISBN 3-540-61037-5

Vol. 218: **Berns, Kenneth I.; Giraud, Catherine (Eds.):** Adeno-Associated Virus (AAV) Vectors in Gene Therapy. 1996. 38 figs. IX,173 pp. ISBN 3-540-61076-6

Vol. 219: **Gross, Uwe (Ed.):** Toxoplasma gondii. 1996. 31 figs. XI, 274 pp. ISBN 3-540-61300-5

Vol. 220: **Rauscher, Frank J. III; Vogt, Peter K. (Eds.):** Chromosomal Translocations and Oncogenic Transcription Factors. 1997. 28 figs. XI, 166 pp. ISBN 3-540-61402-8

Vol. 221: **Kastan, Michael B. (Ed.):** Genetic Instability and Tumorigenesis. 1997. 12 figs.VII, 180 pp. ISBN 3-540-61518-0

Vol. 222: **Olding, Lars B. (Ed.):** Reproductive Immunology. 1997. 17 figs. XII, 219 pp. ISBN 3-540-61888-0

Vol. 223: **Tracy, S.; Chapman, N. M.; Mahy, B. W. J. (Eds.):** The Coxsackie B Viruses. 1997. 37 figs. VIII, 336 pp. ISBN 3-540-62390-6

Vol. 224: **Potter, Michael; Melchers, Fritz (Eds.):** C-Myc in B-Cell Neoplasia. 1997. 94 figs. XII, 291 pp.
ISBN 3-540-62892-4

Vol. 225: **Vogt, Peter K.; Mahan, Michael J. (Eds.):** Bacterial Infection: Close Encounters at the Host Pathogen Interface. 1998. 15 figs. IX, 169 pp.
ISBN 3-540-63260-3

Vol. 226: **Koprowski, Hilary; Weiner, David B. (Eds.):** DNA Vaccination/Genetic Vaccination. 1998. 31 figs. XVIII, 198 pp.
ISBN 3-540-63392-8

Vol. 227: **Vogt, Peter K.; Reed, Steven I. (Eds.):** Cyclin Dependent Kinase (CDK) Inhibitors. 1998. 15 figs. XII, 169 pp.
ISBN 3-540-63429-0

Vol. 228: **Pawson, Anthony I. (Ed.):** Protein Modules in Signal Transduction. 1998. 42 figs. IX, 368 pp. ISBN 3-540-63396-0

Vol. 229: **Kelsoe, Garnett; Flajnik, Martin (Eds.):** Somatic Diversification of Immune Responses. 1998. 38 figs. IX, 221 pp.
ISBN 3-540-63608-0

Vol. 230: **Kärre, Klas; Colonna, Marco (Eds.):** Specificity, Function, and Development of NK Cells. 1998. 22 figs. IX, 248 pp.
ISBN 3-540-63941-1

Vol. 231: **Holzmann, Bernhard; Wagner, Hermann (Eds.):** Leukocyte Integrins in the Immune System and Malignant Disease. 1998. 40 figs. XIII, 189 pp.
ISBN 3-540-63609-9

Vol. 232: **Whitton, J. Lindsay (Ed.):** Antigen Presentation. 1998. 11 figs. IX, 244 pp.
ISBN 3-540-63813-X

Vol. 233/I: **Tyler, Kenneth L.; Oldstone, Michael B. A. (Eds.):** Reoviruses I. 1998. 29 figs. XVIII, 223 pp.
ISBN 3-540-63946-2

Vol. 233/II: **Tyler, Kenneth L.; Oldstone, Michael B. A. (Eds.):** Reoviruses II. 1998. 45 figs. XVI, 187 pp.
ISBN 3-540-63947-0

Vol. 234: **Frankel, Arthur E. (Ed.):** Clinical Applications of Immunotoxins. 1999. 16 figs. IX, 122 pp. ISBN 3-540-64097-5

Vol. 235: **Klenk, Hans-Dieter (Ed.):** Marburg and Ebola Viruses. 1999. 34 figs. XI, 225 pp. ISBN 3-540-64729-5

Vol. 236: **Kraehenbuhl, Jean-Pierre; Neutra, Marian R. (Eds.):** Defense of Mucosal Surfaces: Pathogenesis, Immunity and Vaccines. 1999. 30 figs. IX, 296 pp.
ISBN 3-540-64730-9

Vol. 237: **Claesson-Welsh, Lena (Ed.):** Vascular Growth Factors and Angiogenesis. 1999. 36 figs. X, 189 pp.
ISBN 3-540-64731-7

Vol. 238: **Coffman, Robert L.; Romagnani, Sergio (Eds.):** Redirection of Th1 and Th2 Responses. 1999. 6 figs. IX, 148 pp.
ISBN 3-540-65048-2

Vol. 239: **Vogt, Peter K.; Jackson, Andrew O. (Eds.):** Satellites and Defective Viral RNAs. 1999. 39 figs. XVI, 179 pp.
ISBN 3-540-65049-0

Vol. 240: **Hammond, John; McGarvey, Peter; Yusibov, Vidadi (Eds.):** Plant Biotechnology. 1999. 12 figs. XII, 196 pp.
ISBN 3-540-65104-7

Vol. 241: **Westblom, Tore U.; Czinn, Steven J.; Nedrud, John G. (Eds.):** Gastroduodenal Disease and Helicobacter pylori. 1999. 35 figs. XI, 313 pp.
ISBN 3-540-65084-9

Vol. 242: **Hagedorn, Curt H.; Rice, Charles M. (Eds.):** The Hepatitis C Viruses. 2000. 47 figs. IX, 379 pp. ISBN 3-540-65358-9

Vol. 243: **Famulok, Michael; Winnacker, Ernst-L.; Wong, Chi-Huey (Eds.):** Combinatorial Chemistry in Biology. 1999. 48 figs. IX, 189 pp. ISBN 3-540-65704-5

Vol. 244: **Daëron, Marc; Vivier, Eric (Eds.):** Immunoreceptor Tyrosine-Based Inhibition Motifs. 1999. 20 figs. VIII, 179 pp.
ISBN 3-540-65789-4

Vol. 245/II: **Justement, Louis B.; Siminovitch, Katherine A. (Eds.):** Signal Transduction on the Coordination of B Lymphocyte Development and Function II. 2000. 13 figs. XV, 172 pp. ISBN 3-540-66003-8

# Springer and the environment

At Springer we firmly believe that an international science publisher has a special obligation to the environment, and our corporate policies consistently reflect this conviction.
We also expect our business partners – paper mills, printers, packaging manufacturers, etc. – to commit themselves to using materials and production processes that do not harm the environment. The paper in this book is made from low- or no-chlorine pulp and is acid free, in conformance with international standards for paper permanency.

Printing: Saladruck, Berlin
Binding: H. Stürtz AG, Würzburg